DATE DUE

GAYLORD			PRINTED IN U.S.A.

ADVANCES IN

BIOLOGICAL AND CHEMICAL TERRORISM COUNTERMEASURES

ADVANCES IN

BIOLOGICAL AND CHEMICAL TERRORISM COUNTERMEASURES

EDITED BY
RONALD J. KENDALL
STEVEN M. PRESLEY
GALEN P. AUSTIN
PHILIP N. SMITH

CRC Press
Taylor & Francis Group
Boca Raton London New York

CRC Press is an imprint of the
Taylor & Francis Group, an **informa** business

CRC Press
Taylor & Francis Group
6000 Broken Sound Parkway NW, Suite 300
Boca Raton, FL 33487-2742

© 2008 by Taylor & Francis Group, LLC
CRC Press is an imprint of Taylor & Francis Group, an Informa business

No claim to original U.S. Government works
Printed in the United States of America on acid-free paper
10 9 8 7 6 5 4 3 2

International Standard Book Number-13: 978-1-4200-7654-7 (Hardcover)

Library of Congress Cataloging-in-Publication Data

Advances in biological and chemical terrorism countermeasures / Ronald J. Kendall ... [et al.]. -- 1st ed.
    p. cm.
Includes bibliographical references and index.
ISBN-13: 978-1-4200-7654-7 (alk. paper)
    1. Bioterrorism--Prevention. 2. Chemical terrorism--Prevention. 3. Emergency management. I. Kendall, Ronald J.

HV6433.3.A38 2008
363.325'37--dc22
                               2007047129

Visit the Taylor & Francis Web site at
http://www.taylorandfrancis.com

and the CRC Press Web site at
http://www.crcpress.com

Contents

Preface

This textbook, *Advances in Biological and Chemical Terrorism Countermeasures*, is offered as a contribution to establish the state-of-the-science of research on countermeasures to biological and chemical threat agents. Although the context of this book is heavily focused on the United States, its application should be considered global in nature. Biological and chemical terrorism are continuing threats to the United States and other nations; ironically, as we began writing this book in July 2007, the National Intelligence Council of the United States had just issued a report in which it is stated, "As a result, we judged that the United States currently is in a heightened threat environment." The July 2007 National Intelligence Report went on to state that "we assessed that al-Qa'ida will continue to try to acquire and employ chemical, biological, radiological or nuclear material attacks and will not hesitate to use them if it develops what it deems is sufficient capability." Therefore, we believe that this textbook is timely and will continue to offer strategies and perspectives to assist the United States and other nations to defend themselves from terroristic threats.

Research was begun in 1998 at Texas Tech University to develop countermeasures against biological and chemical threat agents. Subsequently, through The Institute of Environmental and Human Health (TIEHH) at Texas Tech University, the Admiral Elmo R. Zumwalt, Jr. National Program for Countermeasures to Biological and Chemical Threats (Zumwalt Program) was commissioned to further advance the university's and its collaborators' research and development of countermeasures to biological and chemical threats. Support for this research program has primarily occurred through the U.S. Army and, particularly, through the Research Development and Engineering Command, which has challenged our program to develop a multidisciplinary toxicology research program to address countermeasures to biological and chemical threat agents. Indeed, *Advances in Biological and Chemical Terrorism Countermeasures* draws heavily from the funding received through the U.S. Army and incorporates other leading scientists' research and their involvement with countermeasures against biological and chemical threats over the past few years.

Drawing upon the research data developed on countermeasures to biological and chemical threats as well as literature review, this book involved many months of planning and coordination by its authors as well as a meeting in July 2007 in Beaver Creek, Colorado, to bring all the chapters together for the book. Following months of planning, the authors met and discussed these issues for several days to coordinate this book as it relates to advances in countermeasures to both types of threats. Although the book does not attempt to completely implement all aspects of countermeasures to biological and chemical threats, it particularly addresses the research and development that has occurred through the Zumwalt Program at Texas Tech University.

The main authors of *Advances in Biological and Chemical Terrorism Countermeasures* participated fully in a collegial and multidisciplinary perspective at the

Beaver Creek meeting. Additional persons who offered input into individual chapters are noted as contributing authors to an individual chapter, even if they did not participate in the Beaver Creek meeting. The authors did participate in the Beaver Creek meeting and fully support the conclusions reached by the group, particularly those related to the conclusions and research recommendations chapter.

Countermeasures to biological and chemical threats continue to evolve as a national priority issue and we envision this issue will be before us for many years to come. We offer this book as a science-based text to improve our ability to implement countermeasures to biological and chemical terrorism. The authors believe that this book will contribute to developing the science of addressing countermeasures to biological and chemical terrorism that oftentimes challenge environmental toxicologists by virtue of the potential threats that are continuing to emerge with biological and chemical threat agents and that may require more complex experimental designs to evaluate.

We appreciated the opportunity to work together as a team in the publication of *Advances in Biological and Chemical Terrorism Countermeasures*, and we appreciate the Research Development and Engineering Command of the U.S. Army in supporting our research and the ultimate development of this textbook.

Authors

Galen P. Austin, Ph.D.
Senior Research Associate
Texas Tech University
Lubbock, Texas

Chia-bo Chang, Ph.D.
Professor
Texas Tech University
Lubbock, Texas

George P. Cobb, Ph.D.
Professor
Texas Tech University
Lubbock, Texas

Gopal Coimbatore, Ph.D.
Senior Research Associate
Texas Tech University
Lubbock, Texas

Stephen B. Cox, Ph.D.
Assistant Professor
Texas Tech University
Lubbock, Texas

Joe A. Fralick, Ph.D.
Professor
Texas Tech University Health Sciences
  Center
Lubbock, Texas

Ronald J. Kendall, Ph.D.
Professor/Director/Chair
Texas Tech University
Lubbock, Texas

Jeremy W. Leggoe, Ph.D., PE
Associate Professor
Texas Tech University
Lubbock, Texas

Steven M. Presley, Ph.D., BCE
Associate Professor
Texas Tech University
Lubbock, Texas

Seshadri S. Ramkumar, Ph.D.
Assistant Professor
Texas Tech University
Lubbock, Texas

Philip N. Smith, Ph.D.
Assistant Professor
Texas Tech University
Lubbock, Texas

Jean Strahlendorf, Ph.D.
Professor
Texas Tech University Health Sciences
  Center
Lubbock, Texas

Richard Zartman, Ph.D.
Professor
Texas Tech University
Lubbock, Texas

Contributing Authors

Jonathan Boyd, Ph.D.
Senior Toxicologist
Applied Physics Laboratory, Johns
  Hopkins University
Baltimore, Maryland

Prabhjit Chadha-Mohanty, Ph.D.
Research Associate
Texas Tech University Health Sciences
  Center
Lubbock, Texas

Tom Gill, Ph.D.
Associate Professor
University of Texas at El Paso
El Paso, Texas

Munim Hussain
Research Assistant
Texas Tech University
Lubbock, Texas

Guigen Li, Ph.D.
Professor
Texas Tech University
Lubbock, Texas

Eric J. Marsland, Ph.D.
Medical Entomologist
Cincinnati, Ohio

Christopher B. Pepper, J.D., M.S.
Environmental Law
Jackson Walker, L.L.P.
Austin, Texas

Utkarsh Sata, Ph.D.
Research Associate
Texas Tech University
Lubbock, Texas

Howard Strahlendorf, Ph.D.
Professor
Texas Tech University Health Sciences
  Center
Lubbock, Texas

Jia-Sheng Wang, M.D., Ph.D.
Professor
Texas Tech University
Lubbock, Texas

Acknowledgments

We gratefully acknowledge and appreciate the financial support for the development of this book from the United States Army Research, Development and Engineering Command (RDECOM). In particular, we appreciate the encouragement and support of Dr. William Lagna and Dr. John White from RDECOM throughout the years of our association. They have been tireless champions of our program and research efforts and we could not have had better project officers who also supported and encouraged the production of this timely textbook. A major part of the research that was ultimately integrated into the context of this textbook resulted from research conducted through the Admiral Elmo R. Zumwalt, Jr. National Program for Countermeasures to Biological and Chemical Threats (Zumwalt Program), which is operated through The Institute of Environmental and Human Health (TIEHH), an institute within the Texas Tech University System. The Zumwalt Program is comprised of and supported by many individual investigators through out the Texas Tech University System whose research and written contributions to this book are greatly appreciated. We also want to express our sincere appreciation to the administrative staff and support personnel at TIEHH, in particular Ms. Tammy Henricks, Ms. Lori Gibler and Mr. Ryan Bounds, for their professionalism and dedication in the completion of this textbook.

About the Editors

Galen P. Austin is a senior research associate at The Institute of Environmental and Human Health, Texas Tech University. He earned his doctoral degree in animal science with an emphasis on beef cattle production from Texas Tech University in December 2003. As an animal scientist, Dr. Austin's research interests are varied regarding livestock production and the environment. He has utilized GPS/GIS technology to monitor beef cattle movement and behavior and is interested in livestock disease epidemiology, in particular, zoonotic diseases. Additionally, Dr. Austin is concerned with and has research interests in the protection from and detection of agricultural terrorism directed at both on-farm/ranch livestock and confined animal feeding operations.

Ronald J. Kendall serves as the founding director of The Institute of Environmental and Human Health (TIEHH), a joint venture between Texas Tech University and Texas Tech University Health Sciences Center at Lubbock, Texas. He is professor and chairman of the Department of Environmental Toxicology at Texas Tech University. He graduated from the University of South Carolina and received his MS degree from Clemson University and his PhD from Virginia Polytechnic Institute and State University. He received a U.S. Environmental Protection Agency (EPA) postdoctoral traineeship at the Massachusetts Institute of Technology. Dr. Kendall served on the EPA's Science Advisory Panel from June 1995 to December 2002, and was appointed chairman from January 1999 to December 2002. He has served on many other boards, including past president of the Society of Environmental Toxicology and Chemistry, Board of Directors of the SETAC Foundation for Environmental Education, the Endocrine Disrupters Screening and Testing Advisory Committee of the U.S. Environmental Protection Agency, and multiple panels of the National Research Council. He currently serves as editor of terrestrial toxicology for the journal *Environmental Toxicology and Chemistry*. In addition, he has authored more than 200 refereed journal and technical articles and has published or edited many books. He has made more than 170 public and scientific presentations in the field of wildlife and environmental toxicology and has successfully won 136 research grants from federal, state, and foreign governments, industries, and foundations. He has served as advisor for 31 students at the graduate levels, including MS and PhD degrees, and has authored 10 courses in environmental toxicology and wildlife toxicology. He has received numerous awards, addressed the U.N. Committee on Sustainable Development, and has consulted with many foreign countries on environmental issues. Dr. Kendall was awarded a Fulbright Fellowship in 1991.

Steven M. Presley is an associate professor of environmental toxicology in The Institute of Environmental and Human Health, and serves as research coordinator for the Admiral Elmo R. Zumwalt, Jr. National Program for Countermeasures to Biological and Chemical Threats at Texas Tech University. Professionally trained as a medical entomologist, Dr. Presley also served as a U.S. Navy officer and is a graduate of the U.S. Marine Corps Command and Staff College, earning a master's of military studies degree focused on domestic terrorism, and has extensive training and practical experience in various aspects of biological, chemical, and radiological incident detection, response, and mitigation. His operational and research experience has focused upon the surveillance, prevention, and control of biological threats in the environment, and specifically vector-borne infectious diseases in tropical and semi-tropical environments. He has led malaria control operations and research efforts in Africa, Asia, and South America, as well as Rift Valley fever, Crimean Congo hemorrhagic fever, and cutaneous leishmaniasis studies in Africa and Asia. He has published more than 35 scientific and technical manuscripts, and was awarded the Rear Admiral Charles S. Stephenson Award for Excellence in Preventive Medicine for the year 2000–2001.

Philip N. Smith received his doctoral degree in environmental toxicology from Texas Tech University in May 2000, and has since advanced from senior research associate to assistant professor. Dr. Smith is an ecotoxicologist with wide-ranging interests in contaminant exposure and responses among ecological receptors. His research is focused on pathways of contaminant exposure among mammals, birds, aquatic organisms, and trophic transfer of environmental contaminants. Additionally, physiological and population-level responses to contaminant exposure are of particular interest to Dr. Smith. Dr. Smith's research is grounded in ecological relevance and is strategically aligned with his academic emphasis, which is ecological risk assessment. He is a reviewer for ten national and international journals and serves as editorial board member for two international journals, *Environmental Pollution*, and *Environmental Toxicology and Chemistry*.

List of Tables

1 State of the Science
Background, History, and Current Threats

Steven M. Presley, Christopher B. Pepper, Galen P. Austin, and Ronald J. Kendall

Alas America's future enemies may not fight according to these Marquis of Queensberry rules. They might use nuclear, biological, or chemical weapons, not only on the "regional" battlefield that the Pentagon planners assign to them, but also in that unanticipated region of warfare—the United States itself.

—Former Under-Secretary of Defense for Policy Fred C. Iklé (1997)

CONTENTS

1.1 INTRODUCTION

1.1.1 SUMMARY OF WORLD SITUATION AND PERSPECTIVE ON LIKELY FUTURE SITUATION

Western civilization is at war—a multifaceted, asymmetric, global war being fought in a nondelineated, undefined battle space, waged against a faceless enemy that operates from the shadows, utilizing both conventional and unconventional weapons and tactics to achieve its objectives. These 21st-century terrorists do not officially represent nation-states, but often they represent a religious ideology expressed through violence and death. They are driven by hatred and religious fanaticism, with many striving for the destruction of Western society and culture, and ultimately for the establishment of a global theocracy. The employment of unconventional weapons and weapons of mass destruction against civilian noncombatants is not novel or unique to present times. Mankind has exploited diseases, toxins, and poisons since the earliest days of recorded history to wage warfare, commit crimes, and influence or coerce others. However, accessibility to biological and chemical weapon agents, and their enhanced capacity to cause morbidity and mortality, as well as improvement of tactics for their employment, have significantly increased the need for the development of more effective means of detecting and countering such weapons. Additionally, Western society has become considerably more vulnerable to terrorism.

Many advanced sovereign nations have experienced an increase in terrorist activities due to an erosion of the restraints that once limited the terrorist's abilities to engage modern military, law enforcement, and intelligence agencies. Previously effective restraints included political and ideological isolation, prohibitive technical and fiscal requirements for the production of adequate qualities and quantities of terror-based weapons, and complex logistical and organizational obstacles that precluded delivery of such weapons (Stern 1999). Terrorist groups and individuals that Western societies now face, both militarily and in civil society, are not restrained in their actions. They have proven to be innovative and resilient, with a willingness to murder civilians and to martyr themselves without guilt or hesitation. Terrorists capitalize on the critical vulnerability fundamental to most advanced Western societies—openness and freedom. The avowed willingness of terrorists to use any means, including covert biological and chemical weapons against noncombatants, has dramatically impacted the psyche, social norms, and economies of many Western societies, and has instilled a chronic state of awareness of the ever-present threat of terrorism into every aspect of daily life in those societies (e.g., air travel, food and water supplies, public gatherings, etc.). "Noncombatant" is a misnomer in this context; the reality of the "global war on terrorism" is that civilians and civil society are the actual tactical targets for terrorism and are forced into a role as primary combatants.

The citizenry and governmental infrastructures of the West continue to regain their footing following the horrific terrorist attacks upon the United States on September 11, 2001. As a direct result of the September 11 attacks, there was a dramatic realization of the unrestrained, ruthless violence that could be targeted at and perpetrated upon Western civilization by relatively low-tech terrorists. This changed the collective mind-set of Western governments and militaries with regard to terrorists and their threats, and caused reassessment of vulnerabilities and protective capabilities. Western governments and peoples have increasingly recognized the vulnerabilities inherent to a free and open democratic society. Such vulnerabilities are not limited to those overt and covert threats associated with expansive coastlines and borders, or industrial and transportation systems, but include the day-to-day necessities of life such as food production, water supply, and a safe environment within which to work and recreate. The ability to reduce or eliminate these vulnerabilities and respond to terrorism, particularly biological and chemical terrorism, is highly dependent upon innovative and unprecedented mergers and collaborations among academia, engineering, industry, medical arts, and research sciences.

The primary focus and intent of this book is to improve the reader's understanding of the current status of scientific research on countermeasures to biological and chemical threat agents through an enhanced knowledge of the history of their usage, the types and extent of the threats to humans and society at large that they pose, and an awareness of the vulnerabilities within Western societies that exist due to lifestyle and demographics.

Scientific research efforts to develop and employ capabilities to counter biological and chemical threat agents and weapons must address the basic tenets of environmental toxicology and focus upon all environmental compartments, including the air, biological organisms, land, and water. The relevance of the relationship between exposure, dose, and effect, and how toxicants may move through or be retained in the environment is critical to identifying and characterizing the hazards associated with both biological and chemical threat agents.

1.1.2   CRITICAL TERMINOLOGY USED IN THIS BOOK

Throughout the chapters of this book there are critical terms and phrases that are used, many of which may have different specific meanings to scientists from different disciplines. Because of the highly multidisciplinary composition of the authors of the present text, every effort has been made to standardize usage and meaning of such terms and phrases, and a glossary of terminology and phrases is provided following the final chapter. A few critical concepts essential to the understanding of this topic are explained in this chapter.

The terms *chemical weapon* and *biological weapon* are often used collectively in reference to a chemical compound or substance, or a pathogenic organism or toxin derived from a living organism that has been enhanced or modified for use as a weapon to cause morbidity or mortality in a population; whether this agent is specific to humans, animals, or plants is dependent upon the objectives of the user. The enhancement or "weaponization" of the biological pathogen, toxin, or chemical substance may be by means to improve its ease of delivery, longevity in the environment in which it is delivered, toxicological or disease-causing effects upon the intended target population, or speed of action once within the intended target population. Biological weapons may be either living organisms that infect the victim and cause disease (such as bacteria and viruses) or specific toxins derived from bacteria, viruses, and other naturally occurring organisms.

Biological or *chemical terrorism* can be defined as the threat of, or intentional release or delivery of such agents for the purpose of influencing the conduct of a government, or intimidating or coercing a civilian population, which is an expansion on the definition used by the U.S. General Accounting Office (GAO 2001). The term *"international terrorism"* means activities that involve violent acts or acts dangerous to human life that are a violation of the criminal laws of the United States or of any state, or that would be a criminal violation if committed within the jurisdiction of the United States or of any state. The term *"domestic terrorism"* means activities that involve acts dangerous to human life that are a violation of the criminal laws of the United States or of any state and appear to be intended (1) to intimidate or coerce a civilian population; (2) to influence the policy of a government by intimidation or coercion; or (3) to affect the conduct of a government by mass destruction, assassination, or kidnapping; and (4) occur primarily within the territorial jurisdiction of the United States (U.S. Code 18).

The actual use of biological or chemical agents by terrorists to cause disease or debilitate a population can be either overt or covert. The overt use of a biological or chemical weapon, particularly a chemical agent, is an immediately recognizable incident, either through the delivery method (e.g., explosion, motor vehicle, etc.) or the near-immediate physiological effects on the targeted population. Most military use of chemical weapons can be characterized as overt, is typically tactical in scope, and is focused on immobilizing, debilitating, or killing victims within a specific building, location, or area of the battle space. An excellent discussion of the comparative and theoretical differences in the delivery and intended impacts of biological and chemical weapons, both covert and overt, is provided in the first two chapters of Falkenrath and others (2001). Covert use of either weapon type, but more especially biological agents, may be intended to accomplish more strategic objectives. Because of the self-perpetuating capabilities and delayed morbidity or mortality following exposure or infection, most disease-causing organisms used as biological weapons, particularly the zoonoses, can be delivered upon a target population without risk of immediate detection. Strategic objectives that may be sought through the covert delivery of a biological weapon might include the disruption of food production, processing, or delivery, and the disruption of industry or the economy through worker absenteeism.

Only moderate technical skills are required to develop or improvise effective delivery equipment for the covert use of either chemical or biological weapons. An assessment of a population's vulnerability to attack with biological or chemical weapons must consider the potential use of any delivery method, not just highly technical and use-specific delivery systems. An example of diverse and simple low-technology means of agent delivery is the use of the U.S. Postal Service to deliver letters and parcels contaminated with anthrax spores (*Bacillus anthracis*) during October 2001. Further discussion of types of biological and chemical weapons, their deployment and potential effects are provided in Chapter 2. Directly compared, there is a greater likelihood of the surreptitious release of a biological agent than for a chemical agent, for unlike chemical agents, which are often more acute or immediate in their effects, biological pathogen agents must invade and replicate within the host animal or plant tissues before pathology and clinical symptoms of infection present themselves (MacIntyre et al. 2000).

Information available in this chapter will provide a brief overview of the history of biological and chemical weapons and their use for terrorism, briefly discuss the technical aspects characteristic of currently recognized biological and chemical threat agents, and relate the importance of ongoing and needed multidisciplinary research programs to address countermeasures to biological and chemical threats to both military and civilian elements of Western society, economic viability, and political stability.

1.2 HISTORY OF BIOLOGICAL AND CHEMICAL AGENTS AS WEAPONS

It is important for the reader to understand that the use of biological and chemical threat agents against humans and their interests, including crops, livestock, and wildlife, is not a new or novel concept. Numerous references to the use of biological pathogens, toxins, and chemical agents as weapons can be found throughout written history. The modality of these weapons has not significantly changed, but technologies to enhance their effectiveness and capacity to exploit those modalities have improved. The actual delivery and resultant human morbidity or mortality resulting from the use of biological and chemical weapons is only one aspect of their effectiveness. The psychological aspects such as fear and terror produced within a population and society at large that result from the threat of their use can be just as effective, if not more so. References to the use of biological and chemical agents as weapons reach back into the earliest annals of recorded history. Although not an exhaustive or all-inclusive listing of the historic use of biological or chemical weapons, we present an overview of those specific incidents that represent the wide spectrum of technology utilized in the production and delivery or dissemination of such weapons.

Perhaps one of the earliest reported and most simple uses of a biological agent in warfare is from the 6th century BC, when Assyrian armies used a toxin derived from ergot-infected rye to poison the water wells of besieged enemies.* Another unique and innovative use of biological agents is reported from around 400 BC, in which it was the practice of Scythian archers to dip the heads of their arrows into vats of bacteria-rich human excrement and decomposing corpses (Smart 1997). Although not necessarily a distinct and effective biological weapon by modern standards, the bacterial contamination to the wound caused by such an arrow and highly probable secondary infection was most likely very effective in causing increased (however delayed) morbidity and mortality in their enemies. Effective chemical weapons were believed to have been used in warfare as early as 429 BC during the Peloponnesian War. Using hollowed-out wooden beams, Spartan forces and their allies directed smoke from a burning concoction of sulfur and pitch into the Athenians' fort—disabling the defenders with an effective choking agent (Thucydides 431 BC). The tactics used by the Spartans bear a striking resemblance to those of Sadam Hussein's Iraqi military use of chemical agents such as tabun and mustard gas during the Iran-Iraq War and against the Kurds throughout the 1980s nearly 2,500 years later (DOD 1996; Zilinskas 1997).

Warfare and weapon technologies and tactics advanced significantly by the Middle Ages, but the frailty and susceptibility to disease of the warriors had not much improved. During the long siege of Kaffa† by the Tartars, squalid and desperate conditions led to an outbreak of plague (*Yersinia pestis*) among the Tartar forces in 1346 (Deaux 1969; Gottfried 1983; Marks 1971). With death claiming a large portion of the army, the Tartars catapulted corpses of those who succumbed to the disease over the city walls into the Genoese defenders. This caused great terror among the city's defenders who, in an attempt to escape infection, fled by ship back to Genoa and took the plague back to southwestern Europe. Much speculation and discussion on

---

* Ergot is a fungal (*Claviceps* spp.) disease of cereal grains, including rye, from which various water-soluble toxins can be derived that, when ingested, cause abdominal cramps, spasms, and a form of gangrene.

† Kaffa was a Greek colony built on the site of the ancient city of Theodosia and is currently the Black Sea port city of Feodosiya. Genoese traders assumed control in early 13th century and developed Kaffa into a major Black Sea point of commerce. Tartar forces under Mongol control ultimately conquered the city and drove out the Genoese in 1475.

whether the catapulted, infected corpses were truly an effective means of infecting the defenders has been exhaustive; some argue that the fleas would have detached from the bodies prior to being catapulted and thus infection from the corpses could not have occurred. Perhaps dogs and rats fed upon the corpses, became infected, and thus infected the fleas that fed upon them. Subsequently, those rats and fleas then boarded the ships, where the fleas then fed upon the crowded and fleeing Italians (thus completing the zoonotic disease cycle), who then transported the plague and contributed to the establishment of a second epidemic focal point of the Black Death pandemic in Europe. The original focal point of the Black Death epidemic that decimated Europe is believed to have been Constantinople.

Russians under the leadership of Peter the Great exploited the same "biological pathogen" delivery methods that the Tartars used, catapulting of plague-infected corpses, against the Swedes during the Great Northern War (1700–1721). After a long and severe Russian winter, the plague-devastated and severely weakened Swedish army under the leadership of Charles XII was soundly defeated at Poltava in July of 1709 (Smart 1997).

Smallpox has been used throughout history as a very effective biological weapon agent. It is suspected that Francisco Pizarro (circa 1470–1541), in his conquest of Peru, presented the immuno-naive natives blankets and clothing contaminated with the smallpox virus—thus causing a widespread smallpox epidemic and decimating their defenses. A later and controversial suspected use of smallpox as a biological weapon agent occurred during the French and Indian War (1754–1767). English forces were frustrated and suffering extensive losses to the guerilla tactics of the Indians during Pontiac's Rebellion in New England. After trying numerous unorthodox approaches, English forces reportedly distributed blankets soiled with the exudates, excreta, and vomit from smallpox victims at the English Fort Pitt to Indians loyal to the French. An epidemic of smallpox ensued and Fort Carillon fell to the English soldiers (Smart 1997). Numerous historical documents exist to support these stories, including correspondence between Governor General Jefferey Amherst and his field commander Colonel Henry Bouquet that were discovered as part of the British Manuscript Project, 1941–1945, undertaken by the U.S. Library of Congress during World War II. In a letter from Colonel Bouquet to General Amherst, dated July 13, 1763, he suggests the distribution of blankets to inoculate the Indians with the disease: "I will try to innoculate the Indians by means of blankets that may fall into their hands, taking care [illegible] not to get the disease myself."* In General Amherst's reply dated July 16, 1763, he approves Colonel Bouquet's suggested method and encourages him to do whatever is necessary to gain victory: "You will do well to try to innoculate the Indians by means of blanketts, so well as to try every other method that can serve to extirpate this execrable race. I should be very glad your [illegible] scheme for hunting them down by dogs could take effect."†

---

* See letter from Colonel Henry Bouquet to General Jefferey Amherst, dated July 13, 1763, at http://www.nativeweb.org/pages/legal/amherst/34_40_305_fn.jpeg.

† See letter from General Amherst to Colonel Bouquet, dated July 16, 1763, at http://www.nativeweb.org/pages/legal/amherst/34_41_114_fn.jpeg. Also see Francis Parkman, The Conspiracy of Pontiac and the Indian War after the Conquest of Canada (Boston: Little, Brown, 1886).

During World War I, German agents (including Captain Erich von Steinmentz), disguised as women, illegally entered the United States to inoculate horses, mules, and cattle with anthrax and glanders prior to their shipment to France to support the war effort (Smart 1997). The arrival of the 20th century not only brought more covert usage of biological pathogens but also welcomed the large-scale production, stockpiling, and overt use of chemical weapons on the battlefield. The first large-scale use of chemical weapons on the modern battlefield occurred on April 15, 1915, near Ypres, Belgium. Approximately 150 tons of chlorine gas was released from 6,000 cylinders upwind of Allied forces, killing 800 and debilitating 15,000. Although very simplistic in the delivery and dissemination technologies used (gas cylinders and wind), the attack was very effective both physically and psychologically. German forces once again tested new chemical weapon technologies on July 12, 1917, again near Ypres, Belgium, when artillery units delivered sulfur mustard via artillery shells onto Allied infantry and caused more than 20,000 casualties (Smart 1997).

Immediately after witnessing and suffering the horrors of an estimated 530,000–1,300,000 casualties resulting from the use of approximately 125,000 tons of chemical weapons during World War I (Legro 1995), the international community moved to outlaw the use of such weapons through the Geneva protocol of 1925.* The protocol was initially signed by only 38 nations but has since been signed by more than 130 nations. Neither the United States nor Japan was an initial signatory, but eventually the United States did conditionally ratify the protocol in 1975.

During World War I, World War II, and throughout the Cold War, vast quantities of chemical weapons were produced and stockpiled by the Soviet Union and the United States; however, very few were actually employed during World War II. It is estimated that as much as 181,000 metric tons of chemical weapons were produced and stockpiled in the Soviet Union during this period, while some 27,000 metric tons were stockpiled in the United States (Falkenrath et al. 2001; U.S. Office of Technology Assessment [OTA] 1993). The most notable use of biological weapons during wartime, at least to any significant scale, occurred in the 1930s and 1940s. The Japanese Imperial Army established Unit 731 in Beiyinhe, Manchuria, in 1932 to research and manufacture biological warfare agents, including anthrax, glanders, and plague. A full account of Unit 731's activities and the overall efforts of the Japanese army to research, build, and employ biological weapons during WWII can be found in books by Harris (1994) and Williams and Wallace (1989). The facility was moved to Ping Fan in 1937 and large-scale biological weapon production and delivery research was conducted. Various pathogens and delivery methods were refined, but one stands out as an excellent example of a simple method for delivering biological agents. In 1937 Japanese military airplanes dropped plague-infected fleas, some contained in porcelain bombs and some loose, as well as *Yersinia pestis*–saturated rice onto Chinese and Soviet villages, which ultimately caused significant plague outbreaks among civilians and military personnel.

---

* The 1925 Geneva Protocol was actually the League of Nations' *Protocol for the Prohibition of the Use in War of Asphyxiating, Poisonous or Other Gases, and of Bacteriological Methods of Warfare.*

In 1942 the U.S. military began research into the offensive use of biological weapons in response to a perceived German biowarfare threat (U.S. biological weapon efforts were located at Camp Detrick, Maryland). The program was terminated in 1969 by President Richard M. Nixon, and the stockpiles of biological weapons were destroyed in 1971 and 1972. Also in 1972, the Convention on the Prohibition of the Development, Production, and Stockpiling of Bacteriological and Toxin Weapons and Their Destruction (The Biological Weapons Conventions) was signed.* Although the United States used significant amounts of various herbicidal defoliants to gain visual access to enemy actions and supply routes throughout Southeast Asia during the Vietnam War, the use of these defoliants was not targeted at humans. Nevertheless, human exposure to one compound in particular, "Agent Orange" (2,4-D and 2,4,5-T) and dioxin contaminants, has been shown to have devastating long-term health consequences.†

Throughout the second half of the 20th century, particularly after the Biological Weapons Conventions, the use of biological and chemical agents as weapons of war, at least on a large scale, has been limited. However, there have been numerous incidents of biological or chemical agents being used in limited and focused attacks against individuals or small groups. Historical trends related specifically to events involving the use of biological agents ($n = 415$) have been empirically reviewed and were classified according to three general types: terrorist events, criminal events, and state-sponsored assassinations (Tucker 1999). In that article, Tucker concludes that although the historical records may lead to the belief that future incidents of bioterrorism will likely involve hoaxes and relatively small-scale events, the ability to utilize dual-use technologies for the production of bioterrorism agents coupled with the availability of scientists formerly employed by the Soviet Union have actually *increased* the potential for mass casualty terrorism.

One relatively recent incident involving the use of a biological agent on a community scale very clearly fits the definition of bioterrorism and is an excellent demonstration of the difficulty associated with recognizing a covert bioterror attack. In an attempt to sway a countywide election by inhibiting the ability of voters to reach polling stations, members of a cult following of Baghwan Sri Rajneesh contaminated the salad bars of four different restaurants in the Dalles, Oregon, area with *Salmonella typhimurium* in 1984. More than 750 people suffered the ill effects of the exposure, but the knowledge that it was an act of bioterrorism was not revealed until

---

* There were 140 states that ratified the Biological Weapons Convention as of May 1997 according to the Arms Control and Disarmament Agency (ACDA Fact Sheet, May 3, 1997). See also Convention on the Prohibition of the Development, Production, and Stockpiling of Bacteriological (Biological) and Toxin Weapons and on Their Destruction, April 10, 1972, 26 U.S.T. 583, 1015 U.N.T.S. 163. There are now 160 nations that have signed and 143 nations that have ratified the conventions.

† Numerous references are available regarding the use and health effects of herbicides used in Southeast Asia during the Vietnam War, most notably: *Veterans and Agent Orange: Health Effects of Herbicides Used in Vietnam,* Institute of Medicine, National Academy of Sciences, National Academy Press, Washington, D.C. 1994; *Comparison of Serum Levels of 2,3,7,8-Tetrachlorodibenzo-p-Dioxin with Indirect Estimates of Agent Orange Exposure among Vietnam Veterans, Final Report,* Centers for Disease Control, 1989, Atlanta, GA; *Oversite Review of the CDC's Agent Orange Study,* U.S. Congress, Hearing before the House Committee on Government Operations, July 11, 1989.

almost 2 years later when a cult member being tried on unrelated charges confessed to the 1984 attacks. A full report describing this community-focused, politically driven act of biological terrorism is provided by Torok et al. (1997).

A more elaborate and deadly attack, this time a chemical terrorism attack by members of the Aum Shinrikyo, a Japanese apocalyptic cult, was carried out in the Tokyo subway system in March 1995 (Olson 1999). It was suspected by international intelligence agencies that Aum Shinrikyo was working to develop biological and chemical weapons, but not until they killed 12 and severely injured thousands more by releasing sarin gas were they taken seriously.

There are numerous excellent sources of additional information and details regarding the historical use and impact of both biological and chemical weapon agents, in warfare as well as for terrorism, available to the reader through the Internet, particularly the U.S. Army's Textbooks of Military Medicine entitled *Medical Aspects of Chemical and Biological Warfare* (Sidell et al. 1997).

History can be an excellent source of information for planning strategies and developing methods and technologies to prevent and respond to terrorist threats, but we must not limit our assessment of potential threats and countermeasure strategies and technologies to only addressing a repeat of historical events. The historical record regarding the use of biological or chemical weapons by terrorist groups as weapons of mass destruction, particularly by domestic and non-state-sponsored groups and rogue fanatical religious or apocalyptic groups, suggests that technological, organizational, and logistical restraints limit the threat they pose on a national scale to the United States. However, a statement before the House Subcommittee on National Security, Veterans Affairs, and International Relations by John V. Parachini (senior associate, Center for Nonproliferation Studies, Monterey Institute of International Studies) in October 1999, in response to the U.S. Government Accounting Office's report on the threat posed by terrorists' use of biological and chemical threat agents, very clearly identifies the changing threat scenario posed by international terrorist groups such as Osama bin Laden's al-Qa'ida and others with superior organizational structure, near limitless monetary and technological resources, and worldwide reach. However, information gained during the global war on terror regarding the intricate and complex organization and technologies available to groups such as al-Qa'ida and nation-states that indirectly support and sponsor them suggests that those restraints may no longer exist—at least, not to an extent upon which we can rely.

1.3 GENERAL FOCUS AND INTENDED TOPICS

The following section provides a brief description of the intended knowledge and concepts to be conveyed through the various chapters of this textbook. The research efforts and successes resulting from the Admiral Elmo R. Zumwalt, Jr. National Program for Countermeasures to Biological and Chemical Threats (Zumwalt Program) have focused, since its inception, upon four critical areas: (1) modeling, simulation, and visualization of threats; (2) strategies for environmental protection from chemical and biological threats; (3) personal protection and therapeutics; and (4) mechanistic and toxic effects of biological and chemical weapons. These topic areas were used in developing the specific chapter topics used in this book.

Chapter 2 provides an extensive discussion of the threats and vulnerabilities associated with the employment and effects of biological and chemical threat agents by terrorists. The chapter strives to educate the reader on the relationships among risk (potential for exposure), vulnerability (weakness or situation predisposing one to exposure) and threat as they relate to effectively responding to and countering such an attack: Vulnerability + Risk = Threat.

Chapter 3 focuses upon the research findings and technical advances in the modeling, simulation, and visualization of how biological and chemical threat agents disperse and move through the environment and structures. Chapter 4 reports on the state-of-the-science related to the strategies and approaches for assessment necessary for environmental protection from biological and chemical threat agents. Chapter 5 discusses the important mechanistic and toxic effects of chemical weapons on humans. Chapter 6 provides an extensive overview of the challenges faced and successes accomplished in the field of sensor development to detect and identify biological and chemical threats in the environment. Chapter 7 reports on recent advances and remaining opportunities for research in the area of phage display and its applications for the detection and therapeutic intervention of biological threat agent exposures, *in vivo*. Chapter 8 summarizes the need for, and research-based advances in, the development of personal protective capabilities against chemical threats. Chapter 9 provides an overview of recognized biological threat agents and their mechanisms of effect, and summarizes advances and accomplishments of related research. And finally, Chapter 10 offers significant conclusions of the scientists involved in the Zumwalt Program, specific areas identified as needing further research, and how their current and future multidisciplinary research findings may contribute to countering biological and chemical threats.

1.4 CONCLUSIONS

Although the challenges Western civilization now faces both at home and abroad as a result of this global war on terrorism are numerous and daunting, as we have discussed here, the concept of biological and chemical warfare is not new. However, the technologies associated with these tactics and the vulnerabilities inherent in modern Western society have changed immensely. As we recognize and assess potential vulnerabilities that are common within Western societies, such as unrestricted movement and travel within continental borders, food production and distribution technologies and methods, communication systems and electronic essentials, as well as medicine and public health services, it is critical that we design and implement scientific research programs to effectively address and counter the threats. Research and development programs specifically designed to address these threats must integrate multidisciplinary expertise and high levels of experience, and maintain research momentum to ensure that there exists the capability to successfully counter future biological and chemical threats. As we have learned from history, strategic advances gained through applied scientific research will ultimately ensure victory in the war against terrorism.

REFERENCES

Deaux, G., 1969. *The Black Death 1347*, Weybright and Talley, New York, pp. 1, 2, 43–49.

Falkenrath R.A., Newman, R.D., and Thayer, B.A., 2001. *America's Achilles' Heel: Nuclear, Biological and Chemical Terrorism and Covert Attack*, MIT Press, Cambridge, MA, pp. 16–18, 27–164.

Gottfried, R.S., 1983. *The Black Death*, Free Press, New York, p. 35.

Harris, S.H., 1994. *Factories of Death: Japanese Secret Biological Warfare, 1932–45, and the American Cover-Up*, Routledge, NY.

Iklé, F.C., 1997. Naked to our enemies, *Wall Street Journal*, March 10, p. A18.

Legro, J., 1995. *Cooperation under Fire: Anglo-German Restraint during World War II*, Cornell University Press, Ithaca, NY, p. 145.

MacIntyre, A.G., Christopher, G.W., Eitzen, Jr, E., Gum, R., Weir, S., DeAtley, C., Tonat, K., and Barbera, J.A., 2000. Weapons of mass destruction events with contaminated casualties: effective planning for health care facilities, *JAMA*, 283(2), pp. 242–249.

Marks, G., 1971. *The Medieval Plague: The Black Death of the Middle Ages*, Doubleday, New York, pp. 1–5, 29, 45–49.

Olson, K., 1999. Aum Shinrikyo: once and future threat? *Emerg. Infect. Dis.* 5(4), pp. 513–516.

Sidell, F.R., Takafuji, E.T., and Franz, D.R., Eds., 1997. *Textbooks of Military Medicine: Medical Aspects of Chemical and Biological Warfare*, Borden Institute, Walter Reed Army Medical Center, Washington, DC, available at *http://www.bordeninstitute.army. mil/published_volumes/chemBio/chembio.html* (accessed August 15, 2007).

Smart, J.K., 1997. History of chemical and biological warfare: an American perspective, in *Medical Aspects of Chemical and Biological Warfare*, F.R. Sidell, E.T. Takafuji, and D.R. Franz, Eds., Borden Institute, Walter Reed Army Medical Center, Washington, DC.

Stern, J., 1999. The prospect of domestic bioterrorism, *Emerg. Infect. Dis.*, 5(4), special issue.

Thucydides, ~431 BC, *The History of the Peloponnesian War*, Richard Crawley, Trans., available at *http://classics.mit.edu/Thucydides/pelopwar.html* (accessed August 15, 2001).

Torok, T.J., Tauxe, R.V., Wise, R.P., et al., 1997. A large community outbreak of salmonellosis caused by intentional contamination of restaurant salad bars. *JAMA* 278, pp. 389–395.

Tucker, J.B., 1999. Historical trends related to bioterrorism: an empirical analysis. *Emerg. Infect. Dis.*, 5(4), pp. 498–504.

U.S. Code, Title 18, chaps. 2331 and 2332.

U.S. Department of Defense (DoD), 1996. *Impact and implications of chemical weapons use in the Iran-Iraq War*, report from the director of Central Intelligence (July 2, 2007), available at WarGulfLINK, http://www.fas.org/irp/gulf/cia/960702/72566_01.htm (accessed August 13, 2007).

U.S. General Accounting Office (GAO), 2001. *Bioterrorism: Federal Research and Preparedness Activities*, report to U.S. Congressional Committees, September 28, 2001.

U.S. Office of Technology Assessment (OTA), 1993. *Proliferation of Weapons of Mass Destruction: Assessing the Risks*, U.S. Government Printing Office, Washington, DC, p. 83.

Williams, P. and Wallace, D., 1989. *Unit 731: The Japanese Army's Secret of Secrets*, Hodder and Stoughton, London.

Zilinskas, R.A., 1997. Iraq's biological weapons: the past as future? *JAMA* 278(5), p. 418.

2 Threats and Vulnerabilities Associated with Biological and Chemical Terrorism

*Steven M. Presley, Galen P. Austin,
Philip N. Smith, and Ronald J. Kendall*

CONTENTS

2.1 INTRODUCTION

To provide a better understanding and appreciation for the complexities and challenges associated with countering the effective use of biological and chemical weapons against human populations and our interests, it is necessary to clearly define what biological and chemical weapon agents are, how they may be used, how to accurately detect them, and how to protect ourselves against them. The terms *biological threat agents* and *chemical threat agents* may be collectively used as *bio-chem threat agents* and refer to biological organisms or compounds and chemical compounds and substances that have been identified as having a significant potential pathogenic or toxic use against humans, agricultural resources, and other elements of infrastructure. It is necessary to define some of the terminology associated with this topic to provide the reader with a clearer understanding of how biological and chemical threat agents may be employed against humans and our interests. *Terrorism* can best be defined, for the purposes of this book, as the calculated unlawful use of violence or threat of violence to inculcate fear as a means to coerce or to intimidate governments, societies, the civilian population, or any segment thereof, in the pursuit of goals that are generally political, religious, or ideological.*

There are a significant number of toxic and hazardous chemicals, disease-causing pathogens, and toxins that have been enhanced or modified to make their deleterious impact on living organisms more efficient and effective. These enhancements or modifications may include genomic or proteomic engineered changes of naturally occurring microorganism, or changing the physical properties of a chemical to improve volatility or environmental persistence. Thus, enhancing or "weaponizing" biological agents allows them to more effectively gain entry into the targeted organism through inhalation, ingestion, injection, or transdermally. The various means associated with delivering chemical weapon agents represent the full physical spectrum, including aerosols, gases, liquids, solids, and vapors. Biological and chemical weapon agents can be employed by exploiting several primary routes of entry or means of exposure for the targeted population. Routes of entry include inhalation and absorption by any part of the respiratory tract, ingestion and absorption through the gastrointestinal tract, or absorption through the skin, eyes, and mucous membranes. It is necessary to first understand what types of weapons are available and how those weapons can be employed, to be able to correctly identify and effectively assess the areas or elements of our society that make us vulnerable to biological and chemical terrorism.

A more thorough discussion of the unique characteristics and pathogenic or toxic effects associated with various biological and chemical threat agents is provided in a subsequent section of this chapter. However, a brief definition of biological and chemical threat agents, as compared to biological and chemical weapons, and other related terms are presented here. *Biological threat agents* are microorganisms or toxins produced by living organisms that may be intentionally employed to cause morbidity or mortality in other living organisms and include bacteria, mycotoxins, rickettsia, toxins, and viruses. *Biological weapons* are microorganisms, or toxins produced by living organisms that have been enhanced or modified to more effectively and efficiently cause morbidity or mortality in other living organisms and include bacteria, mycotoxins, rickettsia, toxins, and viruses, or to be more effectively delivered. Therefore, biological terrorism, or *bioterrorism*, can be defined as the calculated use of microorganisms, or toxins produced by living organisms, that may have been enhanced or modified to more effectively and efficiently cause disease in, debilitate, or kill other living organisms in an attempt to intimidate or coerce a government, the civilian population, or any segment thereof, in furtherance of political, religious, or social objectives. *Chemical threat agents* are compounds or substances, either produced naturally or synthetically, that can cause significant morbidity or mortality in humans, as well as other organisms. *Chemical weapons* are compounds or substances, either produced naturally or synthetically, that have been designed or modified to maximize exposure through delivery methodologies to cause significant morbidity or mortality in humans, as well as other organisms. These compounds or

---

* This definition is an integration of terminology provided in U.S. Department of Defense Joint Publication 1–02 and U.S. Federal Bureau of Investigation publication, *Terrorism in the U.S.*, 1995.

substances are generally classified as blood agents, choking agents, nerve agents, psychotic agents, and vesicants. Therefore, *chemical terrorism*, or the use of chemical threat agents or weapon agents for terrorism, can be further defined as the calculated use of hazardous toxic compounds or substances that may have been enhanced or modified to more effectively and efficiently debilitate or kill humans in an attempt to intimidate or coerce a government, a civilian population, or any segment thereof, in furtherance of political, religious, or social objectives. Throughout this book, terminology such as biological and chemical terrorism/threat/weapon agents will be used in collective reference to the various agents and threats associated with both biological and chemical pathogenic and toxic substances.

Employment of biological and chemical threat agents, particularly for the purpose of terrorism against civilian populations as well as against targets within the agricultural animal and food production industries, can be further differentiated based upon the actual intent and objectives of the perpetrators. We will broadly categorize two distinctly different employment strategies as being *overt* or *covert* operations or attacks. Overt attacks are high profile, dramatic, and relatively focused upon a specific geographic site or area, or population—more consistent with achieving tactical objectives. The desired result of an overt attack is to cause immediate terror and to illicit an immediate response by emergency responders and law enforcement authorities. A covert attack or operation will be much more strategic in its ultimate purpose. Covert attacks are, by definition, low profile and clandestine, and target a larger area or multiple foci. As a result of its clandestine nature, and particularly with biological weapon agents, the morbidity and mortality resulting from a covert attack may be more widespread and extensive before it is realized or recognized as an attack than in the case of an overt attack. As a result of the delayed recognition of the covert attack and identification of the specific biological or chemical agent(s) used, significant delays in response time and increased efficacy/effectiveness of the attack may be manifested.

2.2 THREATS

2.2.1 BIOLOGICAL THREATS

pestilences methodically prepared and deliberately launched upon man and beast ...
Blight to destroy crops, Anthrax to slay horses and cattle.

—**Sir Winston Churchill, 1925***

Infectious diseases have unquestionably played a significant and defining role in the overall progression of mankind, from the founding and evolution of civil societies, religions, and cultures to the structure and organization of economies and governments. Microscopic organisms and the misery they may cause have been critical factors in defining and influencing human endeavors, likely as critical as sustenance, weather, and warfare have been on mankind's ability to live and thrive. Just as with

---

* From a speech in which Mr. Churchill opined regarding the potential use of biological weapons for economic or strategic goals in future warfare (Harris and Paxman 1982).

any of the critical influences on mankind's success and survival, throughout history and until today, there are individuals and governments that have tried to harness infectious diseases to further their specific objectives. Their objectives may be motivated by greed for power, for ideological, political, or religious domination, or for militaristic conquest.

For purposes of this book, it is necessary for the reader to understand that in the strictest application, biological terrorism is not biological warfare. *Biological warfare* is limited to the employment of biological weapon agents during warfare and combat operations between organized militaries representing sovereign nation-states, and utilizes these weapon agents to erode or deplete the enemy's ability to effectively continue combat operations. This erosion or depletion of combat operational capabilities may be accomplished by inflicting debilitating morbidity among operating forces, interruption of sustenance replenishment and water supplies through contamination, or through massive mortality at the tactical level. Additionally, the erosion or depletion of a military's ability to effectively conduct combat operations may be accomplished at a more strategic level by attacking its most critical vulnerability, its will to fight. The willingness of a nation to continue support for its military in a war can be clearly diminished by massive morbidity and mortality resulting from the horrific effects of biological weapons—which is, incidentally, a principal strategy for terrorism.

For purposes of this discussion, biological weapon agents are categorized as either pathogens or toxins. Pathogens are living microorganisms that can infect and cause disease in other living organisms, including humans, animals, and plants. Bacteria, fungi, rickettsiae, and viruses have been, or potentially can be, used as biological weapon agents. Toxins are substances that are produced as metabolic by-products of microorganisms, plants, or animals, or are synthetically produced, and can be used to poison other living organisms. There are at least 14 disease-causing pathogens that are believed to have been weaponized and could be potentially employed as terror weapons. An excellent complete listing of traditional biological weapon agents and biological agents that have been associated with biological terrorism or biological crimes is provided by Kortepeter and Parker (1999). Table 2.1 provides a listing of select known pathogens and toxins that are considered to be likely biological terrorism agents.

Bacterial pathogens that can be potentially used as biological threat or weapon agents are typically 1–2 μm diameter by 2–10 μm long and are most effective when inhaled, initiating infection through the lungs, or ingested into the gastrointestinal tract. Bacteria other than *Ricksettsia* and *Coxiella* can reproduce outside of living cells, but once inside the victim can rapidly reproduce, overwhelm the victim's immune defenses, and cause serious illness. Most of the naturally occurring bacterial pathogens are susceptible to modern antibiotics and can be controlled medically; however, it is known that several of the bacterial pathogens have been "weaponized" through bioengineering, including a strain of antibiotic-resistant, vaccine-subverting plague (*Yersinia pestis*) and antibiotic resistant tularemia (*Francisella tularensis*) (Alibek and Handelman 1999). Rickettsia are a unique genera of bacteria in that they must multiply within living cells and most cause zoonoses capable of being transmitted by arthropod vectors. Zoonoses are diseases that normally exist and

TABLE 2.1
Biological Pathogens Identified by U.S. Centers for Disease Control and Prevention as Likely to Have Been Weaponized and Likely to Be Used as Biological Terrorism Threat Agents

Disease-Causing Agent	Threat Category[a] (Class of Microorganism)	Natural Occurrence as Anthroponosis, Zoonosis, or Vector-Borne
Bacillus anthracis	A (bacterium)	Zoonosis—may be vector-borne
Brucella (6 strains)	B (bacterium)	Zoonosis—may be vector-borne
Burkholderia mallei	B (bacterium)	Zoonosis
Coxiella burnetti (Q fever)	B (bacterium)	Zoonosis—may be vector-borne
Francisella tularensis	A (bacterium)	Zoonosis—may be vector-borne
Rickettsia prowazekii	B (bacterium)	Zoonosis—primarily vector-borne
Yersinia pestis	A (bacterium)	Zoonosis—primarily vector-borne
California grp encephalitides	B (virus)	Zoonosis—primarily vector-borne
Crimean-Congo hemorrhagic fever	C (virus)	Zoonosis—may be vector-borne
Dengue fever	A (virus)	Zoonosis—primarily vector-borne
Eastern equine encephalitis	B (virus)	Zoonosis—primarily vector-borne
Ebola	A (virus)	Anthroponosis
Hantavirus pulmonary syndrome	A (virus)	Zoonosis—may be vector-borne
Lassa fever	A (virus)	Anthroponosis
Marburg virus	A (virus)	Anthroponosis
Rift Valley fever	A (virus)	Zoonosis—may be vector-borne
South American hemorrhagic fevers	A (virus)	Zoonosis—primarily vector-borne
Variola major (smallpox)	A (virus)	Anthroponosis
Venezuelan equine encephalitis	B (virus)	Zoonosis—primarily vector-borne
Western equine encephalitis	B (virus)	Zoonosis—primarily vector-borne
West Nile virus	B (virus)	Zoonosis—primarily vector-borne
Yellow Fever	C (virus)	Zoonosis—primarily vector-borne

[a] Category A includes high-priority agents that pose a risk to national security because they can be easily disseminated or transmitted from person to person; result in high mortality rates and have the potential for major public health impact; might cause public panic and social disruption; and require special action for public health preparedness. Category B includes the second-highest priority agents, specifically those that are moderately easy to disseminate, may result in moderate morbidity rates and low mortality rates, and require specific enhancements of CDC's diagnostic capacity and enhanced disease surveillance. Category C is the third-highest priority agents and includes emerging pathogens that could be engineered for mass dissemination in the future because of their availability, ease of production and dissemination; and potential for high morbidity and mortality rates and major public health impact.

are maintained in cycles among wild animals but may be transmitted to humans that come into contact with infected animals or, in many instances the ectoparasites associated with the infected animals. Therefore, an array of biological threat agents, including those involving endemic pathogens, could be used as biological weapon agents (Ashford 2003).

Some of the most horrific, virulent, and rapidly infectious biological threat agents are viruses, particularly the viral hemorrhagic fevers. Viral hemorrhagic fever infections may lead to the breakdown of the vascular walls, inhibit blood clotting mechanisms, and allow blood to leak from capillaries and pool throughout the body and beneath the skin. Viruses must reproduce in living cells, taking over individual cells and causing them to produce more viral particles instead of normal cellular components. Viral pathogens are most infective when exposure is through the respiratory tract, gastrointestinal tract, or mucous membranes.

Biological toxins typically are of lesser molecular weight and size, and are thus more soluble and more easily penetrate the skin than chemical weapon agents. Biological toxins can be categorized based upon their mode of action, such as neurotoxins (disrupt nerve impulses) and cytotoxins (disrupt cell respiration and metabolism). Known biological toxins that are warfare or terror agents include aflatoxin, botulinum toxins, ricin, and T2 mycotoxin.

The threat of the use of biological threat agents as terror weapons is not limited to those affecting human health but includes diseases of food crops and livestock. Agriculture-related terrorism is a real and present threat to a country or region's food supplies and economic stability (Khan et al. 2001). Diseases of crops that are considered biological threat agents include rice blast (*Magnaporthe grisea*), rye stem rust (*Puccinia graminis* forma specialis *avenae*), and wheat stem rust (*Puccinia graminis* forma specialis *tritici*). Numerous diseases of livestock and wildlife that are of major concern, including foot-and-mouth disease (*Rhinovirus* group), glanders (*Burkholderia mallei*), louping-ill (*Flavivirus*), melioidosis (*Burkholderia pseudomallei*), rinderpest (*Paramyxoviridae* family), and vesicular stomatitis (*Rhabdovirus* group).

As a further expansion on the targets of bioterrorism, we can define attacks against agricultural food and fiber production capabilities and resources as agricultural terrorism, or *agroterrorism*. Depending upon the scale of the attack and the scope of its effects, agroterrorism may be more strategic than tactical in its purpose, particularly when the weapon or agent used is biological. This can best be illustrated through a scenario in which an attack is perpetrated by infecting a herd of beef cattle with a highly communicable infectious agent such as foot-and-mouth disease virus (*Rhinovirus*). Foot-and-mouth disease (FMD) is highly infectious in cattle and many other domestic and wild animals, and a thorough description of the disease is provided in the *Merck Veterinary Manual* (2005). The infected, asymptomatic animals are then transported to a large regional livestock auction facility during the pathogen's intrinsic incubation period (typically 2–5 days). These infected cattle are mingled with hundreds or thousands of other cattle at the auction facility, thereby potentially exposing and infecting all animals at the facility. Foot-and-mouth disease-infected cattle, both those initially infected and those infected at the auction facility, are then sold and transported to various feedlots or ranches throughout the region, thereby further disseminating the pathogen. These foot-and-mouth disease-infected cattle would begin to be symptomatic, triggering a massive quarantine and depopulation effort by local, regional, and federal agencies. Such an incident would cost large amounts of money for the immediate response, as well as the longer term costs of depopulating infected and suspected cattle herds throughout the region, and would potentially significantly affect the entire cattle industry by interrupting interstate and international trade. Covert introduction of a biological agent into a large population of animals dispersed over a wide geographic area and the resulting long-range devastating economic impacts would truly have strategic implications for any nation-state.

Additionally, there are several significant differences that should be noted between naturally occurring disease outbreaks or epidemics and those that have been intentionally initiated and caused by bioterrorism. The most significant difference across all incidents is the motivation of the perpetrator, which defines the type of biological attack employed, either covert or overt, and the targeted population of humans or agricultural commodities—either morbidity and mortality of humans or economic disruption resulting from contamination of agricultural commodities. Depending upon the motivation and intent of the perpetrator, extensive fast-acting disease pathologies of humans or livestock may be necessary and may be counter to the intent. As very succinctly stated by Dr. David Franz, director of the National Agricultural Biosecurity Center at Kansas State University, "While bioterrorism is about killing humans, agroterrorism is not necessarily about killing cows. It's more an economic assault on our national security and infrastructure."* Economic stability and strength are ultimately the critical determinants of our national security. Franz believes the United States should prepare most specifically for those pathogens that could do the most economic damage to our nation while preparing more broadly for other biological or chemical terrorism threats.

Many of the disease-causing microorganisms that have been identified as likely biological threat agents that may be employed in biological terrorism are zoonoses and occur within natural foci in the United States. Table 2.1 provides a list of pathogenic microorganisms and toxins identified by the U.S. Centers for Disease Control and Prevention (CDC) as likely to be used as biological terrorism agents. The pathogens are categorized by their respective level of threat to human populations, and whether they occur naturally as zoonoses is indicated.

2.2.1.1 Toxins

Toxins that have been identified by the CDC as being likely for use as biological terrorism agents are derived from the following species: *Clostridium botulinum* (category A); *Clostridium perfringens* (category B); *Ricinus communis* (category B); and *Staphylococcus enterotoxin B* (category B). The use of toxins has a replete history for both biological crimes and as biological warfare weapons. Specific, detailed examples of historic incidents involving toxins are provided in Chapter 1. An excellent illustration of the simplicity of delivery and effectiveness of employing toxins as a weapon was that of the Assyrian army's use of a toxin derived from ergot-infected rye to poison the water wells of besieged enemies during the 6th century BCE. Also, more recently, in 1984 in Dalles, Oregon, members of a cult following of the Baghwan Sri Rajneesh contaminated salad bars in four different restaurants with *Salmonella typhimurium*. This attack caused illness in approximately 700 people and was not revealed until 2 years later.

---

* Presentation to the U.S. House Committee on Agroterrorism, Manhattan, Kansas, May 5, 2007.

2.2.1.2 Emerging and Resurgent Diseases

Emerging diseases are generally identified as recognized disease-causing pathogens that are detected in new host species or in new geographic areas, or a disease that was not previously known. Many of the emerging diseases of concern today also have the potential for use as biothreat agents. Additionally there is significant reason for concern related to the increasing number of genetically engineered biothreat agents known to exist or believed to be in development. Military as well as other sources of intelligence gathering have revealed that weaponization of pathogens have included the development of antibiotic-resistant strains of anthrax, glanders, plague, and tularemia. Antibiotic-resistant strains of anthrax, plague, and tularemia are known to exist naturally and may be exploited for weapons (USAMRIID 2005). Additionally, reports have indicated that the former Soviet Union developed encapsulated dual pathogens in which a synthetic hemorrhagic virus was encased inside the plague bacterium (personal communication), as well as recombinant smallpox and Ebola viruses.

There are approximately 1,415 species of infectious agents known to cause human disease, and include species of bacteria, fungi, helminths, prions, protozoa, rickettsia, and viruses. Of those infectious disease-causing agents, 61% (868/1,415) exist naturally as zoonoses, and 13% (175/1,415) of them are considered to be emerging. Of those pathogens identified as emerging diseases, 75% (132/175) are zoonoses. Of the emerging zoonoses, 28% (37/132) are naturally transmitted by vectors (primarily by arthropods) (Cleaveland et al. 2001; Brown 2004). Table 2.2 provides a listing of human and animal diseases that are considered emerging or resurgent.

As reported in Table 2.1, a majority of the known biological threat agents have been derived from pathogens that exist naturally as zoonoses. Zoonoses may be attractive to terrorists for many reasons, chief among them being the potential for zoonoses to become self-perpetuating after introduction through natural host-vector life cycles. Other reasons for their attractiveness to terrorists may be that there are many strains or isolates available in nature, the biology of many are relatively well understood, there are established animal models for virulence testing, there are animal models for phenotypic manipulation, and animals can be used as production vessels. Finally, there would or could be plausible reasons to possess and work with such agents.

2.2.2 CHEMICAL THREATS

Chemical weapons agents are classified based upon their mode of action and include blood agents, choking agents, nerve agents, psychotomimetic agents, and vesicants or blister agents. The following information on chemical weapon agents is compiled from various sources, including the Agency for Toxic Substances and Disease Registry (ATSDR),* Centers for Disease Control and Prevention (CDC),† relevant DoD Field Manual (FM 8–285/NAVMED P-5041/AFM 160–11 1990), National Library

* ATSDR ToxFAQs: Nerve Agents (GA, GB, GD, VX) Web site http://www.atsdr.cdc.gov/tfactsd4.html.
† CDC Emergency Preparedness and Response Web site http://www.bt.cdc.gov/Agent/agentlistchem.asp.

TABLE 2.2
Emerging and Resurgent Diseases of Humans

Disease	Potential as Bioterrorism Agent (None/Low/Moderate/High—CDC Category)	Naturally Exists as Zoonosis (Yes/No)	Transmission by Arthropod Vectors (Yes/No)
Argentine and Bolivian hemorrhagic fevers	High—A	Yes	Yes
Campylobacter	None	Yes	No
Cholera	None	No	(Mechanically)
Creutzfeldt-Jakob disease	None	Yes	No
Crimean-Congo hemorrhagic fever	Low—C	Yes	Yes
Cyclosporiasis	None	Yes	No
Dengue fever	High—A	No	Yes
Diarrhea (viral)	None	No	No
Ebola and Marburg hemorrhagic fevers	High—A	?	?
E. coli O157:H7	Moderate—B	No	(Mechanically)
Drug-resistant gonorrhea	None	No	No
Grp. B & C rotaviruses	None	No	No
Hantavirus Pulmonary Syndrome	Low	Yes	Yes
Hepatitis C, D, and E	None	No	No
HIV1 & HIV2 (AIDS)	None	No	No
Influenza	? None ?	Yes	No
Japanese encephalitis	Moderate	Yes	Yes
Lassa fever	High—A	No	No
Legionaire's disease	None	No	No
Nipah encephalitis	Low—C	Yes	No
Pertusis (whooping cough)	None	No	No
Pneumococcal disease	None	No	No
Polio	None	No	No
Rabies	None	Yes	No
Adenoviral respiratory disease	? None ?	No	No
Rift Valley fever	Low—C	Yes	Yes
Roseola	None	No	No
Ross River virus	None	Yes	Yes
Salmonella	Moderate—B	Yes	(Mechanically)
Scrub typhus	? None ?	Yes	Yes
Staphylococcous aureus	None	No	No
Toxic shock syndrome	None	No	No
Tuberculosis	None	Maybe	No
Venezuelan equine encephalitis	Moderate—B	Yes	Yes
Venezuelan hemorrhagic fever	High—A	Yes	Yes

of Medicine Specialized Information Services (NLM-SIS),* and the U.S. Army Soldier and Biological Chemical Command (SBCCOM).†

Blood agents, or cyanogens, include *arsine* (SA), *cyanogen chloride* (CK), and *hydrogen chloride* (AC). These chemical weapon agents are highly volatile and very fast-acting toxicants, causing seizures, respiratory failure, cardiac arrest, and death.

Choking agents, or lung-damaging agents, cause direct damage to the respiratory tract, resulting in rapid and severe pulmonary edema, and death. Specific choking agents include *chlorine* (CL), *chloropicrin* (PS), *diphosgene* (DP), and *phosgene* (CG). The effects of choking agent exposure are highly dependent upon the dosage inhaled, and their effects may be delayed depending on the agent, but most will culminate in about 4 hours.

Nerve agents inhibit the ability of cholinesterase to hydrolyze acetylcholine that mediates neurotransmitter function in nerve impulses[1] and are thus also known as *anticholinesterase* agents. Chemical nerve agent weapons include *tabun* (GA)—dimethylphosphoramido-cyanidate; *sarin* (GB)—isopropyl methylphosphono-fluoridate; *soman* (GD)—pinacolyl methyl phosphonofluoridate; and *VX* [O-ethyl S-(2-diisopropylaminoethyl) methylphosphonothioate]. Tabun, sarin, and soman are generally referred to as the *G-series* agents and were developed and refined in the mid-1930s through the mid-1940s by the German government‡ for use as insecticides. Depending upon dose and route of exposure, the G-series agents are generally fast acting, typically within 1–10 minutes, with death occurring within 15 minutes. Tabun is a clear, colorless, and tasteless liquid with a slightly fruity odor, and symptoms of exposure appear relatively slowly if by way of dermal (skin) absorption, as compared to ingestion or inhalation routes of exposure. Sarin and soman are gaseous agents and their lethality is more significantly correlated with the environmental conditions during their release. Their vapors are slightly heavier than air, so the gas will remain relatively close to the ground. During wet and humid weather conditions, sarin degrades rapidly, but as the temperature increases, its lethal duration increases, despite the humidity. Doses that are potentially life threatening may be only slightly larger than those producing least effects. VX was developed by chemists in the United Kingdom while searching for new insecticides and proved to be extremely toxic to humans. The British shared the discovery with the U.S. Army in 1953. At that time, a systematic investigation of these new compounds was begun at Edgewood Proving Grounds, Maryland. The Army soon discovered that effects of VX were slower than those of the G-series agents, typically occurring in 4–42 hours, and were more persistent and much more toxic than the G-series agents.§

---

* NLM-SIS Chemical Warfare Agents Web site http://sis.nlm.nih.gov/Tox/ChemWar.html.
† SBCCOM Homeland Defense Information Products Web site http://www.sbccom.army.mil/services/edu/tabun.htm.
‡ Dr. Gerhard Schrader first noticed the effects of nerve agents on humans when he and his lab assistant began to experience shortness of breath and contraction of the pupils.
§ These compounds were designated V-series agents for "venomous." VX is an oily viscous liquid that is clear or amber colored, odorless, and tasteless. Symptoms of overexposure may occur within minutes or hours, depending upon the dose. Severe exposure symptoms progress to convulsions and respiratory failure.

Psychotomimetic or psychochemical agents are chemical compounds that cause symptoms similar to psychotic disorders, debilitating the victim through disorientation, confusion, and hallucination. These psychoactive chemical weapon agents can be classified into four general categories based upon their respective effects, including deliriants, depressants, psychedelics, and stimulants. The incapacitating effects of these psychoactive agents and their historic uses are discussed by Ketchum and Sidell (1997). Substances that are known to have been studied for their potential as psychotomimetic chemical weapons include amphetamines, barbiturates, cocaine, glycolic acid esters, 3-quinuclidinylbenzilate, phencyclidine, and *D*-lysergic acid diethylamide (LSD) (Ketchum and Sidell 1997; Organisation for the Prohibition of Chemical Weapons [OPCW] 2007). Psychotomimetic agents may be delivered by way of ingestion or inhalation; however, transdermal exposure through direct contact may also occur.

Blister agents or vesicants cause severe blistering of the skin, as well as damage to the eyes and mucous membranes, respiratory tract, and internal organs. This class of chemical weapon agent includes the arsenicals/Lewisites (L), phosgene oxime (CX), and sulfur mustards (HD, HN). Sulfur mustards were created in the 1800s and were first used on the battlefields of World War I by the German army in 1917. Symptoms of blister agent exposure typically present within 12–24 hours following exposure. Purified sulfur mustard in liquid form is colorless; normally it is used in an unpurified state in which it resembles a brown oily substance. Although the blister agents may remain a health hazard in the environment for an extended period of time, they are considered nonlethal by the U.S. Army; even so, complications from blister agent exposure can lead to death.

2.3   VULNERABILITIES

It is not difficult to identify elements of our individual lifestyles as well as within our society at large that could be exploited, making us vulnerable to terrorist attack, whether with conventional weapons such as explosives or with biological or chemical weapons. Vulnerabilities do not necessarily indicate a threat, but for a threat to be viable it must exploit an existing vulnerability of the intended target. Falkenrath and others (2001) define vulnerability as a situation of being open to harm, while a threat is the known or suspected presence of an entity with the ability, will, and motive to inflict harm.

Chemical weapon agents were used extensively on the battlefields of Europe during World War I, particularly gruesome agents such as choking agents (e.g., CL) and vesicant agents (e.g., L, HD, and HN). The horrors of such combat inflicted on the person and memories of veterans of that war, as well as upon civilization as a whole, were a pivotal point in modern warfare and ushered in the military use of nonconventional weapons of mass destruction. Subsequently, as the portability and potential covert use of weapons of mass destruction to attack civilian populations was realized and as efforts to create a nuclear weapon were underway, recognition of U.S. vulnerabilities to such weapons was presented to President Franklin D. Roosevelt by physicist Leo Szilard in 1945:

The United States has a very long coastline which will make it possible to smuggle in such bombs in peacetime and to carry them by truck into our cities. The long coastline, the structure of our society, and our very heterogeneous population may make an effective control of such "traffic" virtually impossible. ... So far it has not been possible to devise any methods which would enable us to detect hidden atomic bombs buried in the ground or otherwise efficiently protected against detection.*

Szilard's assessment of American vulnerability is just as applicable today as it was 60-plus years ago, particularly if you replace "atomic bombs" with biological or chemical terrorism weapons—though radiological threats may be another potential terrorist weapon.

There are limited capabilities available to us that can realistically and cost-efficiently be employed with regard to our geography-related vulnerabilities, the most promising being the further development and fielding of biological and chemical threat agent sensor systems. Even with the advantages of high-technology surveillance assets, effective monitoring of every inch of our borders and coastlines would require the institution of near-martial law and deploying a vast number of law enforcement personnel to patrol and respond to perceived transgressors. The potential for a devastating, covert attack by a terrorist group using less sophisticated "homemade or crudely grown" biological or chemical agents instead of high-performance military-grade weapon agents is not believed to be significantly diminished. An excellent discussion of the comparative potential for terrorist use and effect of less-sophisticated biological or chemical threat agents versus high-performance military weapons is provided by Falkenrath and others (2001). An individual or group of three or four people could successfully smuggle into the United States enough weapon-grade biological or chemical agents to effectively attack any city and cause thousands of casualties.

2.3.1 MODERN LIFESTYLE AND CULTURAL PRACTICES

The "structure of our society, and our very heterogeneous population" (see the quotation by Szilard above) remain as significant contributors to our national vulnerability to terrorism; when we consider the many changes that have occurred since 1945, our vulnerabilities have increased in specific areas. We are a much more mobile society with significantly more vehicles on much better roads. There are also much more affordable and faster options for air transportation available. Larger numbers of people congregate in public places, such as large indoor shopping malls, massive theater complexes with 12–18 viewing screens, large enclosed sporting arenas and coliseums capable of seating more than 70,000 people, and an increasing reliance on public mass transit in metropolitan areas. Additionally, America is a much more urban society than it was in 1945. The increased interaction among people as a result of larger numbers congregating in one place is clearly a significant vulnerability,

---

* Leo Szilard, "Atomic Bombs and the Postwar Position of the United States in the World," memorandum for President Franklin D. Roosevelt, March 1945; reprinted in M. Grodzins and E. Rabinowitch, eds., *The Atomic Age: Scientists in National and World Affairs, Articles from the* Bulletin of the Atomic Scientists 1945–1962 (Basic Books: New York, 1963), pp. 13–14.

but it may also improve our collective acquired immunity to naturally occurring contagious diseases.

Advances in aeronautic technologies, both commercial and private, have significantly shortened the amount of time necessary for travelers to accomplish transcontinental and transoceanic movement, while at the same time significantly increasing the volume of people that are transported. As a result of this increased volume of travelers, the numbers and sizes of commercial airports have also increased dramatically. Thus, at any given time there are large numbers of people arriving into airports and awaiting departure from airports throughout the world. These factors combine to create an opportunity for large numbers of people to be exposed to a biological pathogen simultaneously through the most simple of delivery methods—since one or more terrorists intentionally infected and infective with an airborne, pneumonic pathogen could arrive on various transcontinental flights. A recent example that illustrates the potential for such a scenario was the severe acute respiratory syndrome (SARS) epidemic of 2002. Vulnerability to airborne or direct contact-transmitted diseases is demonstrated annually; influenza kills approximately 36,000 people in the United States each year (CDC 2007). Even though the "flu" is a common and often expected seasonal disease in the United States, it significantly impacts the economy in lost workdays and strains medical facilities throughout the country.

As the global human population has increased significantly in the past 50 years, so has our need for food and water. Much more of the food supply is imported from other countries, processed and prepared outside of the home, and eaten in restaurants or quick-food establishments than in 1945. These factors make domestic and international agricultural industries, as well as foodstuffs importers and processors, a vulnerable target for biological or chemical attack.

2.3.2 ASSESSING THE VULNERABILITIES: HUMAN HEALTH AND AGRICULTURE

Vulnerabilities to bioterrorism at the individual level are increased due to the freedom to congregate and to travel uninhibited throughout most of the Western world. However, on a national scale in the United States, vulnerabilities associated with the economic element of our national power result from international trade and market access, as well as increasing international passenger travel. Open market access through free trade agreements such as Global Agreement on Tariffs and Trade (GATT) and the North American Free Trade Agreement (NAFTA) allows potentially contaminated and infected foodstuffs to enter the food supply. Checks and balances associated with the international agreements are somewhat limiting in their abilities to stop importation. Trade in animals and animal products is restricted if validated human or animal health risks to the importing country is identified. To stop trade, the importing country must show, with a scientifically valid analysis, that a risk exists—essentially a "horse is out of the barn" situation.

International passenger travel has been steadily increasing over the past few decades. For example, during 1980 there were approximately 20 million passengers who arrived in the United States via air travel, while in 1995 that number was up to approximately 47 million passengers—an increase of 131%. Current estimates are that worldwide, approximately 2 billion people travel by air annually. At any given

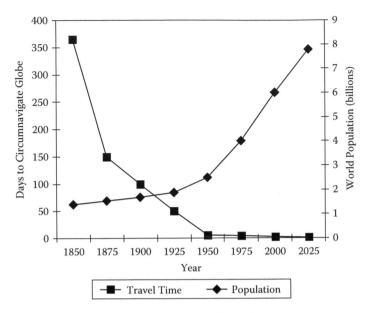

FIGURE 2.1 Relationship between estimated world population growth and amount of time required to circumnavigate the globe (similar to Murphy and Nathanson 1994, with current population estimates from U.S. Census Bureau 2007).

moment, roughly 500,000 people are in transit on airliners globally (Select Committee on Science and Technology [SCST] 2000). Figure 2.1 illustrates the inverse relationship between speed of global travel and world population growth. International travelers may unknowingly transport contaminated animal products from countries known to be infected with dangerous foreign animal diseases (FAD). Contaminated foodstuffs have often served as the source of a FAD in the United States and other countries (U.S. Animal Health Association [USAHA] 1998).

Infectious disease agents and vectors are changing. New disease-causing agents never before a threat to North American agriculture have become an important human health and economic concern. Since the mid-1990s, emerging and reemerging livestock disease outbreaks have cost world economies more than $80 billion, including diseases such as bovine spongiform encephalopathy (BSE), FMD, Avian influenza (H5N1), and others (Karesh et al. 2005). Additionally, the biological threat posed by the importation of live animals is significant. Data reported by the U.S. Fish & Wildlife Service in 2005 on live animal importation from 1997–2003 included: 5,752,168 amphibians; 449,294 birds; 210,947,027 fish; 1,594,415 reptiles; and 63,716 mammals (USFWS 2007).

2.3.3 TARGET-RICH ENVIRONMENT?

The U.S. agricultural industry is capable of clothing and feeding a majority of the world's population—approximately 2% of the population of the United States is capable of feeding and clothing the majority of the rest of the world's population. The U.S. food production and supply infrastructure is a massive and very complex system and accounts for one fifth of U.S. economic activity. There are approximately 2.1 million farms; roughly 600,000 restaurants and food service providers; 235,000 grocery stores and food outlets; more than 880,500 related firms; and more than 1,086,790 individual facilities. A critical contributor to the success and efficiency of this system can be attributed to the fact that it is almost entirely privately owned. However, private ownership also presents challenges to the prevention of tampering/contamination and biosecurity (U.S. Department of Homeland Security [DHS] 2006).

When we consider the multitudes of vulnerabilities to potential terrorists' use of biological and chemical weapons as discussed above, and the rapid expansion of novel production and delivery methodologies increasingly available to terrorists as a result of technology proliferation, it becomes apparent that effective prevention and protection strategies must be multidisciplinary and integrated at the local through national levels, and the research efforts to develop and field effective countermeasures must be interdisciplinary.

REFERENCES

Alibek, K. and Handelman, S., 1999. *Biohazard*, Random House, New York.

Ashford, D., 2003. Planning against biological terrorism: lessons from outbreak investigations. *Emerg. Infect. Dis.* 9(5), pp. 515–519.

Brown, C., 2004. Emerging zoonoses and pathogens of public health significance—an overview, *Scientific and Technical Review*, 23(2), pp. 435–442.

Cleaveland, S., Laurenson, M.K., and Taylor, M.H., 2001. Diseases of humans and their mammals: pathogen characteristics, host range and the risk of emergence, *Philos. Trans. R. Soc. London, B, Bio. Sci.* 356(1411), pp. 991–999.

Falkenrath, R.A., Newman, R.D., and Thayer, B.A., 2001. *America's Achilles' Heel*, MIT Press, Cambridge, MA.

FM 8–285/NAVMED P-5041/AFM 160–11, 1990. *Field Manual, Treatment of Chemical Agent Casualties and Conventional Military Chemical Injuries*, U.S. Departments of the Army, the Navy, and the Air Force.

Harris, R. and Paxman, J., 1982. *A Higher Form of Killing*, Noonday Press, New York, p. 70.

Karesh, W.B., Cook, R.A., Bennett, E.L., Newcomb, J., 2005. Wildlife trade and global disease emergence, *Emerg. Infect. Dis.* [serial on the Internet], available at http://www.cdc.gov/ncidod/EID/v0111n007/05–0194.htm.

Ketchum, J.S. and Sidell, F.R., 1997. Incapacitating agents, in *Medical Aspects of Chemical and Biological Warfare*, Sidell, F.R., Takafuji, E.T., and Franz, D.R., Eds., Borden Institute, Walter Reed Army Medical Center, Washington, D.C.

Khan, A., Swerdlow, D.L., and Juranek, D.D., 2001. Precautions against biological and chemical terrorism directed at food and water supplies. *Public Health Rep.* 116, pp. 3–14.

Kortepeter, M.G. and Parker, G.W., 1999. Potential biological weapons threats, *Emerg. Infect. Dis.*, special issue, 5(4), pp. 523–527.

Merck Veterinary Manual, 2005. Foot-and-mouth disease, 9th ed., Kahn, C.M. Ed., Merck, Rahway, NJ, pp. 507–510.

Murphy, F.A. and Nathanson, N., 1994. The emergence of new virus diseases: an overview, *Semin. Virol.* 5, pp. 87–102.

Organisation for the Prohibition of Chemical Weapons (OPCW), 2007. *Psychotomimetic Chemical Weapons Factsheet*, available at http://www.opcw.nl/resp/html/psyco.html (accessed September 3, 2007).

Select Committee on Science and Technology (SCST), 2000. *Fifth Report: Air Travel and Health*, United Kingdom Parliament.

U.S. Animal Health Association (USAHA), 1998. *The Gray Book: Critical Foreign Animal Diseases Issues for the 21st Century*, available at http://www.vet.uga.edu/vpp/gray_book02/index.php (accessed September 3, 2007).

U.S. Army Medical Research Institute of Infectious Diseases (USAMRIID), 2005. Medical Managmeent of Biological Casualties handbook, 6th ed., Kortepeter, M. et al. Eds., Fort Detrick, MD., avaiable at http://www.usamriid.army.mil/education/bluebookpdf/USAMRIID%20BlueBook%206th%20Edition%20-%205ep%202006.pdf (accessed September 3, 2007).

U.S. Census Bureau, 2007. *Historical Estimates of World Population*, available at http://www.census.gov/ipc/www/worldhis.html (accessed September 4, 2007).

U.S. Centers for Disease Control and Prevention (CDC), 2007. *Key Facts about Influenza and the Influenza Vaccine: Fact Sheet*, available at http://www.cdc.gov/flu/keyfacts.htm (accessed September 3, 2007).

U.S. Department of Homeland Security (DHS), 2006. *National Infrastructure Protection Plan*, available at http://www.dhs.gov/xlibrary/assets/NIPP_Plan.pdf.

U.S. Fish & Wildlife Service (USFWS), 2007. U.S. Wildlife Trade: An Overview for 1997–2003, Office of Law Enforcement Intelligence Unit, available at http://www.fws.gov/le/pdffiles/Wildlife%20Trade%20Overview%20Report.pdf (accessed January 15, 2008).

3 Predicting and Characterizing Threat Transport

Jeremy W. Leggoe, Chia-bo Chang,
Stephen B. Cox, Steven M. Presley,
Richard Zartman and Tom Gill

CONTENTS

3.1 INTRODUCTION

Once hazardous chemical and biological materials have been released, they can be transported through the environment via a variety of mechanisms and media. Although most releases will initially be transported via the atmosphere, the long-term effects of a release will in many instances depend on the subsequent transport of the hazardous materials through the soil, water, or biological populations. Understanding the transport mechanisms in each of these media is essential to the informed deployment of countermeasures in the event of a biological or chemical-threat agent release. Responders need to be aware of the potential extent of the region affected by the release event, the rate at which the material will disperse, the potential for long-term contamination of the affected area, and the potential for any delayed or long-term transport of the material.

Atmospheric transport represents the primary mode of long-range dispersion for hazardous materials. The study of atmospheric dispersion has long been a civilian concern, as scientists have sought to understand the movement of industrial pollutants through the environment. Many of the same principles apply to deliberate releases of hazardous materials, and much of the recent research in the field has been driven by the need to develop countermeasures to terrorist action. The discussion in this chapter focuses on the atmospheric transport of aerosols, as many of the materials of current concern would be released in aerosol form; it should be noted that the mechanisms of transport for gaseous materials in the atmosphere are very similar to those of aerosols.

Advection by the wind field is the dominant mode of transport for aerosols released into the atmosphere. Since atmospheric flows are turbulent, this causes the released material to disperse into a cloud, with the rate of dispersion being governed by the statistics of the turbulent wind field. This chapter first discusses the fundamental science of turbulent dispersion, leading to a description of the techniques used to determine the rates of dispersion and downwind transport. The implementation of these principles in the wide variety of computational models currently being developed to support emergency response operations is described. To illustrate the role that these models play, the operational procedure for developing model predictions using a Gaussian puff-based model is described, highlighting the issues faced by modelers and first responders during the early stages of a release event.

The dominance of advective transport means that the success of predictions of atmospheric dispersion is dependent upon the model being provided with an accurate characterization of the prevailing meteorological conditions. This characterization may be provided through data obtained via observations or, in the absence of sufficient observations, by a meteorological modeling package. The effect of the meteorological data source on dispersion predictions can be profound. This chapter describes the sources and management of observational data, with particular emphasis on the West Texas Mesonet operated by Texas Tech University. The meteorological modeling packages currently available are also described, leading to the presentation of the results of a set of case studies that illustrate the effect of model choices on dispersion predictions. The effect of using four-dimensional data assimilation (FDDA), in which live observational data are integrated to enhance the fidelity of forecast data, on dispersion predictions is discussed.

Particulate and dense gas materials released into the atmosphere will eventually be deposited on the ground surface, either via gravitational settling or precipitation. In addition to the problems caused by biological or chemical contamination, this can be a particular concern in releases of radiological matter. Once deposited, the material may be transported via resuspended blowing dust or through the soil itself via a variety of mechanisms, depending on the interaction between the deposited material and the water, air, and solid matter present within the soil. The models for the different modes of transport through soil are presented here, with emphasis on models accounting for the sorption and degradation of deposited materials within the soil.

For biological threats, environmental transport will result in the exposure of populations of various organisms to the threat. These populations can provide a mode for subsequent long-term transport of the threat. This is a particular concern for zoonoses, diseases that normally exist in wild animals but are transmissible to humans. The populations of interest may include host or reservoir species, such as birds, mammals, reptiles, or humans, and vector species, such as arthropods. This chapter accordingly considers the use of models to understand the interactions between biological threats that have become established in natural host populations and human populations. Approaches for placing these models in a spatially explicit context in order to predict transport are also discussed.

3.2 ATMOSPHERIC TRANSPORT OF AEROSOLS

3.2.1 INTRODUCTION

In deliberate (and in many accidental) releases of hazardous biological, chemical, and radiological species into the atmosphere, the species often enter the atmosphere in aerosol form. An aerosol can be defined as a suspension of solid or liquid particles in a gas (Hinds 1999). Particles in the micrometer size range typically provide optimal delivery properties for weaponized species, with regard to both their potential for long-range atmospheric transport and their potential for inhalation and alveolar deposition.

Common aerosol chemical weapons include the family of nerve agents: tabun, sarin, soman, GF (cyclohexyl methylphosphonoflouride), and VX (*O*-ethyl *S*-diisopropylaminomethyl methylphosphonothiolate). These agents have low vapor pressures and are thus predominantly liquids under normal atmospheric conditions (despite their common misnaming as nerve "gases"). These agents may be absorbed through the skin or by inhalation, leading to muscle paralysis or damage to the respiratory center of the central nervous system (Munro et al. 1999; Benschop and De Jong 1988; Sidell et al. 1997).

The family of "bioaerosols" includes viruses, bacteria, and fungal spores. Viruses, which tend to be on the order of 0.02–0.3 μm in size (Jacobson and Morris 1976), may be carried within a solution for distribution in aerosol (droplet) form. These species may then enter the human body through skin absorption or inhalation; indeed, aerosolization via the sneezing, coughing, or respiration of infected individuals is a common mode of disease spread (Hinds 1999). Bacteria and fungal spores, at 0.3–10 μm and 0.5–30 μm in size respectively (Jacobson and Morris 1976), are often large enough to be transported either singly or attached to carrier particles such as soils (Hinds 1999). *Bacillis anthracis* (anthrax), for example, can be weaponized by attaching spores to carrier particles 1–5 μm in diameter (Inglesby et al. 1999).

Radiological species can similarly be transported tremendous distances if released in aerosol form. Radioactive material released by the Chernobyl reactor in northern Ukraine in 1986 was deposited in dangerous amounts as far away as Europe, and convective weather patterns lifted some of the radioactive matter to altitudes that enabled it to reach the United States within 10 days (Ellis 2003). A particularly dense initial plume known as the "Western Trace" traveled over a hundred kilometers to the west of the plant in the initial wake of the catastrophic event (Chesser et al. 2004). To this day, a residential exclusion zone covering an area of 4500 km^2 is maintained around the reactor.

Although the spread of radioactive materials from reactor accidents has been a rare event, the potential for terrorist use of so-called dirty bombs is a current concern. Dirty bombs would use explosives to aerosolize and spread radioactive material and smoke, contaminating substantial areas. The effects of such weapons are expected to be primarily psychological and economic; although individuals would likely experience only limited radiation exposure as they evacuated the area, the contamination of the area might create significant economic and access issues in the long term (Zimmerman and Loeb 2004). Directing the evacuation and cleanup operations would require detailed modeling and mapping to characterize the dispersion of the aerosolized material.

The dominant transport mechanism for both aerosol and gaseous agents in the atmosphere is advection associated with the bulk motion of the atmosphere. Since airflows in the planetary boundary layer exhibit significant turbulence under most conditions (though turbulence may be suppressed under conditions of temperature inversion), this will cause aerosol releases to disperse into a plume or puff that expands as the distance traveled downwind from the source increases. A continuous release will give rise to a "plume," whose concentration profile can be characterized by a single quasi-steady-state concentration profile that represents an ensemble average of individual temporal realizations of the plume (Arya 1999). Chronic industrial-stack pollution is a classic example of a plume release. A quasi-instantaneous source, in which a fixed mass of aerosol is released over a short period, will give rise to a "puff." This type of source is typical of military or terrorist actions, as well as certain types of catastrophic industrial releases. A mixed Eulerian-Lagrangian approach is often taken to characterizing puff concentration profiles; the motion of the puff centroid is tracked relative to a fixed (Eulerian) coordinate system, while the evolution of the concentration profile is computed in a Lagrangian coordinate system based on the puff centroid.

The dependence of aerosol transport on advective processes means that dispersion is essentially governed by the prevailing atmospheric conditions, meaning that dispersion modeling must combine an understanding of turbulent diffusion processes with a thorough characterization of the prevailing meteorological conditions. Depending on the nature of the release, this characterization can encompass a wide range of length scales. For small releases and for chemical releases in which agent lethality is heavily dependent on concentration, microscale meteorological effects, such as the effects of local topography, buildings, and vegetation, can be critical. In contrast, for radiological events, mesoscale meteorology (covering tens to thousands of kilometers) and even global circulation patterns can become important. For most releases, accurate dispersion predictions will require the provision of time-dependent, three-dimensional (3D) meteorological data over a sufficient range to encompass the area likely to be exposed to toxic effects.

The objective of this section is to describe the fundamental science governing the transport of aerosols in the portion of the atmosphere lying within the planetary boundary layer (which typically extends up to 1000 m above the surface, depending on the time of day and prevailing conditions). Section 3.2.2 discusses those properties of the aerosols that govern their transport and the effects on exposed individuals. In Section 3.2.3, the fundamental processes governing turbulent dispersion in the atmosphere are discussed, examining both theory and experimental data derived from field trials. Section 3.2.4 presents a discussion of models for predicting turbulent dispersion, with detailed discussion of models that are widely used in current emergency management. A discussion of the field experimental programs providing data to be used in the development and validation of models is presented in Section 3.2.5. Finally, Section 3.2.6 presents an overview of the operational protocol for modeling to support emergency response, to highlight the issues faced by modelers and field personnel.

3.2.2 TRANSPORT AND INHALATION PROPERTIES OF AEROSOLS

Particles are transported in an atmospheric airflow via two mechanisms: diffusive transport, associated with the molecular motion of the medium, and advective transport, resulting from the bulk motion of the airflow. For particulate species, diffusive transport arises as a result of the (random) Brownian motion experienced by particles in response to the continual impacts of the molecules of the surrounding gas. For a gaseous agent, the inherent random thermal motion of the molecules of the agent provides the impetus for the diffusion process. In the presence of a concentration gradient, these random motions will result in a net transport of particles from regions of higher concentration to regions of lower concentration.

The origin of this net transport can be illustrated using the following idealized example. Consider a container initially separated into two halves by an impermeable barrier. The gas on the left of the barrier is rich in the particulate aerosol species, while the gas on the right contains no particulates. At time $t = 0$, the barrier is removed. It may be assumed that since the Brownian motion of the particles will be random, there is an equal probability of a particle moving to the left or right. At the newly opened interface between the two halves of the container, there are more particles on the left of the interface than there are right. Thus, there is a greater probability that a particle will move from left to right across the interface than there is of a particle moving from right to left. There must accordingly be a net transfer of particles from left to right (from the region of higher concentration to the region of lower concentration). The process will continue until a steady state is achieved, at which time the concentration of particles will be uniform throughout the container.

The length scale associated with diffusive transport can be characterized in terms of the root-mean-square (rms) Brownian displacement, which is given by (Hinds 1999):

$$x_{rms} = \sqrt{2Dt} \tag{3.2.1}$$

where D is the diffusion coefficient for the aerosol particle (in m^2/s) and t is the time interval. Table 3.1 lists the rms Brownian displacement experienced by spheres of standard density in a second. Even for the smallest diameters, the displacements are on the order of micrometers per second. Atmospheric airflow velocities, in contrast, are typically on the order of meters per second. Thus, while diffusive transport can be critical for short-range processes, such as the deposition of small particles (Hinds 1999), advective transport will dominate the atmospheric transport of aerosols.

TABLE 3.1

Net Displacement (over a Single Second) Due to Brownian Motion of Spheres of Standard Density

Particle Diameter (μm)	RMS Brownian Displacement (μm/s)
0.01	330
0.1	37
1.0	7.4
10.0	2.2

Adapted from Hinds, W.C., 1999, *Aerosol Technology: Properties, Behavior and Measurement of Airborne Particles*, John Wiley & Sons, New York.

TABLE 3.2
Equilibrium Settling Velocities for
Spheres of Standard Density

Particle Diameter (μm)	Settling Velocity (m/s)
0.001	6.9×10^{-9}
0.01	7.0×10^{-8}
0.1	8.8×10^{-7}
1	3.5×10^{-6}
10	3.1×10^{-3}
100	0.25
1000	3.86

The potential range of aerosol transport and the ease of deposition are strongly influenced by the gravitational settling rate. The equilibrium settling velocity of a sphere in a one-dimensional falling motion may be computed by determining the velocity at which the gravitational force exerted in the sphere is balanced by the buoyant and drag forces acting on a sphere. For Reynolds numbers less than 0.1, this may be calculated directly using Stokes' law; for higher Reynolds numbers (typically associated with larger and heavier particles), an iterative calculation is required due to nonlinearity in the drag coefficient at higher Reynolds numbers.

The equilibrium settling velocities for standard density (1000 kg/m^3) spheres under standard atmospheric conditions of 1 atmosphere and 20°C are presented in Table 3.2. It is apparent that a 1-μm particle would take approximately 8 h to fall 1 m; its settling motion would accordingly be insignificant compared to its advective transport by the bulk fluid motion. Particles smaller than 10 μm can accordingly be regarded as moving affinely with the fluid motion for the purposes of dispersion modeling. For larger particles, settling and inertial effects will become significant, and deposition must be accounted for in dispersion models. Indeed, it has been noted that nerve gases are sometimes deliberately deployed in larger droplet sizes to enhance contact deposition (Benschop and De Jong 1988).

Advective transport depends on the interaction between the particles and the airflow, manifested in the drag force exerted on the particle and the rate of gravitational settling. As such, it is governed by the size, density, and shape of the particle. Particle shape will influence advective transport by affecting the drag forces exerted on the particle. Liquid aerosols are confined to roughly spherical forms, but solid aerosols may take on a variety of complex shapes. Many bioaerosols exhibit extremely complex shapes; spores in particular have often evolved shapes designed to maximize their transport range. Radiological particles are often agglomerates of smaller particles, leading to the creation of complex particle forms (Allen et al. 1978). To compute settling rates and model the transport of such particles, an equivalent aerodynamic diameter may be defined, representing the diameter of a standard density sphere that would settle at the same rate as the actual particle (Hinds 1999).

Studies of aerosol inhalation have determined that the optimal particle size range for inhalation (and thus for weaponized species that affect victims primarily via inhalation) is in the range of 0.1–10 μm. Particles larger than around 10 μm will not reach the alveolar ducts but instead will be deposited on the surfaces of the upper airways. Particles smaller than 0.1 μm will not settle once they do reach the alveolar ducts. Particles in the size range of 2–5 μm have been found to most effectively penetrate the nasal and bronchial passages; penetration of these passages is important because the aerosol particles can lodge in the alveolar ducts, which lack the protective mucus layer found in the nasal and bronchial passages (Hinds 1999).

3.2.3 Dispersion of Aerosols in Atmospheric Airflows

The discussion in Section 3.2.2 provides a basis for the assumption that, for particles in the optimal size range for weaponized aerosols, advection is the dominant transport mechanism. Since airflows in the planetary boundary layer exhibit significant turbulence under most conditions, this will cause aerosol releases to disperse into a "cloud" (more rigorously denoted as a plume or puff, depending on the nature of the source) that expands as the distance traveled downwind from the source increases.

To understand atmospheric dispersion, therefore, it is necessary to understand the nature of turbulent flows and the structure of the atmosphere near the Earth's surface. Turbulent flows exhibit apparently random fluctuations in local velocity and pressure (Mathieu and Scott 2000). These fluctuations appear over a wide range of length and time scales. Reynolds (1895) drew an analogy between the behavior of the velocity in a turbulent flow and the velocity of the individual molecules in a gas, leading to the definition of the instantaneous velocity at a point as having a mean and a fluctuating component;

$$\mathbf{U} = \bar{\mathbf{U}} + \mathbf{u} \qquad (3.2.2)$$

The use of bold denotes that all quantities are vectors. The velocity U on the left-hand side is the instantaneous velocity, which is comprised of the sum of the mean velocity at the point (denoted by the overbar) and the fluctuating element of the velocity (denoted by the lowercase u). To permit the use of the Einstein summation convention, the component of the instantaneous velocity in direction i at a point is written as the sum of the mean and fluctuating components of the velocity in that direction,

$$U_i = \bar{U}_i + u_i \qquad (3.2.3)$$

where i takes a value of 1, 2, or 3 according to the coordinate direction under consideration.

It is the fluctuating element of the velocity in a turbulent flow that drives the dispersion process. The foundation for determining the rate of dispersion was set out in papers by G. I. Taylor, who first noted the ability of eddy motion in the atmosphere to diffuse matter in a manner analogous to molecular diffusion (though over much larger length scales) (Taylor 1915), and later identified the existence of a direct relation between the standard deviation in the displacement of a parcel of fluid (and thus any affinely transported particles) and the standard deviation of the velocity (which represents the root-mean-square value of the velocity fluctuations) (Taylor 1923). Roberts (1924) used the molecular diffusion analogy to derive concentration profiles for smoke plumes and puffs. The form of these profiles is essentially the same as the Gaussian profiles commonly used to describe puffs and plumes today.

If it is assumed that the velocities fluctuate randomly, a particle random walk model may be constructed to explain the Gaussian form of the concentration profiles in puffs and plumes (Csanady 1973). Consider the motion of a particle during a small time increment Δt, during which the particle will move distances Δx, Δy, and Δz in the three coordinate directions. Since the average velocity over each time increment in a homogeneous turbulent flow field varies in a random fashion, Δx, Δy, and Δz will also vary randomly from one time increment to the next. Thus the coordinates of any individual aerosol particle position at some point in time is the sum of a set of random numbers; by the central limit theorem, the distribution of positions of a large number of particles must therefore follow a Gaussian distribution in all three coordinate directions. In the years following Taylor's original work, several experiments confirmed that a Gaussian probability distribution described the displacement of a fluid particle at all times after the initial "release" (Batchelor 1949).

The assumption of a Gaussian concentration profile permits the construction of relatively straightforward equations to describe the concentration profile of puffs and plumes. For a continuous point source in a uniform flow having homogeneous turbulence, the plume concentration profile is given by (Arya 1999),

$$c(x_1, x_2, x_3) = \frac{Q}{2\pi \bar{U}_1 \sigma_2 \sigma_3} \exp\left(-\frac{x_2^2}{2\sigma_2^2} - \frac{x_3^2}{2\sigma_3^2}\right) \qquad (3.2.4)$$

where the x_1 direction coincides with the direction of the mean flow velocity $\bar{U}_1$, the x_2 direction is the lateral direction in the horizontal plane, and x_3 represents the vertical direction. Q is the emission rate of the source, and σ_2 and σ_3 are the Gaussian plume-dispersion parameters. Note that while the downstream coordinate x_1 does not directly appear in the equation, the dispersion parameters σ_2 and σ_3 are dependent on x_1, since the plume will expand as it travels downwind.

In practice, most release points are located near a bounding ground surface. If it is assumed that there is no deposition at the surface, a perfectly reflecting boundary condition can be imposed, and a "reflected Gaussian" profile can be derived (Arya 1999),

$$c(x_1, x_2, x_3) = \frac{Q}{2\pi \bar{U}_1 \sigma_2 \sigma_3} \exp\left(-\frac{x_2^2}{2\sigma_2^2}\right) \left\{ \exp\left(-\frac{(x_3 - H)^2}{2\sigma_3^2}\right) + \exp\left(-\frac{(x_3 + H)^2}{2\sigma_3^2}\right) \right\} \qquad (3.2.5)$$

where H is the height of the release point.

The concentration profile within a Gaussian puff is given by (Sykes and Henn 1995):

$$c(x) = \frac{Q}{(2\pi)^{3/2} [\det(\sigma)]^{1/2}} \exp\left(-\frac{1}{2}\sigma_{ij}^{-1} x_i' x_j'\right) \qquad (3.2.6)$$

where $c(x)$ is the concentration at point x, Q is now the total mass of agent released in the puff, x_i' is the coordinate of point x relative to the centroid of the puff (in other words, in the Lagrangian coordinate system based on the puff centroid), and σ is a tensor representing the moments of the Gaussian distribution. Note that summation is implied over repeated indices; and, in the absence of shear distortion, the σ_{ij} may be taken to be zero for $i \neq j$. For a puff whose centroid lies near a reflecting surface, and neglecting shear distortion, the above equation may be modified to yield (Arya 1999)

$$
c(x_1, x_2, x_3) = \frac{Q}{(2\pi)^{3/2} \sigma_{11} \sigma_{22} \sigma_{33}} \exp\left(-\frac{x_1^2}{2\sigma_{11}^2} - \frac{x_2^2}{2\sigma_{22}^2}\right)
$$

$$
\left\{\exp\left(-\frac{(x_3 - H)^2}{2\sigma_{33}^2}\right) + \exp\left(-\frac{(x_3 + H)^2}{2\sigma_{33}^2}\right)\right\}
$$

(3.2.7)

where H is now the height of the puff centroid.

To complete the computation of the concentration field of the puff in Eulerian coordinates, the position of the puff centroid must be updated based on the wind field velocity at the puff centroid; the entire (Lagrangian) puff is then assumed to translate affinely with the centroid. A puff-splitting algorithm may be used to overcome the inaccuracies that arise as the puff dimensions become sufficiently large that the approximation inherent in assuming constant wind velocities throughout the puff becomes invalid (Sykes and Henn 1995).

It is apparent from equations 3.2.4–3.2.7 that the determination of the concentration field is dependent on the values of the Gaussian dispersion parameters σ_i (or σ_{ij} in the fully coupled puff model). Drawing on the fundamental result provided by Taylor (1923), it would be expected that these parameters would relate directly to the statistics of the components of the fluctuating element of the flow velocity. In a neutral atmosphere, the factors affecting these components can be explored by considering the fundamental equations of fluid motion in an incompressible fluid (for airflows less than 70% of the speed of sound, airflows can reasonably be modeled as incompressible); when the temperature of the atmosphere varies with elevation, the fluid must be modeled as compressible (in other words, the density is treated as a variable). The set of equations governing the flow of an incompressible Newtonian fluid at any point at any instant is as follows:

Continuity Equation

$$
\frac{\partial U_i}{\partial x_i} = 0
$$

(3.2.8)

Momentum Equation

$$
\frac{\partial U_i}{\partial t} + U_j \frac{\partial U_i}{\partial U_j} = -\frac{1}{\rho} \frac{\partial P}{\partial x_i} + \frac{\mu}{\rho} \frac{\partial^2 U_i}{\partial x_j \partial x_j} + g_i
$$

(3.2.9)

where U_i is the instantaneous velocity in direction i, P is the pressure, μ is the absolute viscosity, ρ is the density of the fluid, and g_i is the component of gravity in direction i. Once again, summation is implied over repeated indices, which range from 1 to 3 (Equation 3.2.9 thus actually represents three equations, one for each coordinate direction).

In a turbulent flow, the pressure and velocity vary randomly in time and space. It can therefore be more meaningful to average the equations over time or space; for example, if the equations hold at every instant, they must also hold on average over some finite time period. It may be recalled that the instantaneous velocity and pressure may be written as the sum of their mean and fluctuating elements as follows:

$$U_i = \bar{U}_i + u_i \qquad P = \bar{P} + p \qquad \text{(3.2.10 a,b)}$$

Then the Reynolds-averaged Navier-Stokes (RANS) equations are given by

Continuity Equation

$$\frac{\partial \bar{U}_i}{\partial x_i} = 0 \qquad \text{(3.2.11)}$$

Momentum Equation

$$\frac{\partial \bar{U}_i}{\partial t} + \bar{U}_j \frac{\partial \bar{U}_i}{\partial x_j} = -\frac{1}{\rho}\frac{\partial \bar{P}}{\partial x_i} + \frac{\mu}{\rho}\frac{\partial^2 \bar{U}_i}{\partial x_j \partial x_j} + \frac{\partial}{\partial x_j}\left(\overline{u_i u_j}\right) + g_i \qquad \text{(3.2.12)}$$

The averaging process creates a new set of variables, the so-called Reynolds stresses, which are dependent on the averages of products of the velocity fluctuations $\overline{u_i u_j}$ (which for $i = j$ simply represent the standard deviations of the velocity components). This creates a closure problem, which is one of the fundamental issues that has to be addressed in the modeling of turbulent flows. Importantly, Equation 3.2.12 also indicates that the Reynolds stress terms, which in line with Taylor's fundamental result should be related to the dispersion parameters, are coupled to the gradients of the mean flow velocity.

Experimental evidence has indicated that Gaussian concentration profiles can be achieved by releases in the inhomogeneous turbulence typical of near-surface airflows (Barr and Clements 1984; Arya 1999). Recent experimental releases in simulated urban obstacle arrays have, however, found that while approximately Gaussian concentration profiles arose in the crosswind direction, the form of the vertical concentration profiles was more complex than could be predicted by a reflected Gaussian profile (Yee and Biltoft 2004). This is consistent with previous experimental observations, which has prompted some authors to propose the use of non-Gaussian profiles in the vertical direction. A general exponential form for the vertical profile has been proposed for continuous point releases from ground sources (Elliott 1961; Huang, 1979; Brown et al. 1997),

$$c(x) \propto \exp\left(-kx_3^\alpha\right) \qquad \text{(3.2.13)}$$

where k is a constant and the exponent α is dependent on the prevailing atmospheric conditions. Comparisons between the results predicted by equations using this type of vertical profile and wind tunnel experiments have identified reasonable agreement between the experimental and theoretical profiles for surface and elevated plumes, though there can be significant differences in the actual concentration values (Brown et al. 1992).

There are several factors that contribute to the difficulty of predicting the vertical transport of aerosols. Most obviously, the flow field is not homogeneous in the vertical direction. For neutral atmospheric stability conditions, the mean wind velocity follows an approximately logarithmic velocity profile, given by (Wieringa 1980)

$$\bar{U}_{x_3} = \bar{U}_{x_{3ref}} \frac{\ln\left(x_3/r\right)}{\ln\left(x_{3_{ref}}/r\right)} \tag{3.2.14}$$

where $\bar{U}_{x_3}$ is the mean velocity at the height of interest, x_3, $\bar{U}_{x_{3ref}}$ is the mean velocity at the reference height x_{3ref} (usually 10 m above ground), and r is the roughness length for the terrain upwind of the observation point. The profile that results for a wind velocity of 4.44 m/s (10 mph) at a height of 10 m above ground level and a roughness of 0.03 m (consistent with open grassland) is presented in Figure 3.1(a).

The effect of upwind terrain on the velocity profile is accounted for by the roughness length r. The effect of r is to shift the plane of zero velocity, and indeed the entire velocity profile, upwards. Roughness lengths have been compiled for a wide variety of terrain types (Davenport 1960; Wieringa 1980), with the magnitude of the roughness length increasing with the size and number of obstacles. In urban regions, with high densities of solid obstacles of significant height, a profile of the form presented schematically in Figure 3.1(b) develops. The velocity at heights up to the average height of the obstacle array is significantly reduced, as the flow tends to divert preferentially over the array. Above the array, a "displaced" profile of the usual form emerges as the roughness layer flow blends into the free atmospheric flow (MacDonald 2000; Brown et al. 2001).

The nonlinear increase in velocity with increasing elevation affects both the turbulence itself and the production of turbulence through the effect on the mean velocity gradients. The presence of surface obstacles, with the associated distortion of the flow field and the mechanical turbulence that they induce, also significantly affects vertical dispersion. It has also previously been found that although velocity fluctuations are effectively random in the horizontal plane, experimental data have indicated that the vertical fluctuations are not truly random under strongly convective conditions (Arya 1999). Perhaps most important, however, the near-surface atmosphere is not universally neutrally buoyant. Changes in the vertical temperature profile strongly influence vertical transport processes by inducing or suppressing the buoyant (or convective) motion of air near the planetary surface. Since this profile evolves continuously during the day, vertical transport processes will be heavily dependent on both the time of day and the prevailing atmospheric conditions.

In order to understand atmospheric dispersion processes, it is accordingly necessary to understand the structure and characteristics of the planetary boundary layer

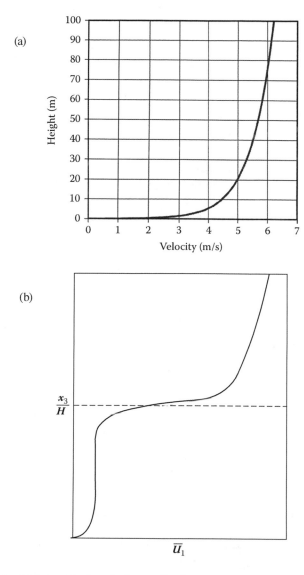

FIGURE 3.1 (a) Logarithmic velocity profile for a neutral atmospheric surface layer for a nominal wind speed is 10 mph (4.44 m/s) at a height of 10 m and a surface roughness length of 0.03 m. (b) Schematic illustration of the effect of an obstacle array in "displacing" the velocity profile upwards; H is the average height of the obstacles in the array (Based on results presented by MacDonald [2000] and Brown et al. [2001].)

(PBL), which forms the lowest layer of the atmosphere. The height of the PBL varies diurnally, ranging from as low as 50–100 m in the night or early morning, to on the order of thousands of meters (corresponding roughly to the height of the cloud base) during the mid to late afternoon; the exact heights will depend on the nature of the surface (land or water, local ground usage, etc.) (Arya 1999). The PBL often consists of a set of sublayers, depending on the nature of the local topography. Nearest the surface there will be a roughness sublayer, where the velocity is affected by the roughness elements (such as buildings or vegetation) present on the ground surface. The roughness sublayer lies within the surface layer, which may be described as a layer in which the atmosphere "adjusts" to the effects generated in the roughness sublayer. The remainder of the PBL is sometimes thought of as being the "free" outer layer, where surface processes become less influential (almost a far field region, unaffected by the bounding surface).

To understand the diurnal variation in PBL height and the mixing processes within the PBL, it is necessary to understand the temperature profile within the PBL and the associated buoyant transport of air and associated aerosol species. In order to account for the effect of declining pressure on the density of air parcels with increasing elevations (and thus obtain a realistic indication of the potential for buoyant motion), the "potential temperature" θ, representing the temperature a parcel of air would have if it were transferred adiabatically to a reference sea-level pressure of 1 bar, is given by (Arya 1999):

$$\theta = T\left(\frac{1}{P}\right)^k \tag{3.2.15}$$

where T is the actual temperature, P is the pressure (in bars) at the original elevation, and k is approximately 0.286. To account for the effect of the water vapor inherently present in the atmosphere, a "virtual" potential temperature θ_v may be defined (Arya 1999):

$$\theta_v = T_v\left(\frac{1}{P}\right)^k \tag{3.2.16}$$

$$T_v = T\left(1 + 0.61q\right) \tag{3.2.17}$$

where T_v is the virtual temperature and q is the specific humidity.

The stability of the atmosphere can be related directly to the "net" gradient of the virtual potential temperature in the vertical direction as follows:

- Stable—$\partial\theta_v/\partial x_3 > 0$: Buoyant motion is suppressed. Under stable conditions, vertical mixing will accordingly be suppressed.
- Neutral—$\partial\theta_v/\partial x_3 = 0$: Buoyant motion is negligible, but vertical mixing is not suppressed. Mixing will occur throughout the full depth of a neutral layer.
- Unstable—$\partial\theta_v/\partial x_3 < 0$: In an unstable layer, strong convective vertical motions (cool air falling from above with warm air rising from below) generate rapid vertical mixing and dispersion.

The height of the PBL, which is also sometimes known as the mixing depth (Arya 1999), is set by the presence of a stable layer that in effect "caps" the PBL, suppressing further vertical transport of dispersing species to greater elevations. In other words, the dispersing species will be essentially confined within the PBL by this cap. This cap may be broken under strongly convective conditions (such as during thunderstorm development) potentially leading to enhanced long-range transport of dispersing species, as was observed during the Chernobyl incident (Ellis 2003). Neglecting "cap breakthrough" could cause local concentrations to be overestimated by dispersion models, which may actually be preferable for the direction of initial emergency operations; for radiological events, or other events where deposition at long range may prove important, the potential for cap breakthrough will have to be taken into account.

Figure 3.2 schematically depicts the evolution of the virtual potential temperature profile. Overnight [Figure 3.2(a)], a near-surface temperature inversion develops as the ground surface cools that increases in depth toward morning. A neutral layer lies above the surface inversion, separating it from the higher inversion that normally caps the PBL. As the day proceeds, the surface inversion is eliminated as the ground surface heats up, first reaching a state where a neutral layer extends from ground level up to the inversion capping the PBL [Figure 3.2(b)]. By late afternoon, heating of the ground surface creates an unstable surface layer, creating a strongly convective environment within the PBL [Figure 3.2(c)]. Obviously, the prevailing weather conditions will influence the extent to which each of these processes occurs on any given day; for example, on days with heavy cloud cover, surface heating may be limited, preventing the formation of an unstable layer, while the presence of snow pack may enhance a surface temperature inversion.

The effect of temperature stratification on the atmosphere can be illustrated by considering the different forms a plume may assume. The form of a plume is determined essentially by the relation of the plume release point to any stable, neutral, and unstable layers that may be present; the basic forms have been summarized by Slade (1968) and Arya (1999). In a stable layer, vertical mixing of the plume will be limited, and the plume will fan out in the horizontal plane. If a plume is released into a neutral layer capped by a stable layer, the plume will mix vertically throughout the entire depth of the neutral layer. If a plume is released into a neutral layer

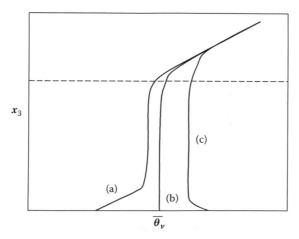

FIGURE 3.2 Diurnal evolution of the potential temperature profile near the surface. (a) Late night/early morning. (b) Late morning/mid-day. (c) Mid/late afternoon.

where there is no immediately adjacent stable layer, the plume will disperse evenly in the horizontal and vertical directions, forming a conical plume. In an unstable layer, the plume will follow a looping pattern, rising and falling as it encounters the up- and downdrafts that arise in a convective atmosphere. If a plume is released into a neutral layer above a stable layer (as may happen for elevated sources under night-time conditions), it will become lofted, dispersing in the neutral layer and remaining above the stable layer. Puffs can be expected to exhibit similar behavior to plumes, allowing, of course, for the translation of the puff along what would be the plume path for a continuous release.

In recognition of the dependence of dispersion on atmospheric structure, the 1960s saw the development of straightforward chart-based schemes for determining the evolution of the dispersion parameters with increasing distance downstream from a ground-level source. The Pasquill scheme (Pasquill 1961, 1974; Gifford 1961, 1976) defines six basic meteorological categories:

A—extremely unstable
B—moderately unstable
C—slightly unstable
D—neutral
E—slightly stable
F—moderately stable

A given set of conditions is assigned to a category according to the mean wind velocity, daytime insolation, or level of nighttime cloudiness in line with tables provided in the original references. For each category, a chart provides values for the horizontal and vertical dispersion parameters as a function of the distance downstream of the source; as would be expected, under unstable conditions, the vertical dispersion parameter rapidly becomes extremely large. These parameters may then be used in Equations 3.2.13 to 3.2.17 to compute plume and puff spread.

This straightforward approach to dispersion predictions, which can be used in the absence of detailed meteorological data, was particularly popular prior to the widespread availability of computing platforms, and was widely adopted for calculations by regulatory agencies (see also Turner 1970). A similar scheme was developed by Brookhaven National Laboratory (BNL) based on categories defined in terms of the range of fluctuations of the wind direction in the horizontal plane (as monitored using a wind direction trace), based on measurements of a release from a 108-meter-tall tower (Gifford 1976). The dispersion parameters could then be determined using a chart provided for each category (or a corresponding power-law fit to the chart). Cramer (1957) also proposed a method based on the fluctuation in wind direction, determining categories based on the standard deviations of the wind directions in the horizontal and vertical planes. Briggs (1974) developed an interpolation scheme to reconcile the effect of source height on downwind dispersion in calculating ground-level concentrations, which tends toward the Pasquill-Gifford and BNL schemes in the respective extremes (Gifford 1976).

As will be discussed in Section 3.2.4, as computational models have increased in popularity, sophisticated approaches have been developing for computing the values of dispersion parameters. At this point, it is important to understand the nature of the data provided by the fundamental Gaussian plume- and puff-based models. The Gaussian profile represents a probability distribution: if a large number of aerosol particles is released from a point, then a Gaussian profile provides the probability of a particle having a given lateral or vertical displacement after the time in question. As such, the Gaussian concentration profile represents the expected ensemble average of a sufficiently large number of instantaneous realizations of the puff or plume, *not* the actual concentration profile that will arise at any instant.

If a plume were being photographed, the ensemble average is the picture that would be obtained if the shutter were left open over a long period. An instantaneous snapshot (realization) of the plume, however, would follow a sinuous path, as the plume is transported by the turbulent eddies that happen to be passing through the region at the instant in question. The velocities within individual eddies are strongly correlated, so the path taken by the particles comprising the plume within a given eddy will also be correlated, giving rise to the sinuous plume observed at any instant. The reader may easily visualize the effect by observing the smoke plume from an extinguished match. Rather than a smooth plume of gradually declining density, a sharp, sinuous plume with regions of alternately very high and very low concentrations is observed. From a practical viewpoint, for a plume, a Gaussian concentration profile will provide valid estimates of the long-term dosage of the transported species at a particular point. To understand the range of instantaneous concentrations, the potential range of variation in local concentrations must be determined. This challenge is particularly important for puff releases, since a puff will transit through an area in a time that may be too short for ensemble averaging to become meaningful.

The need to account for potential variability becomes increasingly important in environments that contain significant obstacles to ground-level flow, such as buildings and vegetation. The effect of obstacles on ground-level flows is largely neglected in the dispersion models discussed thus far. Given that the majority of human exposures to hazardous species will occur at ground level, understanding the microscale meteorological phenomena associated with such obstacles is essential to understanding the true impact of releases in urban environments.

The characteristics of the flow field around isolated solid obstacles have been summarized by Hosker (1984). The flow field exhibits three key features:

- A displacement zone exists upwind of the body, where the incident fluid is slowed and redirected under the influence of pressure gradients.
- The boundary layers that form along the exposed surfaces of the obstacle separate from the obstacle at locations determined by geometry or aerodynamic effects. When geometry does not fix the locations (sharp edges being a key separation trigger), the locations tend to fluctuate with variations in the characteristics of the flow field. This is typically the case for rounded obstacles, where the separation point moves as the Reynolds number of the flow increases.
- The separated boundary layers extend downstream into the surrounding flow field as free shear layers. With increasing downstream distance, the separated layers reapproach the wake axis or the bounding surface, enclosing a

recirculation zone immediately downwind of the body. The recirculation zone is characterized by low-mean velocities, high-turbulence intensities, recirculation, and long residence times for entrained particles.

The recirculating wake zone is of particular importance for dispersion prediction. The dispersing species can become trapped in the wake region, remaining present at elevated concentrations well after a puff might be expected to have passed.

The difficulty of developing universal parameterizations to describe the effects of individual obstacles for use in dispersion models has been highlighted by research undertaken within the Zumwalt Program (Eastepp 2006). The study considered the flow fields associated with isolated individual vegetative canopies (trees). In the development of dispersion models for the urban environment, attention has generally concentrated on modeling the effect of buildings on "street-level" dispersion. Outside the "downtown" environment, however, vegetation elements often represent the dominant physical feature of the environment. For individual trees, it was often found that a conventional recirculating wake failed to form, as illustrated in Figure 3.3. The downstream velocity and turbulent kinetic energy (TKE) values for canopies without a recirculation zone were significantly reduced compared to upstream values. This was significantly different from the behavior observed for solid cylinders attached to a surface plane, for which distinct recirculation zones and very high turbulence levels are observed in the near wake. The flow fields associated with tree canopies must accordingly be regarded as being very different from those associated with solid obstacles of similar dimension.

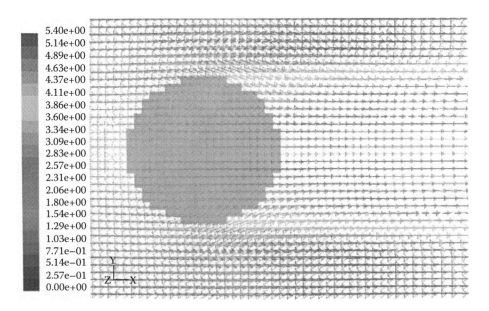

FIGURE 3.3   (See color insert following page 46.) Velocities in the wake region for a cylindrical vegetative canopy with a 5 m ground clearance immersed in a logarithmic upstream velocity profile with effective foliage element diameter = 0.001 m and a porosity of 0.99995.

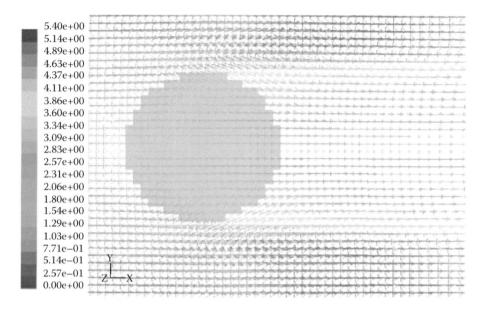

COLOR FIGURE 3.3 Velocities in the wake region for a cylindrical vegetative canopy with a 5 m ground clearance immersed in a logarithmic upstream velocity profile with effective foliage element diameter = 0.001 m and a porosity of 0.99995.

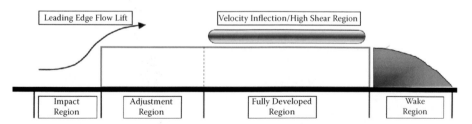

COLOR FIGURE 3.4 Schematic illustration of the evolution of a turbulent flow upon encountering an obstacle array.

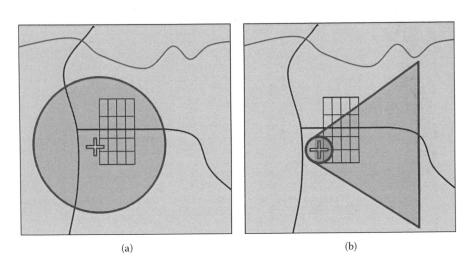

(a)   (b)

COLOR FIGURE 3.9 Schematic illustrations of ATP-45 estimates of the hazard zone (the release is located at the center of the cross). (a) Nominal wind speeds less than 5.8 mph (5 knots). (b) Nominal wind speeds greater than 5.8 mph (5 knots).

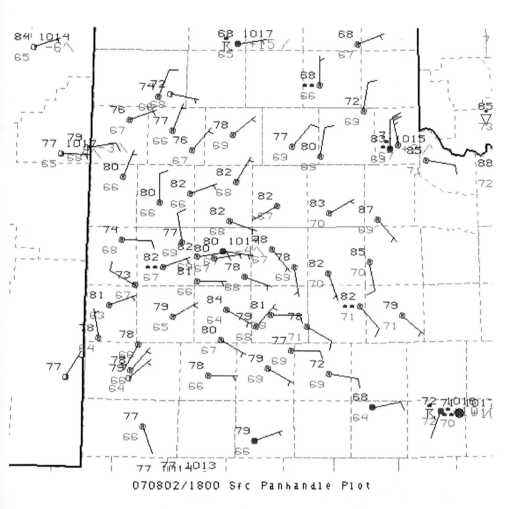

COLOR FIGURE 3.11 West Texas Mesonet site locations and identifiers as of March 2007.

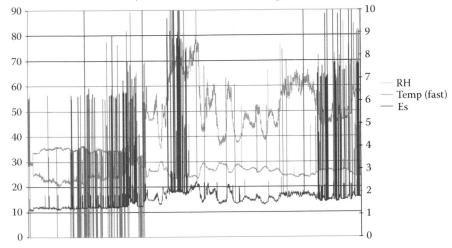

COLOR FIGURE 3.12   Raw data example.

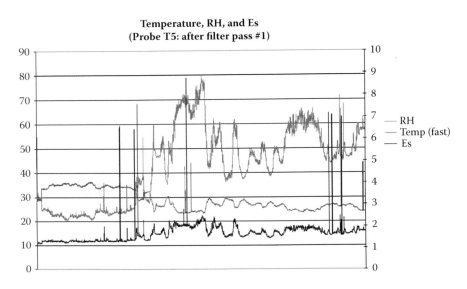

COLOR FIGURE 3.13   First quality control pass.

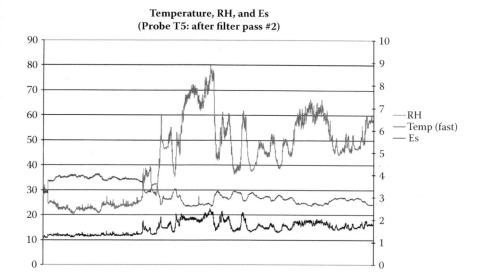

COLOR FIGURE 3.14   Second quality control pass.

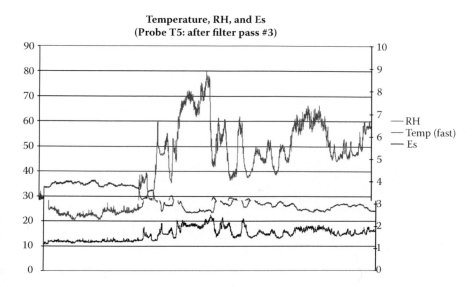

COLOR FIGURE 3.15   Third quality control pass.

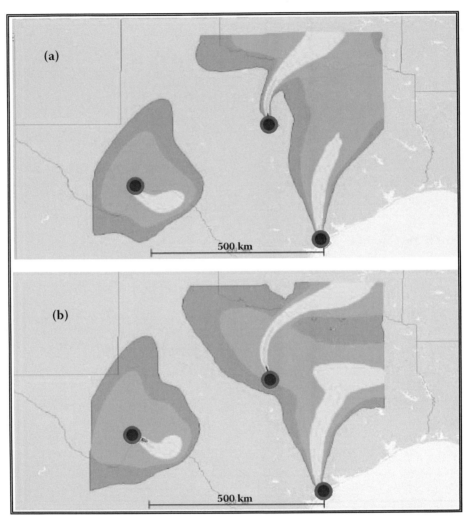

COLOR FIGURE 3.20 HPAC simulated 24-h surface GB plumes at 1200 UTC 9 May 2005; (a) control run and (b) FDDA run.

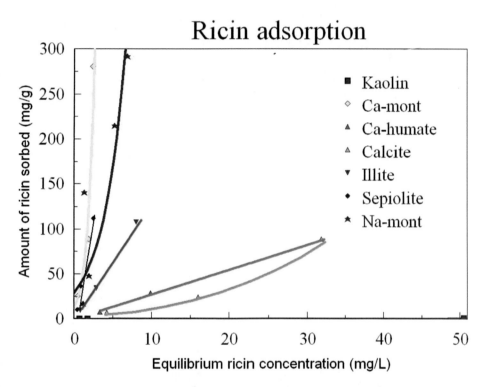

COLOR FIGURE 4.2 Ricin sorption isotherms for several different clay minerals. (Zartman, R E., Green, C.J., San Francisco, M.J., Zak, J.C., Jaynes, W.F., and Boroda, E., 2002, Mitigation of ricin contamination in soils: Sorption and degradation, Joint Services Scientific Conference on Chemical and Biological Defense Research, Hunt Valley, MD, November.)

COLOR FIGURE 4.5 Dust storm along a U.S. highway. (Courtesy of Dr. Tom Gill, UTEP.)

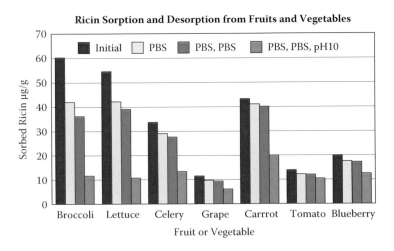

COLOR FIGURE 4.8 Amount of ricin sorbed by various fruits and vegetables before and after rinsing. (Zartman, R.E., Green, C.J., San Francisco, M.J., Zak, J.C., and Jaynes, W.F., 2003, Food sorption of ricin and anthrax simulants, Joint Services Scientific Conference on Chemical and Biological Defense Research, Towson, MD, November.)

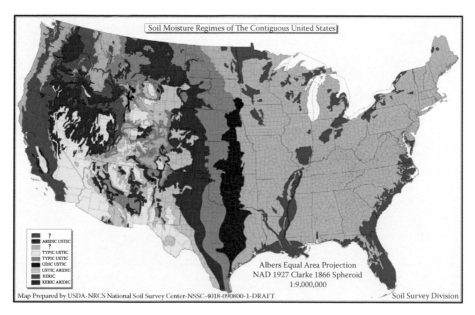

COLOR FIGURE 4.10 Soil moisture regimes for the contiguous United States (http://soils. usda.gov/use/thematic/moist_regimes.html).

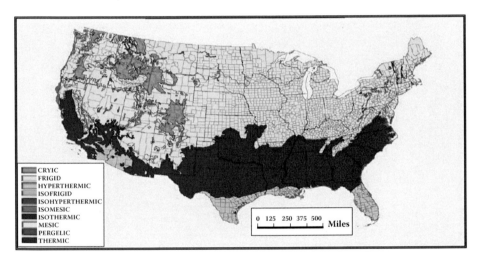

COLOR FIGURE 4.11 Soil temperature regimes of the contiguous United States (http://soils.usda.gov/use/thematic/temp_regimes.html).

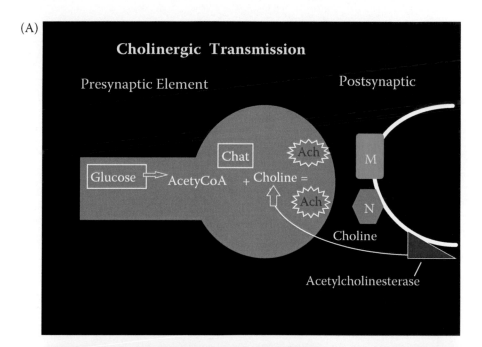

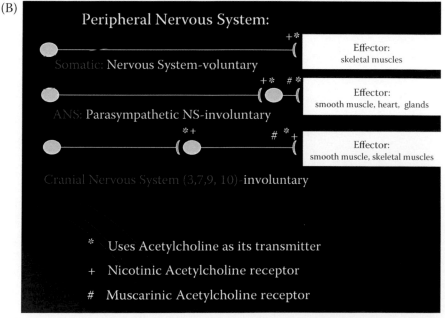

COLOR FIGURE 5.2 Presynaptic and postsynaptic regions of the acetylcholine neuron, emphasizing the synthesis and degradation of acetylcholine and the cholinergic receptor subtypes (Panel [A]); summary of the peripheral nervous system that utilizes acetylcholine as the transmitter (Panel [B]).

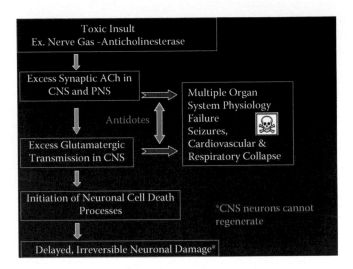

COLOR FIGURE 5.3 Role of excitotoxicity in mediating neuronal cell death induced by nerve gas exposure.

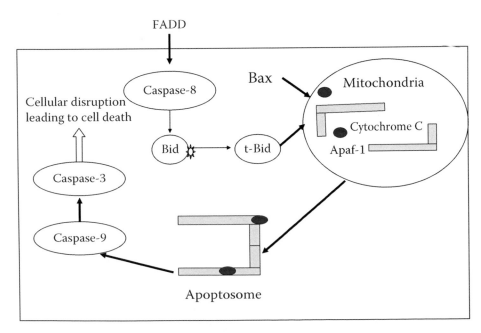

COLOR FIGURE 5.4 Cellular processes underlying the activation of caspases and their role in cellular disruption associated with apoptosis.

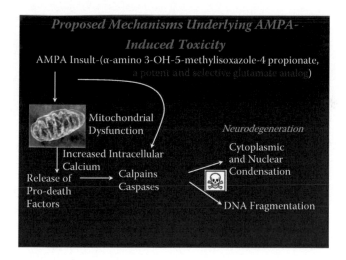

COLOR FIGURE 5.7 Proposed involvement of caspases and calpains in mediating AMPA-induced excitotoxicity.

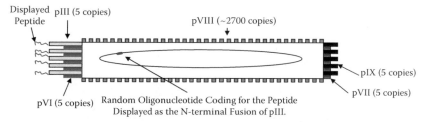

COLOR FIGURE 7.1 M13 display phage. Cartoon of an M13 filamentous bacteriophage depicting the five capsid proteins (pIII, pVI, pVII, pVIII, and pIX), the single-stranded DNA genome, and the random oligonucleotide, cloned into the 5' end of gIII, coding for the sequence of the displayed peptide.

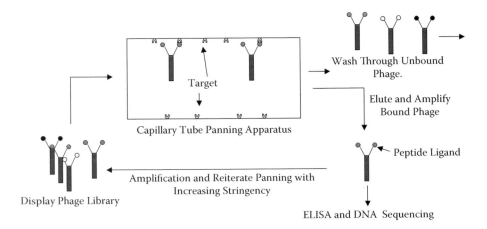

COLOR FIGURE 7.2 Affinity selection. Cycles of selection for phage displayed peptides (display phage) that bind to target ligands.

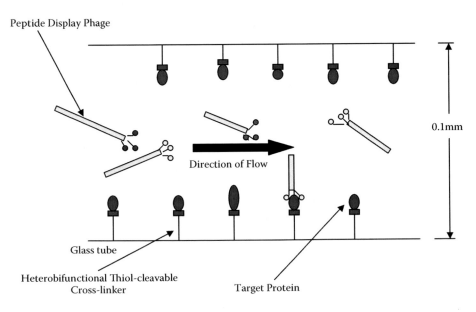

COLOR FIGURE 7.3 Capillary tube panning apparatus. A cartoon of our panning apparatus indicating the luminal surface of the glass capillary tube, the target affixed to the tube via a thiol-cleavable cross-linkage agent, and the display phage flowing through the tube.

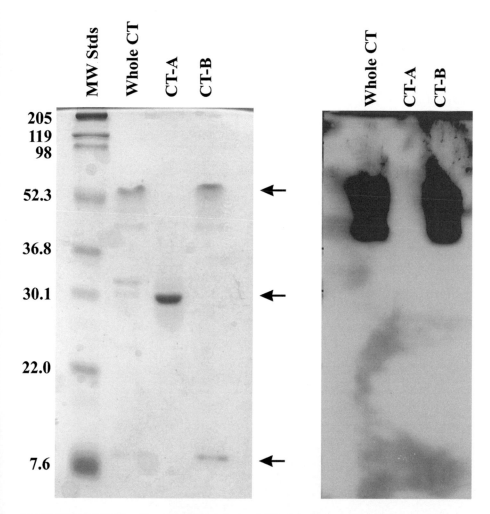

COLOR FIGURE 7.6 Western blot analysis of CT with display phage CT7. Cholera toxin and its subunits were separated by nonreducing SDS-PAGE and transferred to a nitrocellulose membrane. CT7 and anti-M13 HRP-tagged Ab were used to probe and develop the blot. The arrows indicate the single CT-B subunit (bottom arrow), CTA (middle arrow), and pentameric CT-B (top arrow).

COLOR FIGURE 8.1 Typical activated carbon–based CBRN suit. (Photo courtesy of Neal Hinkle, TTUHSC.)

COLOR FIGURE 8.10 M291 applicator pad with particulate resin mixture. (Utkarsh R. Sata, PhD Dissertation, Texas Tech University.)

The situation is further complicated by the fact that the flow-field morphology in the vicinity of the canopy was also strongly influenced by the nature of the velocity profile imposed at the upstream boundary of the model domain. The standard logarithmic profile (Equation 3.2.14) is inherently stable and relatively well behaved, and does not evolve significantly over open terrain. In contrast, profiles based on urban environment data are very unstable, and when perturbed by a significant roughness element—for example, an isolated individual vegetative canopy or even when simply exposed to open terrain—the subsequent evolution of the flow field becomes very complicated. In these circumstances, the natural evolution of the flow field around the isolated canopy becomes subject to the simultaneous evolution of the upstream velocity profile, significantly altering the velocity and turbulence profiles in the vicinity of the canopy compared to those arising in the presence of stable logarithmic profiles. Given this, it appears improbable that universal parameterizations can be developed to describe the effect of individual solid or vegetative obstacles on dispersion; rather, as detailed computational fluid dynamic models (CFD) become more practical, individual obstacles will have to be modeled as part of a detailed model of their immediate environment.

For extensive arrays of obstacles, Belcher et al. (2003) propose that there are three stages in the adjustment of a turbulent boundary layer flow upon encountering the canopy:

- The *Displacement Region*, immediately upstream of the canopy, where the flow is first affected by the canopy, and at least partially deflects around the canopy
- The *Adjustment Region*, arising in the region immediately downstream of the leading edge of the canopy, throughout which the velocity profiles within and immediately above the canopy continuously evolve
- The *Fully Developed Region*, in which the mean velocity and turbulence profiles within and above the array are fully developed

A fourth stage that assumes increasing importance for shorter obstacle arrays may also be defined:

- The *Wake Region*, immediately downstream (and sometimes extending slightly within) the trailing edge of the array.

A schematic illustration depicting the arrangement of these regions for an extensive continuous canopy is provided in Figure 3.4.

The flow field in each of these regions is extremely complex. The fully developed region presents the simplest case, as the flow-field statistics (ideally) become invariant with position once the canopy and adjacent external flow become fully developed, and as such has been the most extensively investigated of the four canopy-flow regions. Depending on the nature and arrangement of canopy elements, the extent of the adjustment region has been estimated at between 5 and 25 times the canopy height, with the wake region extending over distances of the same order. Deep within the canopy, turbulence features on the length scale of the obstacle spacing develops as the large-scale eddies of the free upstream flow are broken up by the

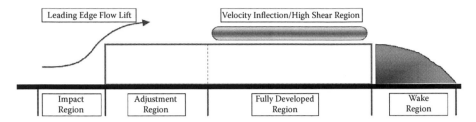

FIGURE 3.4 **(See color insert following page 46.)** Schematic illustration of the evolution of a turbulent flow upon encountering an obstacle array.

interactions with the obstacles. This can promote relatively vigorous mixing of dispersing species within the array, dramatically reducing the temporal fluctuation in concentration levels within a dispersing plume at points within the array of obstacles (Yee and Biltoft 2004).

Understanding the evolution of turbulence within an obstacle array is essential to understanding ground-level dispersion within the array. The difficulty of arranging detailed instrumentation of obstacle arrays has promoted interest in using high-resolution CFD to model flows within obstacle arrays (see also Hanna et al. 2002). Research by the authors of this chapter (Rendon 2005) into the effect of tree spacing and arrangement on airflows within and around forest stands of limited extent found that the nature of the arrangement in particular strongly influenced turbulence statistics within the canopy. A Reynolds-averaged Navier-Stokes (RANS) approach was used for this investigation, with closure of the flow equations achieved by the k-ε turbulence model. Flow within the tree canopies was modeled by including a momentum source term based on Forchheimer's law for a porous medium. The Sanz canopy turbulence model (Sanz 2003) was implemented to account for the transfer of energy from the large scales of turbulence to the small scales and the resultant enhanced dissipation of TKE.

For all ground-coverage area fractions, the averaged velocity profiles for staggered and random arrangements of obstacles were virtually identical (Figures 3.5 and 3.6). This indicates that a staggered array of trees will provide a good approximation of the velocity profile for a random array of trees given a suitable distance for the velocity profiles to develop. Importantly, however, the TKE profiles for staggered and random arrays proved to be significantly different (Figure 3.7). Given the importance of turbulence in governing dispersion, it appears that the nature of the stand arrangement must be characterized and accounted for in dispersion models.

3.2.4 COMPUTATIONAL MODELS FOR THE PREDICTION OF AEROSOL DISPERSION

The chart-based plume models developed during the 1960s, as exemplified by the Pasquill-Gifford and BNL models, have given way to computational models of ever increasing complexity and variety as the availability and power of computational devices has increased. In a recent review of the national capacity for tracking and predicting atmospheric dispersion produced by the National Research Council (2003), models sponsored by four distinct federal agencies were described; furthermore, each of the agencies involved was sponsoring the development of more than one type of model.

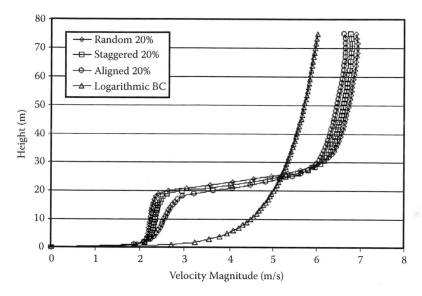

FIGURE 3.5 Averaged velocity profiles at a distance of 80 m downstream from the leading edge of the porous zone for finite forest stands having a 20% ground coverage area fraction. The logarithmic velocity profile at the inlet velocity boundary is provided for comparison (Rendon 2005).

Since the dispersion of aerosols is driven primarily by advection, the sophistication and accuracy of the wind-field characterization provided to a model is the principal factor in determining the accuracy of the dispersion predictions provided by the model. This may range from:

- Basing estimates on a wind field described by a single mean wind vector and an estimate of the atmospheric stability (in effect, the modeler in this situation would simply be using a computer to implement a Pasquill-Gifford–style prediction).
- Using live sensor data from the vicinity of the release to provide a characterization of the actual wind field in the area. In practice, there are currently very few locations in which a dense meteorological data sensor network is available to provide this data.
- More often, three-dimensional gridded meteorological forecast data provided by mesoscale meteorological models such as MM5 will be available. The resolution of the grid will typically be on the order of 1 km, which is too large to capture microscale surface features. Mesoscale model results can be coupled with fine-scale models by enforcing conservation of mass to provide enhanced fine-scale detail.
- Using detailed CFD simulations to capture the microscale wind-field features associated with obstacles and local terrain. Detailed CFD simulations presently take too long to be useful in rapid emergency response situations but can be useful in advance preparation and to develop understanding of local dispersion patterns, especially within urban areas (Hanna et al. 2006).

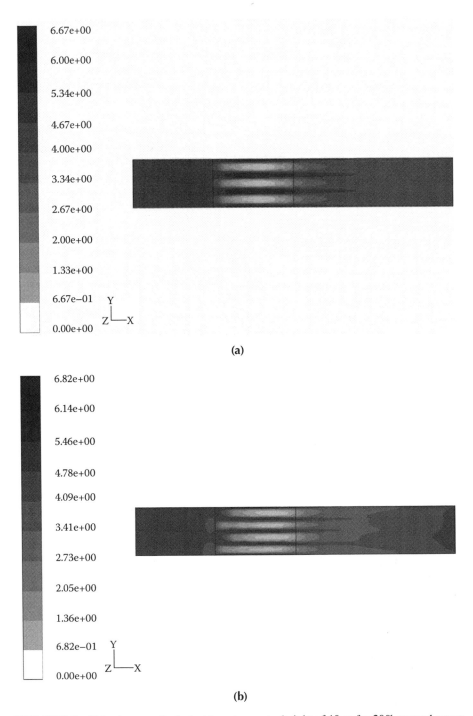

FIGURE 3.6 Velocity magnitude (m/s) contours at a height of 10 m for 20% ground cover-age forest stands. (a) Aligned array. (b) Staggered array. (c) Random array. The black box represents the boundary of the vegetated zone (Rendon 2005).

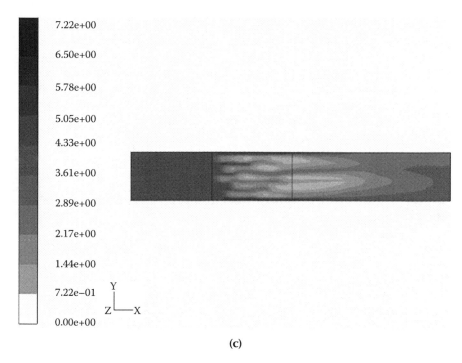

(c)

FIGURE 3.6 (continued).

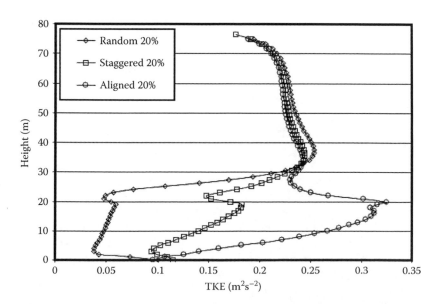

FIGURE 3.7 Averaged TKE profiles at a distance of 80 m (4H) downstream from the leading edge of the porous zone for finite forest stands having a 20% ground coverage area fraction (Rendon 2005).

One approach that is the subject of current interest is to obtain meteorological data from ensemble simulations. There are currently several mesoscale meteorological models available (as described in Section 3.3), and each one of these offers the analysts a variety of options for modeling atmospheric physics. Rather than taking the output of a single model with a single set of assumptions, the predictions of multiple models using multiple internal physical models and parameters are averaged to provide an ensemble prediction of the wind field in the region of interest.

The dispersion models themselves are distinguished by the following characteristics:

- The speed of execution
- The source and complexity of the meteorological (wind) field characterization
- The nature of the dispersion model employed
- The level of expertise needed to develop viable analyses

Models used in emergency response operations generally need to execute rapidly (within minutes) and need to be packaged in a manner that permits nonexpert users to develop viable analyses with only basic training. More sophisticated, longer running models are primarily used in pre-event preparation and postevent analysis, and as such tend to be operated more often by expert users. In any situation, a key capability is for models to be able to produce both an ensemble-averaged concentration field and a prediction of the range of variation in local concentrations.

Gaussian plume- and puff-based computational models are popular for emergency response due to their speed of execution and their suitability for packaging in a format that permits nonexpert users to obtain rapid predictions. Puff-based models are increasingly popular, as a plume may be described as a series of puffs emitted over time. The motion of each puff centroid is updated according to the wind-field velocity at the puff centroid, and the entire puff translates affinely with the centroid. A puff-splitting algorithm may be used to overcome the inaccuracies that arise as the puff dimensions become sufficiently large that the approximation inherent in assuming constant wind velocities throughout the puff becomes invalid (Sykes and Henn 1995); this capability becomes meaningful in computational models that employ detailed gridded wind-field data. Gaussian plume- and puff-based packages currently in use include:

- Hazard Prediction and Assessment Capability/Second-Order Closure Integrated PUFF (HPAC/SCIPUFF), sponsored by the Defense Threat Reduction Agency (DTRA) and the Department of Defense (DOD). This package is designed to use meteorological data in a wide variety of formats and resolutions. It also includes a comprehensive package of source terms representing a wide variety of military, industrial, and potential terrorist releases.
- Computer-Aided Management of Emergency Operations/Areal Locations of Hazardous Atmospheres (CAMEO®/ALOHA®), sponsored by the Environmental Protection Agency (EPA) and the National Oceanic and Atmospheric Administration (NOAA).
- CALPUFF, an air quality dispersion model, sponsored by the U.S. Environmental Protection Agency (EPA).

In their simplest form, these models provide results essentially equivalent to those provided by the Pasquill-Gifford approach. With more sophisticated meteorological data, increasingly sophisticated predictions can be obtained. In the case of the second-order closure integrated puff (SCIPUFF) algorithm used in HPAC, predictions of both the ensemble-averaged puff profile and the range of variation can be provided, as will be discussed in Section 3.2.6.

In stochastic Lagrangian particle models, the evolution of the concentration field is computed in a two-step process. First, the Eulerian velocity field in the region of interest must be calculated, either by solution of the Navier-Stokes equations or via an approximate method that satisfies mass consistency. The solution must also provide the local statistics of the velocity field. Individual particles are then released, and their position is updated over a time increment dt using an equation of the form (Wilson and Sawford 1996)

$$dU_i = a_i\left(\mathbf{X},\mathbf{U},t\right) + b_{ij}(\mathbf{X},\mathbf{U},t)d\omega_j \qquad (3.2.18)$$

where $\mathbf{U}$ is the velocity, $\mathbf{X}$ is the particle position, and U_i represents a component of the velocity; the bold quantities are vectors. The first term on the right-hand side is known as the "drift" term, and the second term, which contains the Gaussian white-noise term, $d\omega_j$, is the diffusion term. The coefficients a and b are functions of the local velocity and the statistics of the local velocity. Once a large number of particles have been released and tracked (simulating the release of interest), the statistics of the concentration field can be computed (Naslund et al. 1994)

The Department of Energy (DOE) has sponsored the development of two Lagrangian-particle-based dispersion models:

- ADAPT-LODI, developed at Lawrence Livermore National Laboratory. The ADAPT model assimilates meteorological data provided by observations and models (in particular, by Coupled Ocean/Atmosphere Mesoscale Prediction System [COAMPS®]) to construct the wind and turbulence fields. Particle positions are updated using a Lagrangian particle approach that uses a skewed (non-Gaussian) probability density function (Nasstrom et al. 1999; Ermak and Nasstrom 2000).
- QWIC-Urban, developed at Los Alamos National Laboratory. This model is designed specifically to operate on local and urban length scales, and combines a set of empirical equations for the flow fields around solid obstacles with a mass consistency requirement to rapidly estimate the velocity field. A simplified solution to the Fokker-Planck equation is used in the updating of particle positions.

CFD-based models use high-resolution grids to develop a detailed representation of the wind field and are typically used to investigate wind fields within urban (obstacle-rich) environments. The models incorporate detailed representations of each obstacle in the environment and provide the only possible approach to the computation of flows deep within urban canyons (a capability not offered by mesoscale meteorological models). A recent field study of dispersion in New York City provided impetus for a study in which five different CFD models were employed to characterize airflows in the vicinity of Madison Square Garden (Hanna et al. 2006). Two different types of CFD model are currently in use for dispersion studies:

- RANS, under which the Reynolds-averaged Navier Stokes equations are solved using some type of closure assumption to account for the Reynolds stress terms. RANS provides the values of the mean wind velocity and estimates of the turbulence statistics within the model domain.
- LES, or Large Eddy Simulation. Under LES, spatial averaging is used to account for the fine scales of the turbulence, and the larger eddies (those above the length scale of the grid) are resolved explicitly. In contrast to the ensemble-averaged results provided by RANS, LES provides individual realizations that can be used to estimate the statistics of both the turbulent flow field and the concentration field of the dispersing species.

One of the advantages of CFD approaches is that data can be provided at any point within the computational domain, providing a level of detail that cannot currently be approached by sensor networks; this of course carries the caveat that CFD models include a number of assumptions in both the physics and the boundary conditions, all of which affect the accuracy of the predicted wind fields. A third CFD approach, Direct Numerical Simulation (DNS), in which the Navier-Stokes equations are solved directly, is presently not feasible due to the excessive computational requirements. At present, all CFD models are too slow for use in emergency operations; they are best suited to detailed postevent studies or preparatory studies to understand the character of the local wind field in complex or urban terrain.

3.2.5 FIELD STUDIES TO SUPPORT MODEL VALIDATION

All numerical models incorporate significant assumptions and approximations, and their predictions must always be regarded as estimates. Solution of the RANS equations, for example, requires some form of closure assumption dealing with the Reynolds stress terms. Since the Reynolds stress terms and the mean flow terms are coupled by the equations, inaccuracies in the closure approximations can affect the predicted mean flow field. Furthermore, the boundary conditions imposed on the model require the assumption of velocity profiles and momentum transport rates, which may themselves be approximated. Similar approximations are inherent in any of the various techniques used to compute the wind field, with further assumptions being present in each of the dispersion models.

Any dispersion model must accordingly be evaluated by comparing its predictions with observations recorded in field experiments. Wind tunnel experiments, while useful for studies of specific terrain, cannot faithfully reproduce complex meteorological conditions. Field studies of dispersion, especially within urban areas, are major undertakings, requiring cooperation and coordination between local, state, and federal authorities in addition to the deployment of an extensive array of meteorological and release sampling equipment. Major field experiments have accordingly been coordinated largely by the federal agencies that have sponsored the development of the various dispersion models. The resultant databases are then typically made available to members of the modeling community for use in model validation.

The Project Prairie Grass field experiment provides an early example of such a study. In this experiment, diffusion of passive (neutrally buoyant) tracers from a point source over a flat field was measured using ground samplers placed on an arc ranging up to 800 m from the source (Hanna et al. 2004). A set of towers placed on arc 100 m from the source provided samplers at varying heights, enabling evaluation of the vertical profile of the plume (Elliott 1961). The fundamental data set (Barad 1958) still serves as the standard database for testing plume models over flat terrain (Hanna et al. 2004; ASTM 2000).

In recent years, attention has focused on experiments designed to validate the performance of models in urban areas. One approach has been to conduct experiments based on idealized arrays of obstacles. For the MUST (Mock Urban Setting Trial) program (Yee and Biltoft 2004), a regular 12×10 array of 120 rectangular obstacles (shipping containers) was set out on flat terrain at the U.S. Army's Dugway Proving Ground (Utah). A series of neutrally buoyant tracer release experiments was then undertaken to collect plume concentration data under a variety of conditions. One of the important results of this study found that while a Gaussian concentration profile was generally achieved in the horizontal plane, a reflected Gaussian profile failed to capture the vertical profile of the plume. It was also noted that the environment within the canopy of solid obstacles became dominated by fine-scale wake turbulence and became very highly mixed. The resultant vigorous mixing dramatically reduced the temporal fluctuation in concentration levels within a dispersing plume at points within the array of obstacles. The resultant smoothing of the concentration data has important implications for the design of sensors for use in the urban environment, as it appears that shorter sampling times may be used to obtain reasonable estimates of the evolution of mean concentration at points within an urban canopy. Yee and Biltoft also determined that turbulence within the obstacle array is dominated by length scales on the order of the obstacle dimensions. Considered in conjunction with the mean flow velocities, this provides a framework for establishing time scales for the design of urban sensor sampling regimes.

Instead of releasing passive tracers, in the Kit Fox field experiment a dense gas plume (CO_2) was released over an array of obstacles set out on flat terrain at the Department of Energy's Nevada Test Site (Hanna and Chang 2001). The obstacle array consisted of an inner core of large (2.4 m^2) plywood billboards, on 6.1 m lateral and 8.1 m downwind spacing, surrounded by an outer rectangle of 0.2 m high plywood billboards on a 2.4 m lateral and longitudinal spacing. The array was designed to simulate the environment in a refinery or chemical process plant, so that the study would provide data to understand the motion of dense gas releases through a plant environment. The data has subsequently been used to validate and tune the performance of models for the dispersion of dense gases in plant environments (Hanna and Chang 2001; Hanna et al. 2004)

In recent years, a series of experiments has been undertaken in specific urban environments within the United States, with the specific objective of providing datasets for the evaluation of dispersion models being developed specifically for the urban environment. The first experiment of this series, Urban 2000, consisted of a series of short-duration sulfur hexafluoride (SF_6) releases within and around the downtown area of Salt Lake City (Utah). In addition to addressing a variety of issues associated with dispersion in an urban area, the study also permitted evaluation of the effects of diurnal weather patterns driven by the area's unique topography (mountains to the east, with open water to the west) on dispersion (Allwine et al. 2002). The data has subsequently been used in a variety of model evaluation studies; for example, the predictions provided by HPAC when coupled with meteorological data sources ranging from single-point observations taken during the study to mesoscale model forecasts have been compared with the dataset (Warner et al. 2004).

For Joint Urban 2003, a field experiment was conducted in Oklahoma City, Oklahoma (Allwine et al. 2004). To give an idea of the scale of this type of field experiment, the experiment had over 150 individual participants drawn from organizations associated with the Department of Defense, the Department of Energy (with participants from five national laboratories), National Oceanic and Atmospheric Administration (NOAA), eight universities, and assorted state and private entities. CFD simulations of the airflow in the downtown area were undertaken to identify optimal locations for sensors (Lee et al. 2003). Once again, SF_6 tracer was released, with the release location varying according to the prevailing wind and whether the objective of the particular release was to maximize the transit of the cloud through the Oklahoma City downtown or to explore dispersion within a specific street canyon.

The Department of Homeland Security (DHS) has established a multiyear Urban Dispersion Program to understand the flow and dispersion in the deep street-canyon environment of New York City. The first experiment under this program, MSG 05, dealt with a limited number of releases in the vicinity of Madison Square Garden; distribution of the results is currently restricted, though plume footprints without quantitative information are available (Allwine and Flaherty 2006). The experiment has also provided the basis for a comparison of the performance of CFD models in predicting flow patterns in this unique environment (Hanna et al. 2006).

3.2.6 FORMULATION AND EXECUTION OF A GAUSSIAN PUFF-BASED MODEL FOR EMERGENCY RESPONSE

With the current level of computational capabilities, Gaussian puff and stochastic Lagrangian particle tracking models provide the primary tools for predicting dispersion to support emergency response. To illustrate the issues facing modelers and emergency response directors, the process of formulating a Gaussian puff model and delivering results is described here. The procedures were developed for the Texas Emergency Analysis and Response Program (TEARP) at Texas Tech University, and have been tested in a series of simulated exercises. The flowchart for the operational procedure that has been developed is provided in Figure 3.8.

The first step in the process is for the modeling center to be notified of the event and provided with an initial description of the source. To formulate a dispersion model, the modeler needs the following information about the source:

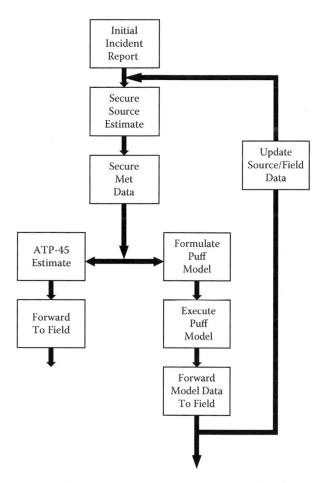

FIGURE 3.8 Operational procedure for formulation of dispersion models to support emergency response.

- The release location (ideally in latitude/longitude [lat/lon] coordinates, though street addresses may be easily converted to lat/lon using inexpensive commercial software)
- The release elevation
- The nature of the released material
- The amount of material released
- The time at which the release occurred
- The duration of the release
- The release mode (instantaneous, continuous, moving source, etc.)

It should be noted that initial reports will often be provided by responders or civilians with little or no understanding of either dispersion or biological and chemical threats (and this state may continue until trained state or federal personnel can be deployed). The modeler will often be forced to make a best estimate of the nature of the release; in software packages such as HPAC, a set of predetermined source terms characterizing specific types of release event are available to provide the initial source estimate. It must be recognized that this estimate will significantly affect the predicted extent of any region of toxic effects, and modelers and responders must be aware of this unavoidable limitation.

The next step is to begin the process of procuring meteorological data for the model. For an initial Allied Technical Protocol -45 (ATP-45) estimate, a single-point observation providing the near-surface wind speed and direction in the vicinity of the release will be sufficient. Such observations can be secured from a variety of freely available meteorological Web sites (such as National Weather Service [NWS] Web sites). For the Gaussian puff model, a more detailed time-dependent, three-dimensional forecast dataset should be secured. For the TEARP program, MM5 forecast models were run for the state of Texas, and software was developed to extract forecast data for specified regions and time periods for use in dispersion modeling. This software formatted the data into a file format suitable for the dispersion program being used.

For chemical releases (in particular), the majority of toxic exposures will occur within the first 30 min. after the initial release (this can vary depending on the amount and nature of the release). Allowing for the time taken for the information regarding the event to reach the modeling center, even with streamlined and practiced operating procedures, it will generally not be possible to formulate and execute a puff model in much less than 30 min. To provide a rapid initial estimate to support first responders, an ATP-45 estimate can be provided while the dispersion model is being formulated. For wind speeds less than 5.8 mph, a circle extending 6.25 miles from the source is constructed; for wind speeds greater than 5.8 mph, a triangle is constructed extending 45° either side of the wind direction and 1.67 times the wind speed downwind of the source. Examples of each type of estimate are provided in Figure 3.9. While this type of estimate is crude, it rapidly provides first responders with a conservative initial estimate of the threatened area.

While the ATP-45 estimate is being formulated, formulation of the Gaussian puff model should proceed. The details of the process will depend on the dispersion software package being used but will always include definition of the source characteristics and integration of the meteorological data. The HPAC package uses the second-order closure integrated puff (SCIPUFF) model, which provides both an ensemble-averaged concentration field prediction and a prediction of the variance in the concentration field (Sykes et al. 1998).

In SCIPUFF, the evolution of the components of the moment tensor for the Gaussian distribution as the puff translates and evolves is calculated using a second-order closure model. The closure model also permits the formulation of equations to compute the variance in the concentration value (Sykes et al. 1998). As puffs expand, the assumption that the entire puff translates affinely with the centroid becomes increasingly suspect; accordingly, a puff-splitting scheme divides the puffs when they become too large in a particular direction. The moments of the new puffs are defined in a manner that conserves the overall moments of the puff and attempts to minimize the local changes in predicted concentration as a result of the splitting operation (a separate algorithm identifies overlapping puffs that can be merged

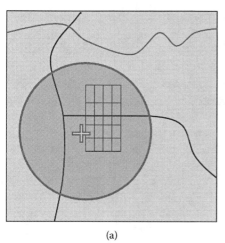 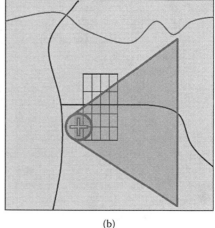

(a)   (b)

FIGURE 3.9 **(See color insert following page 46.)** Schematic illustrations of ATP-45 esti-mates of the hazard zone (the release is located at the center of the cross). (a) Nominal wind speeds less than 5.8 mph (5 knots). (b) Nominal wind speeds greater than 5.8 mph (5 knots).

(Sykes et al. 1999). SCIPUFF also assumes that the ground surface and the inversion capping the PBL act as reflecting surfaces.

A first dispersion estimate can typically be generated using a Gaussian puff model within 5–15 min., depending on the scale of the dispersion model. The funda-mental result provided is the evolution of the ensemble-averaged concentration field. The variance in the concentration field is also critical, as it provides an indication of the extremes that may be encountered due to the potentially sinuous nature of the actual puff or plume. The cumulative dosage field and the contour map of surface deposition concentrations are also of practical utility. Once these images are gener-ated, they can be posted to secure Web sites to enable first responders to access the data from the field using wireless data-reception devices.

Once the initial estimates have been delivered, updated information from the field may be used to formulate updated dispersion estimates. This process would be expected to continue as the response operation proceeds; although in many cases the initial toxic effects may subside rapidly, the long-term effects of deposition and low-level exposure assume increasing importance as the event continues. The model-ing operation would usually continue until a full postevent analysis can be completed, using updated or even actual meteorological data and the full postevent assessment of the characteristics of the source.

3.3 COMBINING METEOROLOGICAL OBSERVATIONS WITH MODEL PREDICTIONS TO ENHANCE THE FIDELITY OF DISPERSION PREDICTIONS

3.3.1 INTRODUCTION

Meteorological measurements and predictions provide the foundation of any suc-cessful method for assessing the consequences of biological or chemical agent release into the lowest layer of the atmosphere known as the planetary boundary layer (PBL). This section describes how observational and modeling techniques can be coordinated to evaluate the accuracy and effectiveness of predicting the airborne dispersion of hazardous materials near the surface. The objective of such an analysis is to improve the ability of first responders to use the available dispersion modeling packages in conjunction with locally collected weather data to gain a tactical advantage, whether it is on the battlefield or in a civilian emergency event.

Texas Tech University (TTU) provides an example of a program where scientists and engineers have been developing and field-testing a variety of instrumentation platforms capable of gathering high-frequency full-scale data with which to challenge the numerical weather prediction (NWP) models. The program has demonstrated the benefit of using real-time field data in the operational forecast of PBL transport of chem-bio agents, and has provided unique data sets for in-depth model validation. The central component of this program is the West Texas Mesonet (WTM), which consists of a grid of dozens of meteorological stations reporting real-time weather data to a central hub. Other facilities include portable Mesonet towers of 10-m height, transportable platforms used for sensing a wide variety of weather data, mobile Mesonets, and the Vehicular Instrumentation Platform for Emergency Response (VIPER) used to collect high-resolution meteorological observations from a moving vehicle.

Numerical weather prediction models currently provide the foundation in generating dispersion predictions in support of an emergency response. The models may encompass length scales ranging from synoptic scale (weather features > 1,000 km in spatial extent) to mesoscale (10–1,000 km) and microscale (~1 km). The U.S. National Oceanic and Atmospheric Administration (NOAA) National Weather Service (NWS) is responsible for regular synoptic-scale and large mesoscale (> 100 km) weather prediction. For many chem-bio releases, the health-threatening impacts occur primarily on the micro- or small mesoscale (meso-β scale) of 10–100 km in length. Meso-β conditions (and thus dispersion) are strongly influenced by fast-changing local conditions such as cloud cover and surface heating, and are heavily influenced by local terrain. These conditions may cause localized variations in weather over small scales not reflected in the data provided by widely spaced weather-reporting stations or model predictions based on larger length-scale conditions. Despite the challenging nature of meso-β and microscale weather prediction, high-frequency three-dimensional meteorological data derived from mesoscale forecasts is often used as a basis for PBL diffusion prediction. Figure 3.10 depicts this coupling approach of using mesoscale weather and dispersion models in tracking the flow of agents near the surface in an urban or battlefield setting.

In an attempt to capture fast-changing local events (and thus maximize model performance), state-of-the-art mesoscale NWP often makes use of four-dimensional data assimilation (FDDA), in which high-frequency observations are used to update the model state during the time integration step (Daley 1991). Examples of suitable data sources include the WTM, where data is reported every 5 min., and NOAA ACARS (Aircraft Communications Addressing and Reporting System) observations, which are taken by commercial aircraft about every 10 min. A better understanding

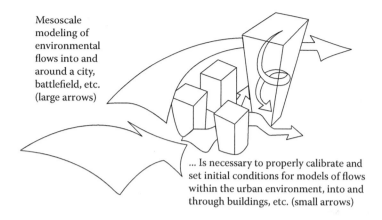

Mesoscale modeling of environmental flows into and around a city, battlefield, etc. (large arrows)

... Is necessary to properly calibrate and set initial conditions for models of flows within the urban environment, into and through buildings, etc. (small arrows)

FIGURE 3.10   Integrated vision of the multiscale nature of airflow modeling and simulation projects.

of how to take advantage of these high-resolution datasets can significantly advance our expertise in meso-β NWP and enable us to use the most up-to-date environmental conditions for PBL dispersion prediction. Several real-data case studies have been conducted to demonstrate the impact of FDDA on mesoscale simulations.

In the development and applications of PBL dispersion models, the following issues must be addressed:

- How does the horizontal grid resolution associated with mesoscale models impact hazardous agent forecasting near the surface?
- How does uncertainty in the lateral boundary conditions associated with mesoscale models impact hazardous agent forecasting near the surface?
- How does the high-frequency mesoscale FDDA impact the model surface flow and consequently hazardous agent forecasting?
- Are existing modeling techniques and observations adequate in providing useful real-time guidance in case of an airborne emergency event?

This work involves both field and modeling initiatives (Gill, Doggett, et al. 2002; Gill, Chang, et al. 2003). However, here we will focus more on modeling than field work. Six real-data case studies are employed to address the abovementioned key issues. Section 3.3.2 describes the WTM and ACARS, highlighting the data-processing procedures needed to provide real-time observational data. Section 3.3.3 describes mesoscale weather models. Section 3.3.4 synthesizes and discusses the real-data simulation experiments, including an overview of HPAC and the manner in which high-frequency mesoscale data assimilation using the WTM and ACARS observations is accomplished. Findings of the effect of different approaches to NWP on HPAC dispersion predictions and suggestions for future research are summarized in Section 3.3.5.

3.3.2 WEST TEXAS MESONET (WTM) AND THE AIRCRAFT COMMUNICATIONS ADDRESSING AND REPORTING SYSTEM (ACARS)

The WTM is a gridded array of meteorological observational platforms. The average spacing of the platforms is about 30 km, providing a much higher horizontal resolution than the National Weather Service surface network (which typically has stations spaced on the order of 150 km apart). The WTM currently consists of 50×10-m surface towers, each providing observations of wind speed and direction, temperature, humidity, dew-point temperature, surface heat flux, and precipitation. The data from all surface stations is distributed in real time via the Internet to users accessing the WTM homepage (http://www.mesonet.ttu.edu), as well as directly from the NWS office in Lubbock. The Mesonet online data is updated in real time every 5 min.

Figure 3.11 provides a map showing the location of the Mesonet sites. Each square block of approximately 50-km × 50-km represents a county. A detailed technical overview of the WTM surface stations can be found in Schroeder et al. (2005). The WTM headquarters are located in the Reese Center (REES), which is located

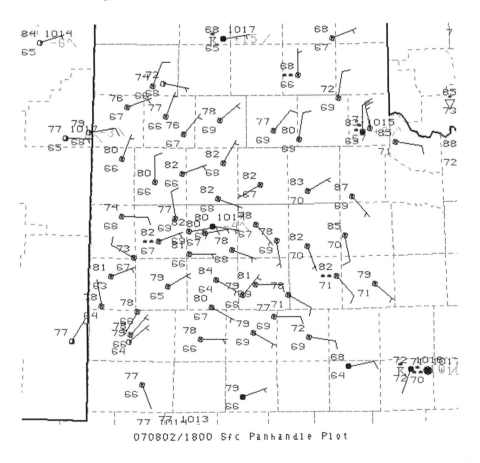

070802/1800 Sfc Panhandle Plot

FIGURE 3.11   (See color insert following page 46.) West Texas Mesonet site locations and identifiers as of March 2007.

in the box of the fourth row from the bottom and third column from the Texas/New Mexico border. The NWS has been incorporating WTM observations into its hourly weather roundup product, available to the public and news media, since 2001. The redundancy of data sources will ensure that data users have a backup in the event of a server failure at Reese Center.

3.3.2.1   Quality Control of WTM Data

A fundamental element of data-collection activities is the development of robust quality control (QC) procedures for the data. In developing a robust QC algorithm, care is taken to remove the erroneous data while maintaining the integrity of the valid data as much as possible. For the WTM data, the first step in the QC procedure is to replace any missing data with interpolation from the nearest two data points. Since missing data points are infrequent, this replacement has proven adequate. Small subsets of the time series are then examined to identify spikes in the data. This is done by moving a 7-point data window through the data, computing mean and standard deviation data for that window, and looking for data points that are significantly out of line with those statistics. The filtering pass is done three times. The first pass has high thresholds for data variability and only removes the most pronounced, short-lived anomalies. The second pass has lower thresholds for data variability and removes smaller but longer lasting anomalies. The final pass is designed to remove the longest lasting of the anomalies.

This procedure has been found to successfully remove bad data without altering the valid data points. It is important to note that the filtering process does not smooth or average out the time series; it only looks for and corrects spurious data points. An example of the procedure is shown in Figures 3.12–3.15 taken from Gill et al. (2002).

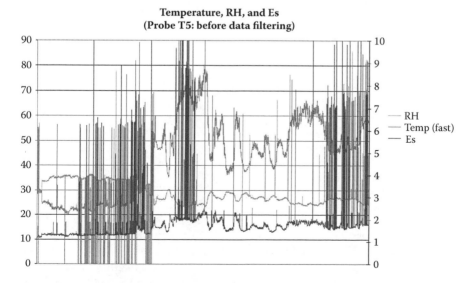

FIGURE 3.12   (See color insert following page 46.) Raw data example.

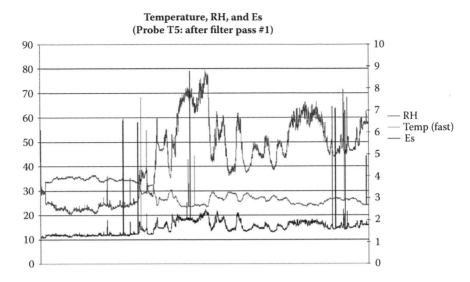

FIGURE 3.13 (See color insert following page 46.) First quality control pass.

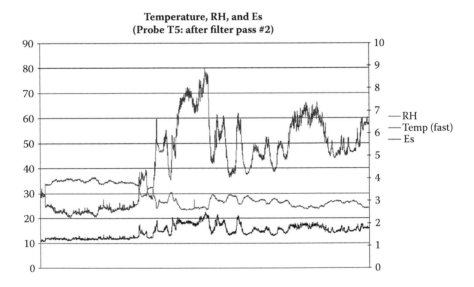

FIGURE 3.14 (See color insert following page 46.) Second quality control pass.

As can be seen in Figure 3.12, the raw data can be quite noisy (though not all datasets exhibit this level of noise). Figures 3.13–3.15 show the improvement following each of the three 7-point-window QC filtering passes. As can be seen, the data after the first and second passes show a great deal of improvement. The final pass removes the longest duration spikes, which account for only a small percentage of the problem data. It should be mentioned that these figures contain more than 10,000 data points, so while the raw data looks unreliable at first glance, there is still a large percentage of valid data that allows the corrections to be applied.

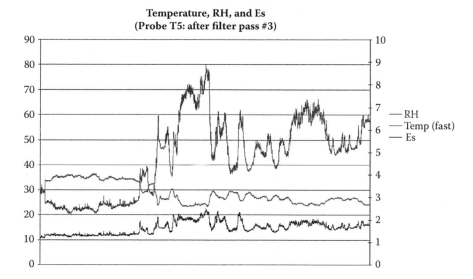

FIGURE 3.15 (**See color insert following page 46.**) Third quality control pass.

3.3.2.2 Quality of ACARS Data

ACARS data are routed by several cooperating airlines to the FSL (Forecast Systems Lab), NOAA Global System Division, for quality control. General information about ACARS and how to access the datasets can be found on the Web site http://www.acweb.fsl.noaa.gov. ACARS wind and temperature data are collected by commercial aircraft during both the en-route and ascent/descent modes of their flights, providing data at locations in between the NWS rawinsonde sites (Mamrosh 1998) at very high frequency. At flight altitudes of about 23,000 ft, data are generally taken every 5–6 min. Near the airports, the data spacing is decreased by some airlines. Below 18,000 ft, a vertical resolution of 1000 to 2000 ft is commonly provided. More than 150 aircraft provide data with a vertical resolution of 300 ft for the first minute after takeoff. ACARS can provide valuable up-to-date weather data for short-range forecasts of PBL winds.

The quality of ACARS data had been examined by many researchers, for example, Mamrosh (1998), Schwartz and Benjamin (1995), and Lord et al. (1984). Estimated wind vector accuracy was about 1.8 m/s and estimated temperature accuracy was about 0.50°C. When ACARS was compared to radiosondes, root-mean-square (RMS) deviations were 7.4° in direction and 3.3 m/s in speed. In comparing ACARS ascent/descent winds and temperatures with radiosondes, it was found that temperature differences were less than 20°C on 94% of all occasions, and less than 10°C greater than 68% of the time. Wind speed rms deviations were 4.1 m/s, while direction rms differences were 35° (mostly due to light and variable wind situations).

3.3.3 Mesoscale Meteorological Models

Evaluation of the suitability of mesoscale NWP models for use in dispersion modeling is based on following criteria:

- Portability: The system should be able to run on various computer platforms, ranging from a personal computer (PC) to a mainframe supercomputer.
- Reliability (robustness): The system must be reliable under various weather and PBL scenarios and not overly sensitive to any variations in model initial conditions.
- Flexibility: The system should possess a modular, easily adjustable structure. Flexibility and portability (above) should allow the system to be adapted to the various computer platforms (e.g., laptop PCs, workstations) that may be required by first responders in a civilian capacity or war fighters.
- Nested grid capability: This is required for multiscale simulations with a very fine model resolution over the areas of interest.
- Full physics: The model system must have an advanced (full) treatment of PBL processes and related physics.
- Nonhydrostatic dynamics: This capability permits modeling of very small-scale airflows on the order of a few kilometers or smaller in the horizontal dimension. Modelers are often initially interested in phenomena occurring on spatial scales on the order of a few kilometers where motion is generally nonhydrostatic, since boundary-layer transport will be at heart of urban airflow modeling.
- Friendliness: The software must be well documented and not too time-consuming to run.
- Terrain capability: If possible, the model should be able to simulate processes in complex terrain.
- FDDA capability: The system ideally should be capable of assimilating high-frequency observations collected from various platforms.

Table 3.3 lists mesoscale meteorological modeling packages that are currently available and in widespread use. COAMPS includes an ocean component for modeling air-sea interaction, which may be of secondary importance for urban boundary-layer modeling. The Advanced Regional Prediction System (ARPS) is similar to MM5 and Regional Atmospheric Modeling System (RAMS) in many aspects and has not been chosen for recent model intercomparison studies. The Operational Multiscale

TABLE 3.3
Mesoscale Numerical Weather Prediction Models

Model	Originating Organization/Agency	Reference
COAMPS	Naval Research Laboratory (NRL)	Hodur et al. 1997
ARPS	University of Oklahoma (OU)	Xue et al. 1995
OMEGA	Science Application International Corporation (SAIC)	Bacon et al. 2000
MM5	National Center for Atmospheric Research (NCAR) and Pennsylvania State University (PSU)	Grell et al. 1994
RAMS	Colorado State University (CSU)	Snook et al. 1995
HOTMAC	Yamada Science and Art (YSA) Corporation, New Mexico	Yamada and Bunker 1988
WRF	NCAR, NOAA, DOD, and a consortium of universities	Klemp 2004

Environment Model with Grid Adaptivity (OMEGA) has a unique triangular grid structure and many attractive features. Its model grid-mesh size can be adjusted dynamically during time integration according to prespecified criteria. OMEGA is, however, privately owned and not commercially available at this time.

Gross (1994) carried out statistical evaluation of several mesoscale models using data from Project WIND (Wind in Nonuniform Domains) over an area of 80 × 80 km² (Cionco 1994). The models considered were MM4 (a precursor of MM5), RAMS, and HOTMAC® (Higher Order Turbulence Model for Atmospheric Circulation). The finest model grid size used in the simulations was 5 km. Two 24-h forecast periods started at 1200 UTC, 27 June 1985, and 1200 UTC, 1 February 1986, respectively. In the first period, the synoptic-scale setting represented an undisturbed environment, while during the second period a cold front passed through the model domain. The model simulations were verified every hour at the surface. For the first period, the forecasts revealed mean errors on the order of 1 m/s for wind speed, 50° for wind direction, and 30°C for temperature. For the second period, the mean errors ranged from over 2 to 6 m/s for wind speed, from 10° to 30° for wind direction, and were on the order of 20°C for temperature. HOTMAC showed slightly higher skill than the other two except in nighttime cooling prediction.

There are many other mesoscale model intercomparison studies (e.g., Pielke and Pearce 1994; Cox et al. 1998; Henmi 2000). The two latter cases, MM5 vs. Battlefield Forecast Model (BFM) over White Sands Missile Range (Henmi 2000) and MM5, NORAPS6, RAMS, RWM (RelocaTable Window Model, USAF) over five theaters (Cox et al. 1998) were carefully considered and were a key to our evaluation of the models. The overall rank suggested by Cox et al. was RAMS, MM5, NORAPS6, and RWM. Also, RAMS showed slightly better performance in temperature and low-wind-condition forecasting. Hanna and Yang (2000) showed very similar verification statistics of surface wind predictions using MM5 and RAMS.

Based on the results of intercomparison studies, HOTMAC, MM5, and RAMS appeared best suited for case studies of dispersion. Each is highly reliable, is portable, and meets the required criteria. These systems are used by various institutes worldwide as a research tool as well as an operational forecasting model. HOTMAC (of YSA Corporation [2000]) has many desirable features, including a multiple (up to five) grid framework, model resolution ranging from 100 m to 10 km, and a high-order turbulence closure scheme. The earlier version of HOTMAC exhibited similar skill to MM5. In addition to HOTMAC, the YSA system provides a compatible diffusion model called RAPTAD® (Random Puff Transport and Diffusion) and lively three-dimensional graphics and animations. The latest HOTMAC/RAPTAD system is, however, relatively costly. The MM5 system is well documented, free of charge, and widely used in many research and operational centers. MM5 was accordingly adopted for this work.

MM5 is a multiscale (10–1000 km) weather prediction system consisting of data analysis and initialization, dynamical prediction, and postprediction diagnosis and verification codes. The MM5 model (Grell et al. 1994) system source codes and documentation are in the public domain (i.e., the Internet) for use by any person. The MM5 model has demonstrated high skill in mesoscale simulation of tropical cyclone-orography interaction (Wu et al. 1999) as well as various mesoscale weather phenomena. The model is composed of nonhydrostatic dynamic framework, terrain-following vertical coordinates, nested grid and FDDA capabilities, cumulus parameterizations, and PBL processes. The PBL scheme over land comprises two basic regimes: nocturnal and free convection (Hong and Pan 1996). They are built upon the Blackadar high-resolution structure (Blackadar 1979).

For the model output display, NCAR graphics and Vis5D were employed. Vis5D, developed at the Space Science and Engineering Center (SSEC) at the University of Wisconsin, is a multidimensional visualization display and animation software. Both Vis5D and NCAR graphics are user friendly and have a wide range of graphical capabilities. Particularly attractive is that Vis5D permits the user to conduct a "data probe" at any location and supports parcel tracking within the model domain. The data-probe capability may be used to locate points (e.g., population centers) prone to high concentrations of biological and chemical agents under a particular attack scenario. It also permits evaluation of the temporal variations of the biological and chemical threat agent concentration at the selected points. The forward and backward tracking tool is most useful in projecting the spread of biological and chemical threat agents and revealing the possible sources.

3.3.4 CASE STUDIES

The results of six real-data case studies are presented here. The case studies considered conditions including dry-line perturbations, tranquil conditions, weak disturbances with light precipitation, mild synoptic forcing, and severe weather environment. The full physics and nonhydrostatic version of MM5 was used in the case studies. The six cases were as follows:

(A) 30–31 July 2001: No significant weather
(B) 6–7 April 2002: Mild synoptic circulation
(C) 14–16 April 2002: Quiescent dry line
(D) 23–24 November 2002: Tranquil conditions
(E) 21–22 February 2003: Weak disturbance with light precipitation
(F) 8–9 May 2005: Severe storm in central Texas

Table 3.4 lists some of the key MM5 model case-study parameters. Case A involved three-grid MM5 simulations, while in Cases B and C two grids and in Cases D, E, and F single grids were used. There were 24 terrain-following levels in the vertical in all cases. For the coarse (outer) grids, the model initial and lateral boundary conditions were derived from the NWS Eta model (Black 1994), which had 50 vertical levels and a horizontal grid mesh size of 22 km. For the fine (inner) grids, the initial states were obtained from the interpolation of the coarse grid model. All model integrations were performed on an SGI Octane-2 workstation.

MM5 was first tested using the three-grid domain (Case A) as shown in Figure 3.16(a). The model appeared to be quite reliable and robust. Close examination of model output did not reveal any indication of computational instability during the 24-h integration. Figure 3.16(b) shows the two-grid model domain and 9-km terrain for use in Case B. The inner grid, centered near Lubbock, Texas, was located

TABLE 3.4

Key MM5 Case Study Parameters. Columns 2 to 4 Show the Numbers of Horizontal Grid Points and Spacing

Case	Outer Grid	Inner Grid	Third Grid	Length of Integration	Start Time UTC/DY/MN
A	$(41 \times 41)/27$ km	$(73 \times 53)/9$ km	$(73 \times 73)/3$ km	24 h	12/30/07
B	$(67 \times 67)/9$ km	$(61 \times 61)/3$ km	NA	24 h	12/06/04
C	$(67 \times 67)/18$ km	$(67 \times 67)/6$ km	NA	24 h	00/14/04
D	$(67 \times 67)/15$ km	NA	NA	24 h	00/23/11
E	$(67 \times 67)/15$ km	NA	NA	24 h	00/21/02
F	$(67 \times 67)/20$ km	NA	NA	36 h	00/08/05

in the middle of the WTM area. The outer grids of Cases C, D, and E were also centered near Lubbock and covered a bigger area than that of Case B. HPAC diffusion computations were performed within the general area of DØ3 in Figure. 3.16(b) in Cases B, D, and E. Because of the relatively flat surface in West Texas terrain, forcing did not have a major role in the HPAC simulations. The FDDA experiments were performed on Cases C, D, and E using the WTM observations, and on Case F using the ACARS observations.

3.3.4.1 MM5/HPAC and WTM/HPAC Simulations

The PC version of HPAC 4.0 and 4.0.1 from DTRA (Defense Threat Reduction Agency) was adopted for case studies. HPAC is well documented and has many desired features for predicting the concentration field and displaying exposure information for the population in the vicinity of accidents. The capabilities of HPAC have been demonstrated by many users (Cox et al. 1998; Chang et al. 2003; Chang and Gill 2005). We have developed an MM5/HPAC conjugation for modeling diffusion.

The basic components of HPAC (*HPAC User's Guide* 2001) are: (1) an editor module for specifying project, map, and materials released; (2) incident models, for example, chem-bio and nuclear weapons; (3) an atmospheric diffusion/dispersion model known as SCIPUFF (Sykes et al. 1998); (4) a weather module; and (5) a display module. The weather module can work with data from various sources (e.g., observations or models). HPAC accepts both gridded meteorological data on the prespecified isobaric surfaces (e.g., 500 hPa or mb) and profile (vertical column) data at the individual model grid points. Two diagnostic wind models, the terrain influenced SWIFT (Stationary Wind Fit and Turbulence) and Mass Consistent (MC) SCIPUFF, are provided in HPAC to derive mass-consistent gridded wind fields from observations or model output for dispersion computation over complex terrain. Caution must be taken in applying the gridded format data over the areas of high terrain. Some lowest isobaric surfaces could be located beneath the ground. For example, in West Texas, where surface elevations are on the order of 1 km above sea level, both 1000 and 900 hPa levels are beneath the ground. This can result in erroneous wind fields being used in the HPAC computation, as was identified in a case study and will be discussed.

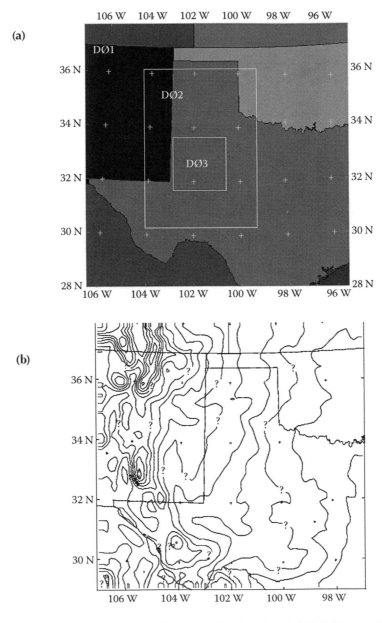

FIGURE 3.16 (a) MM5 three-grid domain for Case A and (b) MM5 two-grid domain for Case B.

MM5/HPAC and WTM/HPAC simulations were performed on Cases B, D, and E. The simulations focused on the WTM area and consisted of a large number of runs initialized with various datasets, including the WTM, MM5 surface, MM5 profile, and MM5 grid-point data. The Mesonet provided high-quality datasets for MM5 PBL forecast verification. Data for a direct quantitative verification of the diffusion

TABLE 3.5
Verification Statistics Based on the WTM Surface Data

CaseTime/Date		# Sites	α_o	α_m	$\alpha_o - \alpha_o$	σ
Case B	T	24	12.04	13.38	−3.34	1.63
00 UTC 7 April 2002	T_d	24	11.42	12.54	−1.13	1.33
Case D	T	28	4.71	9.96	−3.35	2.44
12 UTC 23 November 2002	T_d	27	0.59	0.78	−0.19	1.30
Case E	T	24	3.75	4.13	−0.38	1.01
12 UTC 21 February 2003	T_d	23	3.26	2.73	0.52	0.79

Note: α and σ in °C are the mean and standard deviation, respectively. The subscript o denotes observation, while m denotes model.

Source: Chang, C.-B. and Gill, T.E., 2005, MM5 and HPAC experiments, *Meteorol. Atmos. Phys.*, 90, 127–138.

computations were not available. However, WTM/HPAC simulations based solely on the WTM observations provided a benchmark for qualitative assessment of the MM5/HPAC simulations. Also, the role of MM5 grid resolution (3 km vs. 9 km) and lateral boundary conditions in the performance of HPAC was evaluated.

The details of experimental design and simulated results for these three case studies can be found in the paper by Chang and Gill (2005). Some significant findings of the paper are summarized in this section. Table 3.5 shows the 12-h MM5 verification statistics computed at the WTM sites for temperature (T) and dew point (T_d). The model overpredicted T and T_d in Case B. In Case D, the model did less well in T but the best job for T_d. Case E was much better predicted overall by the model than the other two cases. Also, the MM5 12-h simulated surface flow patterns were in good agreement with the WTM observations for all three cases. The mean square errors of T have about the same size as those of 24-h values for typical regional model forecasts summarized by Anthes (1983).

The GB (sarin) agents delivered as a 500-kg bomb at several prespecified sites were selected as an incident model in all HPAC runs. Comparison between the HPAC simulated 12-h GB plumes using the hourly WTM reports and the hourly MM5 surface winds in Case B revealed no significant differences in location, shape, and concentration level. Also, there was almost no difference between the HPAC computed 12-h GB plumes using the MM5 coarse-mesh (9 km) and fine-mesh (3 km) simulated hourly profile data, respectively. However, a shift in the GB plume location was noted when the MM5 surface winds were replaced by the MM5 profile data to drive HPAC. One of interesting and unexpected outcomes was the HPAC runs with the MM5 grid-point data in Cases D and E. The major features of the GB plumes did not resemble their counterparts in the WTM/HPAC runs. The plume pathways in the grid-point data experiments were not consistent with the model surface flow: the surface GB plumes veered far away from the prevailing winds. As in Case B, the MM5/HPAC runs based on the model profile data performed fairly well, but the similar runs with the model surface data appeared to result in lower forecast skill in both cases.

The sensitivity of HPAC diffusion model in response to different lateral boundary conditions was also tested in Case E. This involved an MM5/HPAC run using the Eta model forecasts instead of analyses as in all simulations described above to generate the lateral boundary conditions for MM5. In principle, the Eta analyses continuously updated by observations should provide more accurate boundary conditions for the MM5 model. But to aid development of a real-time operational prediction and warning system, the ultimate goal of this research, we must invoke large-scale model forecasts to close the system. The question being investigated here is whether altering the boundary conditions would significantly alter the MM5 forecasts and subsequent HPAC diffusion computations in the central portion of the domain. The results suggest the GB plumes produced using the MM5 profile data from two different boundary conditions are almost identical in the two runs.

3.3.4.2 MM5/FDDA Simulations

A method known as observational nudging (Newtonian relaxation) is employed in FDDA. There are other more sophisticated FDDA methods involving the Kalman filter or three-dimensional variational (3DVAR) procedures (Kalnay 2003), but these advanced treatments are much more computationally intensive and require background information generally not available in a fast-response scenario. The Newtonian relaxation technique represents a good balance between complexity, timeliness, and accuracy, which is a guiding principle in U.S. Army "nowcasting" efforts (Dumais et al. 2003).

Several case studies of MM5/FDDA using the WTM surface tower data were carried out. In conjunction with Dr. Dumais of ARL (Army Research Laboratory) at White Sands Missile Range (WSMR), New Mexico, trials were also conducted in MM5/FDDA using ACARS data. Computer programs for processing ACARS data to provide the input files for MM5/FDDA were also developed, and two case studies were completed. In the 5–6 January 2005 case, 24 ACARS vertical profiles over north West Texas were used in FDDA. In the second case of 5–6 May 2005, FDDA with more than 90 profiles over a bigger MM5 domain was conducted in simulating the severe weather environment in Central Texas. The results of the second case are presented.

3.3.4.2.1 FDDA with WTM Data

The MM5 observational nudging package was used in the FDDA experiments. This empirical approach was first proposed by Hoke and Anthes (1976). The MM5 observational nudging parameters selected in this study were as follows:

- Nudging factor for wind, temperature, and humidity is 4×10^{-4} s^{-1}
- Horizontal radius of influence is 100 km from the observation site
- Vertical radius of influence is 0.002 (about 20 m) from the level of $\sigma = 0.995$
- Time window is 60 min. centered at the observation time

In Case C, FDDA failed to produce notable improvement in dry-line forecasting. In Case D, MM5 performed equally well in reproducing the observed surface flow with and without FDDA. In Case E, MM5 overpredicted the surface wind speeds but did very well for T and T_d forecasting (Table 3.6). All case studies revealed that

TABLE 3.6

Surface *u*, *v*, and *T* Differences (CNTR-FDDA) at Every 3 h

Time (h)	3	6	9	12	15	18	21	24
u mean	0	−0.05	−0.11	−0.09	0.01	0.01	0.00	0.00
SD	0	0.17	0.34	0.32	0.08	0.04	0.02	0.03
v mean	0	−0.02	0.01	0.09	0.03	0.00	−0.01	0.00
SD	0	0.09	0.13	0.31	0.09	0.06	0.03	0.03
T mean	0	0.00	0.00	0.00	0.00	0.00	0.00	0.00
SD	0	0.01	0.07	0.07	0.05	0.03	0.02	0.02

Note: *u* and *v* are in m/s, and *T* is in °C.

FDDA with the WTM surface had very minor impacts on the MM5 model performance. As an example, the results of Case D are presented below.

Figures 3.17 and 3.18 show the 12-h simulated surface wind vector and temperature differences between the control run (CNTR) without FDDA and the run with 6-h FDDA between 6-h and 12-h model time. As expected, at the end of FDDA the most notable discrepancies with the highest magnitudes close to 3.5 m/s and 0.50°C in the surface winds and temperatures, respectively, occur over the WTM area. However, the differences dissipate rapidly after FDDA is turned off. In 3 h, the maximum differences are less than 1 m/s in wind speed, while 0.30°C in temperature (not

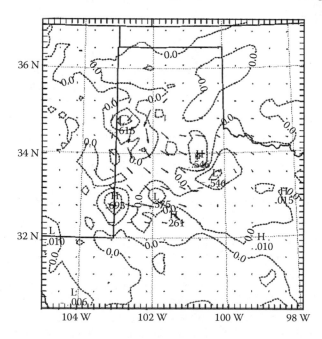

FIGURE 3.17 Surface vector wind and *T* differences (CNTR-FDDA) at 12 UTC 23 November 2002 (Case D). The contour interval for *T* is 0.2°C. The maximum speed difference is 3.2 m/s.

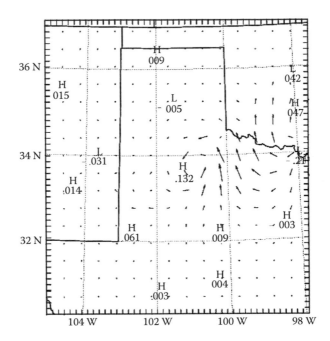

FIGURE 3.18   Surface vector wind and T differences (CNTR-FDDA) at 18 UTC 23 November 2002. The maximum speed difference is 0.4 m/s.

shown). Increasing the length of FDDA to 9 h (from 3 h to 12 h) resulted in minor changes.

Table 3.6 shows the means and standard deviations (SDs) of the surface wind (u, v), and T differences averaged over the model domain as a function of time. The means show temporal oscillations, while SDs increase monotonically after FDDA is invoked at 6 h. SDs reach their highest values around the end of FDDA at 12 h, drop precipitously in 6 h, and become quite small at 18 h. Also, the surface SDs are one order of magnitude larger than their counterparts in the model upper layers (not shown). Clearly, the insertion of WTM tower data has little effect on the MM5 forecast, and any impact on the model surface flow diminishes rapidly after the termination of FDDA. The indicator is that the effect of FDDA with WTM data on MM5/HPAC prediction would be negligible.

3.3.4.2.2   FDDA with ACARS

The nudging parameters were similar to those mentioned earlier except that the temporal window was reduced to 30 min., centered at the observation time. Several MM5/FDDA model simulation experiments were carried out on Case F using ACARS data, and the sensitivity of the model forecasts to changes in the nudging parameters was examined. Results of two experiments are presented. They are (1) a 36-h, 0000 UTC 8 to 1200 UTC 9 May, control run (CNTR) and (2) a 36-h FDDA run with 12-h observational nudging between the 12 h and 24 h of the model integration. The ACARS observations obtained for the case study initially consisted of 1928 data points of wind and temperature fields. The number dropped to 1213 after averaging observations within the same proximity, meaning that they were reported in the same time slot and at the same vertical level while within 10 km (half the grid size of the individual model grid points). Also, in this case study many more observations were inserted into the model upper layers than the middle and lower layers.

Figure 3.19 shows the 36-h (CNTR-FDDA) surface vector wind and T differences. There are clear differences between the two runs at 12 h after the termination of observational nudging. The wind field shows over 5 m/s in Central Texas, while the T field reveals several maximum and minimum centers with a magnitude close to 60°C near the U.S.-Mexico border. Table 3.7 shows the means and SDs of the u, v, and T differences in the model lower troposphere ($\sigma = 0.68$ or $p \approx 700$ hPa) as a function of time. FDDA causes temporal oscillations in the means between 12 and 36 h but not in the SDs. The SDs increase rapidly as FDDA starts at 12 h and approach their highest values around 24 h. They remain quite high and close to those at 24 h at the end of simulations. The similar features are observed in the middle and upper levels (not shown). The relatively small amplitudes of the temporal oscillations suggest that the model was only undergoing mild adjustments to the data insertion, while the SDs reveal lasting impact of the ACARS profile data on the MM5 forecasts throughout the entire model atmosphere 12 h after the termination of FDDA.

Figure 3.20 shows the 24-h HPAC prediction using the MM5 simulated profile data in (1) the control and (2) FDDA runs. GB agents were released as a 500-kg bomb at 1200 UTC, 8 May 2005, at three separate locations indicted by the incident icons. The meteorological data was updated every 2 h during the 24-h HPAC computations. The GB plumes spreads along the direction of the prevailing low-level flow

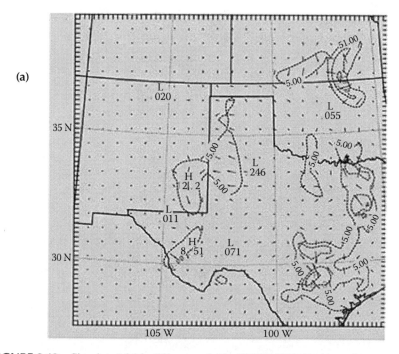

FIGURE 3.19 Simulated 36-h difference fields (CNTR-FDDA). (a) surface vector winds, the contour interval is 5 m/s; and (b) surface T, the contour interval is 1°C.

(b)

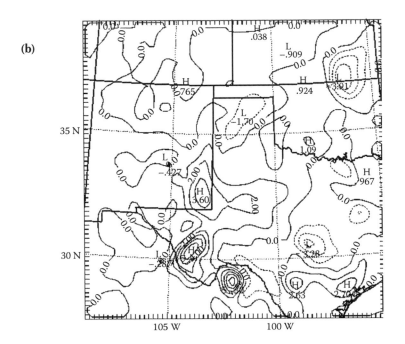

FIGURE 3.19   (continued).

TABLE 3.7
u, v, and T Differences (CNTR-FDDA) at Every 3 h at $\sigma = 0.68$ (~ 700 hPa)

Time	9	12	15	18	21	24	27	30	33	36
u mean	0.00	−0.04	0.06	0.15	0.09	−0.04	−0.02	−0.11	0.03	0.09
SD	0.00	0.19	0.58	1.09	1.37	1.78	1.92	1.71	1.87	1.65
v mean	0.00	0.02	0.15	0.17	0.23	0.33	−0.05	−0.24	−0.14	−0.18
SD	0.00	0.12	0.69	1.42	1.78	2.39	2.09	1.87	1.92	1.65
T mean	0.00	−0.01	−0.03	0.02	0.04	0.04	−0.01	0.00	0.02	0.01
SD	0.00	0.05	0.21	0.29	0.37	0.47	0.48	0.49	0.41	0.39

Note: u and v are in m/s, and T is in °C.

(not shown). Here we are interested in assessing the impact of ACARS on the plume location rather than the absolute values of concentration. There are similarities in the general patterns of the GB plumes between the two runs. There are minor differences in the first 6 h. However, as the plumes spread further away from the release point, differences are quite notable over the areas with relatively large changes in the surface winds shown in Figure 3.20 in Central Texas and Oklahoma.

3.3.5   SUMMARY AND DIRECTIONS FOR FUTURE RESEARCH

This publication describes a research project founded upon a combination of state-of-the-art meteorological field and modeling techniques to predict PBL dispersion

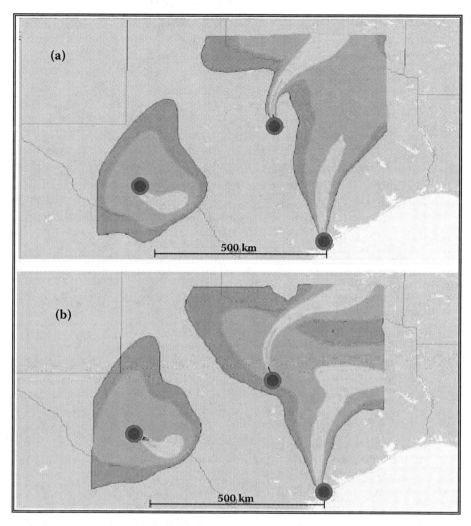

FIGURE 3.20 (See color insert following page 46.) HPAC simulated 24-h surface GB plumes at 1200 UTC 9 May 2005; (a) control run and (b) FDDA run.

of hazardous substances. The research is carried out based on real-data simulation experiments using the MM5 weather model and HPAC diffusion model in conjunction with high-frequency meteorological observations. Results of six case studies are presented. The experiments include multiple-grid MM5 simulations, MM5/HPAC diffusion simulations, MM5/FDDA simulations with the WTM and ACARS data, and sensitivity tests of HPAC to various forms and sources of weather data input. We have performed in-depth intercomparisons of the model results to highlight strengths and weaknesses of the joint MM5 and HPAC modeling approach and to demonstrate the potential for significant improvement of PBL dispersion forecasting through data assimilation. The major findings are:

(1)   Minor differences in the HPAC computations are found between experiments using the 3-km mesh and 9-km mesh MM5 data. This is likely due to the relatively smooth terrain of West Texas, such that its effects on the surface flow of the two scales are not very distinguishable. It appears that a 10-km grid mesh weather model would be adequate for an operational warning system over rural areas of relatively consistent terrain. However, this would probably not be the case in a rough urban setting, in which detailed urban terrain with substantial variations within a distance of 10 km must be considered.

(2)   The use of MM5 simulated surface flow alone to run the HPAC model worked reasonably well in some cases but not so in other cases. The failure has a profound impact on 12-h plume prediction. The most desirable built-in HPAC input mode appears to be the 3D profile form of MM5 model data in all cases.

(3)   The HPAC grid mode option of creating weather data on isobaric surfaces should probably be avoided. It is not clear why such a format led to an improbable dispersion pattern of chem-bio plumes inconsistent with the flow patterns. This might be caused by the use of isobaric coordinate and the domain-averaged geopotential heights in defining data locations. Over WTM on the "high plains" of the central United States, several isobaric surfaces specified lie beneath the surface.

(4)   We evaluated the impact of lateral boundary conditions imposed on the MM5 model. The MM5 domain is about ten times bigger than the HPAC domain. The ratio appears to be sufficiently "large" to prevent any serious contamination of 12-h diffusion computations due to the lateral boundary conditions. For a longer HPAC run (e.g., 24 h) or a strong forcing case, the ratio might need to be increased. However, when doing so, one would need to consider the additional time and computational resources potentially demanded by a much larger domain.

(5)   Our studies show that the MM5/HPAC conjugation can provide useful prediction of airborne transport of hazardous materials near the surface. It also demonstrates that the accuracy of HPAC computation strongly depends on the performance of MM5. The forecast skill of mesoscale models is likely to be a function of weather scenarios and the terrain over which the models are being run. Because the numerical techniques are different and the model physics (e.g., PBL, surface, and moist processes) vary considerably among different mesoscale meteorological models, we anticipate that there would be discrepancies between the predictions of individual models.

(6)   We have determined that nudging the models with the WTM surface observations alone appeared to have very limited impact on the mesoscale weather and dispersion forecasts. It is basically three-dimensional (two horizontal dimensions plus time) data assimilation over a small area. The impulses caused by the data insertion dissipated rapidly within 3 h after the turnoff of data nudging regardless of the length of relaxation time scale for model adjustment and nudging period. However, impulses caused by FDDA

with vertical profile data such as ACARS can have long-lasting effects on PBL wind forecasting.

(7) The experiments suggest that ACARS data has high potential to advance our expertise in mesoscale NWP and supporting the need to rapidly and accurately adjust high-resolution meteorological model forecasts to real-time observations. The experiments have also revealed the potential impact of ACARS on MM5/HPAC prediction. More case studies of diverse events (e.g., oscillating dry lines, rainstorms, wind/dust storms) are desired. We are particularly interested in those occurring in the vicinity of the WTM domain. This will enable us to carry out comprehensive model validation with the Mesonet temperature, wind, and rainfall observations, enabling identification in which diffusion predictions can be enhanced through the use of composite forecast statistics derived from a large number of diverse mesoscale weather events.

Based on these results, it is reasonable to hypothesize that FDDA with ACARS data could significantly improve the quality of mesoscale NWP and the subsequent dispersion prediction in the atmospheric boundary layer. The hypothesis needs to be systematically tested.

An advanced mesoscale model may perform well in some cases but not in others, and the model responses to data assimilation are likely to vary with weather scenarios. There would be a potential benefit of using several model winds separately to run HPAC. A composite result of the HPAC runs would give a more complete depiction of the potential surface-based transport of hazardous agents. Also, the latest version of HPAC allows us to display dispersion output within a geographic information system (GIS) environment. We should explore all possible GIS applications in dispersion modeling and damage assessment.

The MM5 model will eventually need to be replaced by the new-generation Weather Research and Forecasting (WRF) model (e.g., Klemp 2004) to provide meteorological data inputs to HPAC. WRF has many improved features over MM5 and will enhance our ability to generate more timely and accurate near-surface flow for dispersion computation in an urban setting. Also, we would like to evaluate the impact of ACARS data on WRF/HPAC simulations. An interesting experiment would be to assimilate both WTM and ACARS data simultaneously into the WRF model.

From the standpoint of emergency response to toxic dispersion in the urban PBL, the microscale prediction is a most important task because of the life-threatening nature of plumes of hazardous materials acting at small spatial scales where their concentrations would be highest. Here, our main concern is the movement of plumes and their vertical and horizontal dispersion. Large eddy simulation (LES) may be employed in the design of microscale prediction. LES is an efficient and consistent computational technique available for simulating turbulent flow in the atmospheric PBL. In LES the larger scales of motion (plumes of hundred meters) are resolved explicitly and the smaller ones (subgrid scales) are parameterized. This approach requires fewer assumptions and uses internally generated PBL parameters to represent the subgrid-scale motion. Over recent years, LES has enabled researchers to probe various turbulent flows by generating unprecedented high-resolution, four-dimensional turbulence data (Meoeng and Sullivan 1994). As a consequence, scientists start to gain a better understanding of some of the complex PBL phenomena (e.g., plume-generation mechanism in a buoyancy-driven PBL). A possible approach is using the LES technique jointly with a mesoscale model and diffusion modeling systems (e.g., HPAC) to predict the movement of toxic plumes.

Hopefully, this work will encourage further research on how to capitalize state-of-the-art modeling techniques and high-frequency meteorological measurements to improve our real-time ability in response to atmospheric releases of hazardous materials from an industrial accident or terrorist act. An urgent and challenging issue that needs to be addressed is diffusion model validation. Without comprehensive validation, even a well-designed prediction system cannot be accepted for operational application.

3.4   TRANSPORT OF BIOLOGICAL AND CHEMICAL THREAT AGENTS IN SOIL

3.4.1   INTRODUCTION

While the movement of chem-bio agents within the atmosphere has previously been discussed, their transport differs in the soil. The subsequent transport and fate of chem-bio agents in this soil will be discussed in this section. Soil is a dynamic, natural three-phase system comprised of solids, liquids, and gases. The transport and fate of biological and chemical threat agents depends upon the state of the agent and how the agent reacts with the soil. In general, gases will mix and move through the gaseous soil phase while liquids will move through the liquid phase or sorb to the solid phase. Biological and chemical threat agents applied as solids can be mixed with soil solids or dissolve in the aqueous phase surrounding the solids. Once the agent comes in contact with the soil, it may sorb to the soil, move with the air or water in the soil, or be degraded within the soil.

A biological and chemical threat agent that is soluble in water and does not exert a vapor pressure will generally exist in the soil in two phases: as dissolved solute in soil water and as a sorbent to soil particles or organic matter. Mathematically, the partition of a specific chemical can be written as

$$C = \rho_b C_a + \theta_v C_l \qquad (3.4.1)$$

where C is the chemical content in the soil (M/L^3), ρ_b is the soil bulk density (M/L^3), C_a is adsorbed chemical concentration expressed as mass of sorbent per mass of dry soil (M/M), θ_v is the volumetric water content (L^3/L), and C_l is dissolved chemical concentration expressed as mass of solute per volume of soil solution (M/L^3) (Jury et al. 1991). Please note that the fundamental units M = mass, L = length, and T = time are used. If the chem-bio agent is also present as a gas, an additional term must be added

$$C = \rho_b C_a + \theta_v C_l + f_a C_g \qquad (3.4.2)$$

where the f_a is the air concentration of the soil (L^3/L^3) soil and C_g is gaseous chemical concentration (M/L^3) (Scott 2000). Biological and chemical threat agents move by diffusion, convection, or by both modes.

3.4.2 DIFFUSION

Diffusion is the movement by a gradient from an area of biological or chemical threat agent concentration to an area of low concentration. It is generally considered conceptually to be the differences in the biological or chemical threat agent concentrations divided by the distance between the two concentration areas. Diffusion always goes from an area of high concentration to an area of low concentration. Molecular diffusion for steady-state transport is generally written as

$$J = -D \frac{\partial C}{\partial z} \tag{3.4.3}$$

where J is the solute flux density $(M/L^2/T)$; D is the molecular diffusion coefficient (L^2/T), and z is distance (L). For transient-state conditions, the conservation of mass equation in one-dimension without generation or consumption is

$$\frac{\partial C}{\partial t} = -\frac{\partial J}{\partial z} \tag{3.4.4}$$

where t is time. Combining Equation (3.4.3) with (3.4.4) gives what is generally considered to be Fick's second law,

$$\frac{\partial C}{\partial t} = \frac{\partial}{\partial z} \left(D \frac{\partial C}{\partial z} \right) \tag{3.4.5}$$

For solute diffusion coefficients independent of solute concentration, Equation (3.4.5) becomes

$$\frac{\partial C}{\partial t} = D \frac{\partial^2 C}{\partial z^2} \tag{3.4.6}$$

If the initial concentration of the solute in the soil is negligible, only diffusion occurs, and there is no uptake of the solute; the solute will diffuse downward, and the solute concentration distribution will be as follows:

$$C(z,t) = C_0 \, erfc \left[\frac{z}{2\sqrt{Dt}} \right] \tag{3.4.7}$$

where C_0 is the initial solute concentration and *erfc* is the complementary function. Note in this instance z is depth (L).

3.4.3 CONVECTION

Convection (also known as advection) is a more passive movement of biological and chemical materials, in that the agent moves with the water or air within the soil. One-dimensional solute flow rates in the z dimension can be represented by

$$J_z = -q_z C_l \qquad (3.4.8)$$

where q_z is the water flow rate (L T⁻¹). Equation (3.4.8) can be rewritten as

$$J_z = -v\theta_v C_l \qquad (3.4.9)$$

where v is the average water velocity (L/T) in the z direction.

3.4.4 TRANSIENT-STATE CONVECTION

Combining (3.4.4) and (3.4.9) gives the one-dimensional conservation equation for convective flow

$$\frac{\partial C}{\partial t} = -\frac{\partial}{\partial z}\left(v\theta_v C_l\right) \qquad (3.4.10)$$

from Scott (2000). Application of Equation (3.4.10) to real-world problems would indicate a sharp concentration front with the high solute concentration, C_0, abruptly changing to a low initial value. This type of flow has been called "piston flow" and seldom, if ever, occurs in the environment. The diffusion portion of the diffusion-convention advances or retards the flow and precludes the sharp line of demarcation. Due to dispersion, preferential flow, and fingering, solute flow is unstable and presents diffuse flow boundaries (Hillel 1998). Examples of this type of breakthrough curve will be shown in the Zumwalt accomplishment section later on in this chapter.

There is not only mixing by diffusion, but there is also a mechanical dispersion. Mechanical dispersion occurs when the two solutions of differing chemical compositions meet. Mixing along the direction of flow path is called longitudinal dispersion, while dispersion perpendicular to the flow path is called transverse dispersion (Scott 2000). The mathematical equation for the flux density of solutes by hydrodynamic dispersion is

$$J_h = -D_h \frac{\partial C_l}{\partial z} \qquad (3.4.11)$$

where D_h is the hydrodynamic dispersion coefficient (L³/T).

3.4.5 Solute Transport Equations

3.4.5.1 Steady-State Transport

The steady-state solute transport equation can be written as follows:

$$J_s = J_l + J_g + J_a \tag{3.4.12}$$

where s is for soil, l is for liquid, g is for gas, and a is for solid or adsorbed phases, respectively. Using Fick's law as before

$$J_g = -f_a D_g \frac{\partial C_g}{\partial z} \tag{3.4.13}$$

where f_a is the aeration porosity (L/L) and C_g is the solute concentrations of the gas phase (M/L). The total solute flux density in the liquid phase can be written as follows:

$$J_l = -\theta_v D \frac{\partial C_l}{\partial z} + v \theta_v C_l \tag{3.4.14}$$

Since J_a is the flux density of the adsorbed phase, it can be neglected. Therefore,

$$J_s = -\theta_v D \frac{\partial C_l}{\partial z} + v \theta_v C_l - f_a D_g \frac{\partial C_g}{\partial z} \tag{3.4.15}$$

as shown in Scott (2000).

3.4.5.2 Transient-State Transport

The general mass balance equation for one-dimensional solute flow can be written as follows:

$$\frac{\partial}{\partial t} \left(\theta_v C_l \right) = -\frac{\partial J_s}{\partial z} \tag{3.4.16}$$

For a conservative solute with minimal diffusion in the gas phase, Equations (3.4.14) and (3.4.16) can be combined to produce

$$\frac{\partial C}{\partial t} = \frac{\partial}{\partial z} \left(\theta_v D \frac{\partial C_l}{\partial z} - v \theta_v C_l \right) \tag{3.4.17}$$

This can be simplified to

$$\frac{\partial C_l}{\partial t} = D\frac{\partial^2 C_l}{\partial z^2} - v\frac{\partial C_l}{\partial z} \qquad (3.4.18)$$

This is known as the convective-dispersive equation (CDE) (Scott 2000). For chem-bio agents, Equation (3.4.16) can be written as follows:

$$\frac{\partial}{\partial t}\left(\theta_v C_l\right) = -\frac{\partial J_s}{\partial z} \pm r \qquad (3.4.19)$$

where r is a source/sink term. For biological and chemical threat agents, the r is a sink term related to the sorption of agent to the soil and can be expressed as follows (Scott 2000):

$$r = -\rho_b\frac{\partial S}{\partial t} \qquad (3.4.20)$$

where S is the amount of solute in the adsorbed phase (M solute/M soil). Equation (3.4.2) can now be revised as follows:

$$C = \rho_b S + \theta_v C_l \qquad (3.4.21)$$

Differentiating (3.4.21) with respect to t, assuming the bulk density does not change and substituting the result into (3.4.18) gives the following:

$$\frac{\partial}{\partial t}\left(\theta_v C_l\right) + \rho_b\frac{\partial S}{\partial t} = \theta_v D\frac{\partial^2 C_l}{\partial z^2} - v\theta_v\frac{\partial C_l}{\partial z} \qquad (3.4.22)$$

Applying the chain rule, assuming that θ_v does not change with respect to z or t, and collecting the $\partial C/\partial t$ terms yields the following:

$$R = 1 + \left(\frac{\rho_b}{\theta_v}\right)\left(\frac{dS}{dC_l}\right) \qquad (3.4.23)$$

where R is the retardation coefficient. For cations and neutral molecules, the R will be greater than 1; for no sorption, R becomes 1; and for anion exclusion, R becomes less than 1 (Scott 2000). The CDE that accounts for sorption by solid surfaces can be written as follows:

$$R\frac{\partial C_l}{\partial t} = D\frac{\partial^2 C_l}{\partial z^2} - v\frac{\partial C_l}{\partial z} \qquad (3.4.24)$$

For chem-bio agents that undergo decomposition, the r is represented as follows:

$$r = -kC \tag{3.4.25}$$

where k is the first-order decay constant (T^{-1}). Adding the retardation term (3.4.25) to (3.4.24) gives the CDE for a chemical and biological threat agent that not only sorbs to the soil solid surfaces but also undergoes first-order decomposition,

$$R\frac{\partial C_l}{\partial t} = D\frac{\partial^2 C_l}{\partial z^2} - v\frac{\partial C_l}{\partial z} - k\theta_v C_l \tag{3.4.26}$$

If one defines an apparent diffusion-dispersion coefficient, D_e, as follows:

$$D_e = \frac{D}{R} \tag{3.4.27}$$

and the apparent pore-water velocity, V_e as follows:

$$V_e = \frac{v}{R} \tag{3.4.28}$$

and the apparent degradation rate coefficient, B, as follows:

$$B = \frac{k}{R} \tag{3.4.29}$$

the final CDE becomes

$$\frac{\partial C_l}{\partial t} = D_e\frac{\partial^2 C_l}{\partial z^2} - V_e\frac{\partial C_l}{\partial z} - B\theta_v C_l \tag{3.4.30}$$

This transient-state equation connects the one-dimensional flow of soil-water with the temporal and spatial transports of solutes as influenced by sorption and degradation (Scott 2000).

3.4.6 Prediction of Biological and Chemical Threat Solute Transport in the Soil

The first form of flow to be considered would be the steady-state case. In this case $\partial\theta_v/\partial t$ is zero. Transient-state conditions occur when $\partial\theta_v/\partial t$ is not zero. If a biological or chemical threat agent were applied to the soil surface, the mathematical solution to (3.4.18) would be as follows:

$$C(z,t) = \left[\frac{C_0 Z}{2\sqrt{\pi D t^3}} \right] \left[\exp\left\{ -\frac{(z-vt)^2}{4Dt} \right\} \right] \tag{3.4.31}$$

given the appropriate boundary conditions of Jury et al. (1991). Van Genuchten and Wierenga (1986) present four analytical solutions with differing inlet and exit boundary conditions.

3.4.7   PREDICTION OF BIOLOGICAL AND CHEMICAL THREAT SOLUTE TRANSPORT COEFFICIENTS IN THE SOIL

3.4.7.1   Molecular Diffusion Coefficients

For nonvolatile, nonsorbed biological and chemical threat agents, the apparent diffusion coefficients can be expressed as follows:

$$D_e = w D_0 \tag{3.4.32}$$

where D_0 is the solute molecular diffusion coefficient (L^2/T) and w is an empirical coefficient that is related to tortuosity (Scott 2000). For biological and chemical threat agents that are sorbed to soil surfaces, the apparent diffusion coefficient is further retarded. This apparent diffusion coefficient can be represented as follows:

$$D_e = \left(\frac{w}{R} \right) D_0 \tag{3.4.33}$$

3.4.7.2   Hydrodynamic Diffusion Coefficients

Once the mechanical dispersion associated with water is accounted for, the dispersion coefficient, D, is a function of two components—molecular (D_e) and mechanical (D_h) dispersion. D can be represented as follows:

$$D = D_e + D_h \tag{3.4.34}$$

Equation (3.4.34) can be altered by making assumptions as to the influence of molecular diffusion and rate of water flow. For example, if the molecular diffusion is negligible and the dispersion carries with the pore-water velocity,

$$D_h = \varepsilon v \tag{3.4.35}$$

where ε is the dispersivity. For unsaturated flow systems, molecular diffusion is not negligible and

$$D = w D_0 + \varepsilon v \tag{3.4.36}$$

where D_0 is the solute molecular diffusion coefficient and w is a tortuosity factor. At high pore-water velocities, the molecular diffusion coefficient can be neglected and

$$D_h = \varepsilon v^n \tag{3.4.37}$$

where n is an empirically determined constant. Most often, D incorporates both molecular diffusion and pore-water velocity terms and is written as follows:

$$D = wD_0 + \varepsilon v^n \tag{3.4.38}$$

The values for w, n, and D_e all depend on the soils and their water contents (Scott 2000).

3.4.8   SORPTION BY SOIL SURFACES

Previous discussions within this section have been concerned only with the movement of biological and chemical threat agents within the soil system. There has been no discussion about the loss of biological and chemical threat agent concentration from the soil due to sorption onto the solid surfaces that are the fabric of the soil. Sorption of biological and chemical threat agents does occur in real-world situations.

3.4.8.1   Sorption Models

Selim (1992) proposed a multirelational approach based upon the types of sorption sites:

$$S = S_e + S_k + S_{ir} \tag{3.4.39}$$

where the subscript e assumes rapid equilibrium, k is a kinetic reaction, and ir refers to irreversible retention.

3.4.8.1.1   Equilibrium Sorption Models

The simplest form of model for the sorption of biological and chemical threat agents to the soil is a linear model in which the rate of sorption is dependent on the concentration of the agent. A linear equilibrium model can be represented as follows:

$$S_e = K_d C_l + b \tag{3.4.40}$$

where K_d and b are empirical constants representing the slope and intercept, respectively. The K_d represents the equilibrium distribution coefficient or the slope. Often the intercept, b, is forced to be zero, reflecting the physical reality of the system. Additionally, the equilibrium sorption of organic molecules is dominated usually by the soil organic fraction. To account for this, the K_d is divided by the fractional organic matter content, f_{oc}. The organic carbon fraction equilibrium constant, K_{oc}, is represented as

$$K_{oc} = \frac{K_d}{f_{oc}} \tag{3.4.41}$$

For the equilibrium constant models, the S_e takes on the mathematical structure of the model. For example,

$$S_e = K_f C_l^N \tag{3.4.42}$$

represents a Frundlich equation model where K_f is the Frundlich equilibrium constant (L^3/M) and N is an exponent. For most pesticides, N is approximately 0.9 (Scott 2000).

For example,

$$S_e = \frac{aQC_l}{\left(1+aC_l\right)} \tag{3.4.43}$$

represents a Langmuir equation model where S_e represents the number of adsorbed molecules per unit mass of soil, C_l is the solute concentration in solution, and Q is the total number of adsorption sites available (Scott 2000).

3.4.8.1.2 Kinetic Sorption Models

Nonlinear sorption of chem-bio agents can be based on the different kinetics of sorption. Chemical nonequilibrium must also be considered in real-world problems. One of the early kinetic sorption models used the CDE first-order reactions (Scott 2000). This is represented as follows:

$$\rho_b \frac{\partial S_k}{\partial t} = k_f \theta_v C_l - k_b \rho_b S_e \tag{3.4.44}$$

where S_k and S_e are the kinetic and equilibrium sorbed solute and k_f and k_b are the forward and backward sorption rate coefficients (Scott 2000). For large values of k_f and k_b, sorption is rapid and approaches quasi equilibrium. For small values of k_f and k_b, sorption is slow and strong kinetic dependence is anticipated. At long times, Equation (3.4.42) reduces to

$$S_e = K_d C_l \tag{3.4.45}$$

and this kinetic equation reduces to a linear sorption form.

3.4.9 DEGRADATION

Biological and chemical threat agents not only move through the soil and sorb to the soil, but they are also degraded within the soil. These degradations may be due to abiotic means such as degradation by ultraviolet light and chemical volatilization. More often, however, the degradation is due to microbial breakdown by indigenous microflora. The indigenous microflora can change due to competitive advantage of the microorganisms if there are persistent quantities of a chem-bio agent within the rhizosphere.

3.4.9.1 Rate Models

A zero-order degradation rate model can be represented as follows:

$$\frac{\partial C_l}{\partial t} = -k_0 \tag{3.4.46}$$

where k_0 is the zero-order rate constant (M/L/T). Integrating (3.4.44) with respect to time gives

$$C_l = C_{l0} - k_0 t \tag{3.4.47}$$

where C_l is the solute concentration in solution and C_{l0} is the solute concentration in solution at $t = 0$. A first-order kinetic model mathematically becomes

$$\frac{\partial C_l}{\partial t} = -k_1 C_l \tag{3.4.48}$$

where k_1 is the first-order degradation rate coefficient. Equation (3.4.46) can be integrated to produce the following two working equations:

$$C_l = C_0 \exp\left(-k_1 t\right) \tag{3.4.49a}$$

$$\ln\left(C_l\right) = \ln\left(C_0\right) - k_1 t \tag{3.4.49b}$$

The half-life of the solute in the soil system is computed by (Scott 2000)

$$t_{\frac{1}{2}} = \frac{\ln\left(2\right)}{k_1} = \frac{0.693}{k_1} \tag{3.4.50}$$

3.4.10 ZUMWALT ACCOMPLISHMENTS

Sorption depends on the type and amount of the charge on the biological or chemical threat agent. Soils are typically negatively charged due to the clay mineralogy and the organic matter present. Most cationic agents readily sorb to the soil. Some soils with low pH values that contain kaolinite clay posed a positive charge and sorb anionic agents. The isoelectric point of the organic molecules also determines how tightly the agents are sorbed.

These principles will be illustrated using the toxin, ricin, peanut lectin (a surrogate for ricin), and aflatoxin B_1. One method of biological and chemical threat agent movement in the soil is for the agent to flow through the soil in a liquid state. Two aspects of this movement are: (1) When does the first quantity of the agent pass through the soil? and (2) When does the bulk of the agent pass through the soil?

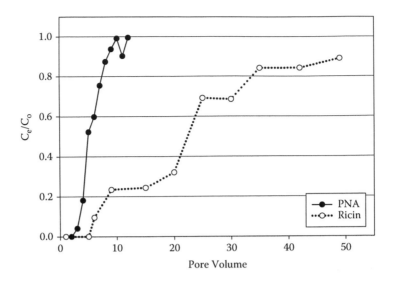

FIGURE 3.21 Breakthrough curves for peanut seed lectin and ricin solution moving through a Brownfield fine sand. (Basinger 2003)

Using a flow-through method similar to that described by Skaggs et al. (2002), peanut seed lectin (PNA) or ricin were passed through columns. The soils used were Amarillo fine sandy loam (fine-loamy, mixed, superactive, thermic Aridic Paleustalfs) or Brownfield fine sand (loamy, mixed, superactive, thermic Arenic Aridic Paleustalfs). The Amarillo has 63% sand, 28% silt, and 9% clay while the Brownfield has 84% sand, 13% silt and 3% clay. After the columns were saturated and at a steady-state flow rate, solutions containing PNA or ricin were added using a constant head pump. The concentration of solute in the effluent will be dependant on the adsorption of the solute on clay and soil organic matter. A plot of effluent concentration as a function of pore volume is called a breakthrough curve. Figures 3.21 and 3.22 are breakthrough curves for PNA and ricin for two soils. The Brownfield soil was used in Figure 3.21, while the Amarillo soil was used in Figure 3.22.

The difference between these two figures represents the components of the above-described equation. PNA and ricin are cations and are sorbed to the clay fraction. The greater the content of clay, organic matter, or sesquioxides (Fe, Al), the greater the sorption of these lectins. The lectins are sorbed at lesser pore volumes for the Brownfield soil than the Amarillo soil. This reflects the lower clay content of the Brownfield soil than the Amarillo soil. Additionally, differences were noted in how tightly the two lectins were bound to the soil. The PNA was less tightly sorbed and moved more quickly than the ricin. These differences have protection implications for safety. To sorb more biological or chemical threat agent, the clay fraction must be increased.

3.5   BIOLOGICAL THREAT TRANSPORT

The above sections describe the utility of models in predicting the transport of biological and chemical threat threats in the atmosphere (Section 3.2), and under given

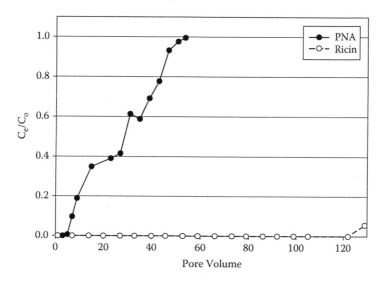

FIGURE 3.22 Breakthrough curves for peanut seed lectin and ricin solution moving through an Amarillo fine sandy loam. (Basinger 2003)

meteorological conditions (Section 3.3). In Section 3.4, the fate of those threats in the soil is discussed and approaches for modeling transport through the soil are presented. However, for biological threats, consideration must be given to further transport once populations of organisms are exposed. Moreover, in many cases, the use of populations of host organisms can prove to be an effective transport mechanism. This section focuses on the use of models to understand the interactions between biothreats (that have become established in natural host populations) and human populations. In addition, approaches for placing these models in a spatially explicit context in order to predict transport will be discussed.

Transporting biological and chemical threats to targeted populations is a significant obstacle. However, there are natural routes of transport via biological populations that can be exploited. For example, it is known that many of the disease pathogens identified as probable biological terrorism agents and suspected to be available to terrorist organizations are zoonoses (diseases that normally exist in wild animals but are transmissible to man). Many such diseases occur and cycle naturally within wild animal populations in the United States and are called enzootic foci. As a result, weaponized or enhanced strains of zoonotic pathogens could potentially be employed as biological threat agents and could be disseminated using naturally occurring or bolstered populations of endemic wild-animal host and vector species. Such attacks potentially require much less technical expertise than is required to develop, weaponize, and disseminate some of the more publicized biological agents, such as anthrax. In particular, zoonoses that have been identified as having the potential for use as biological terrorism agents include plague (*Yersinia pestis*), tularemia (*Francisella tularensis*), and several viral hemorrhagic fevers and encephalitides (Western equine encephalitis, Saint Louis encephalitis, West Nile encephalitis) (Childs et al. 1998; Gubler 1998; O'Toole et al. 2002; Rotz et al. 2002; Chang, Glynn et al. 2003). Furthermore, the bacteria that cause plague and tularemia are identified as Category

A biological threat agents by the Centers for Disease Control and Prevention (Rotz et al. 2002). Finally, it is also known that plague, tularemia, and at least one hemorrhagic fever virus have been weaponized through genomic or proteomic engineering to be antibiotic resistant or vaccine subverting.

In the majority of human exposures to zoonoses, the bite of an arthropod vector or direct contact with infective fluids or tissue from an infected host or reservoir animal is involved. Thus, there are two potential dissemination mechanisms that could be exploited to intentionally introduce either pathogenic organisms into a human or domestic livestock population. The first is a focused terrorist incident in which an aerosolized form of the pathogen is released with the intent of initiating a primary pneumonic plague or tularemia epidemic in a clustered population. In this case, models from Sections 3.2 and 3.3 can be used to predict transport of the agent. However, this form of attack requires a level of funding and technical expertise probably available only to state-sponsored or supported terrorist organizations. Alternatively, arthropod vectors infective with a weaponized, more pathogenic, or antibiotic resistant strain of *F. tularensis* (in ticks), *Y. pestis* (in fleas), or an arboviral encephelatide (in mosquitoes) could be seeded into an area to induce an epizootic near human population centers (Galimand et al. 1997). This method of biological threat agent delivery was effectively used by the Japanese during World War II (~1938–1940) in China to cause a plague epidemic (Tomilin and Berezhnoi 1985; see Chapter 1 for more historic information). Furthermore, this latter scenario, being less expensive and technologically less sophisticated, is a far more realistic bioterrorism threat to communities throughout the southwestern United States, for example, than a large-scale aerosolized pathogen attack.

The introduction of zoonoses is particularly effective when natural hosts are present. For example, outbreaks of plague within colonies of black-tailed prairie dogs (*Cynomys ludovicianus*) are relatively common in the southwestern United States. Moreover, prairie dogs are increasingly associated with urban and suburban residential areas in the southwestern United States (Pepper et al. 2004), and the canine and feline predators and other animals and arthropods associated with prairie dog colonies may readily harbor *F. tularensis*. Thus, the intentional introduction of epizootic plague or tularemia into areas with populations of black-tailed prairie dogs increases the likelihood of disease crossover into human populations. Dengue hemorrhagic fever, on the other hand, can be found wherever potential mosquito vectors (*Aedes aegypti* and *Ae. albopictus*) are naturally occurring or have been introduced, whether in rural or urban areas (Mitchell 1991). These mosquito vectors are peridomestic, often breeding in artificial containers in and around human dwellings (e.g., old tires, flowerpots, water storage containers). Additionally, these mosquitoes prefer to feed during daylight hours and are most active in the early morning and late afternoon (Moore and Mitchell 1997).

3.5.1 Modeling the Threat of Zoonotic Introductions

Although a thorough review of all models that have been developed for zoonotic pathogens is beyond the scope of this chapter, we will illustrate the general approach for applying models to assist in the understanding of zoonotic disease dynamics in

TABLE 3.8

Parameters in a Possible SIR Model of Plague within Prairie Dog Colonies

Parameter	Meaning
r_{PD}	Prairie dog's reproductive rate
p	Probability of inherited resistance
K_{PD}	Prairie dog's carrying capacity
d_{PD}	Death rate of prairie dogs
β	Transmission rate
m	Infectious period
g	Probability of recovery
μ_F	Movement of fleas
a	Flea searching efficiency
r_F	Flea's reproductive rate
d_F	Death rate of fleas
K_F	Flea's carrying capacity

Note: The model is based on a model of bubonic plague proposed by Keeling and Gilligan (2000a).

light of potential bioterrorism attacks using *Y. pestis*. One approach to predicting risk is to use statistical models that are based on empirical data gathered from previous plague outbreaks. An alternative approach to understanding the potential threat of an introduced zoonosis is to develop spatially explicit models of potential host populations. Although these models do not require as much prior data as statistical models, they do require enough data or other knowledge to be available to derive estimates of model parameters (e.g., see Table 3.8). Such models can be utilized for variety of purposes. For example, simulation of models across a range of parameter values is useful for discerning the factors that are important determinants of human exposure. In addition, simulation of empirically based versions of models can be used to conduct experiments involving a variety of intentional disease-introduction scenarios. Such experiments are particularly useful in the case of zoonotic introductions because of the importance of environmental drivers of potential host populations.

3.5.1.1 Statistical Models

Statistical models can be used to predict the risk, or probability, of a disease outbreak based on data from previous outbreaks. The most common statistical model used in this context is the logistic model

$$\ln\left(\frac{p}{1-p}\right) = \beta_0 + \beta_1 x_1 + \beta_2 x_2 + \cdots + \beta_k x_k$$

where p is the probability of plague, x_i are predictor variables, and β_i are model parameters that are estimated based on previous data. When the predictor variables vary spatially, the logistic model can be integrated into a GIS to create spatially explicit maps of risk. For example, Eisen et al. (2007) used data from the Four Corners region of the United States, collected from 1957 to 2004, to identify landscape features associated with human cases of plague. They then predicted where future human cases were likely to occur. Because of their reliance on past data, statistical models will not necessarily provide insight about the potential implication of an intentional introduction of plague. However, they can provide information about the likelihood of an introduction becoming established in a natural host population. Moreover, they also can be used to distinguish intentional introductions from naturally occurring plague outbreaks.

3.5.1.2 Simulation Models

Ultimately, models can provide information on the potential number of human cases that would result from an intentional introduction. However, the ability to successfully design and develop a realistic and effective stochastic simulation model of a biological terrorism attack, which can then be utilized to generate dynamic patient-flow rates, requires the consideration and integration of five critical components. These critical components include: (1) characteristics of the specific biological pathogen to be utilized; (2) methods or means of effectively delivering the biological agent to the target population; (3) accurate estimates of the initial and subsequent number of people exposed and susceptible to infection over a defined time period; (4) environmental factors that might negatively or positively influence the survivability, longevity, reproductive potential, and virulence of the biological agent following release; and (5) availability and effectiveness of various intervention or control approaches to mitigate the impact of the biological terrorism attack (e.g., vaccination of human and domestic animal populations, control approaches to preempt potential wild animal reservoirs, and vectors that could maintain and transmit the pathogen).

Models at multiple levels of detail (individual, population, metapopulation) can be used to predict the effects of epizootic plague die-offs of black-tailed prairie dogs, prairie dog colony population dynamics, vector population trends, and the resultant threat of *Y. pestis* to human populations related to these combined factors. The level of modeling detail that is necessary to adequately describe plague dynamics is currently unknown. However, by examining a range of levels one can: (1) discern the relative importance of individual, population, and landscape determinants of plague threat to humans; and (2) develop a modeling framework that would allow for simulation experiments of various intentional plague introduction scenarios in the most computationally efficient manner. Three distinct modeling approaches, containing varying levels of model detail and complexity, can be used to capture disease dynamics (Figure 3.23). Although each of these approaches differs in its hierarchical focus, they all may be used to predict prairie dog and plague dynamics among multiple colonies at the regional or landscape scale. Thus, in the general sense, they all could be considered metapopulation approaches (Grenfell and Harwood 1997).

The metapopulation concept has been cited as one of the most important influences on modern biology because it provides a way of using spatial as well as demographic data to predict population persistence and extinction (Gilpin and Hanski

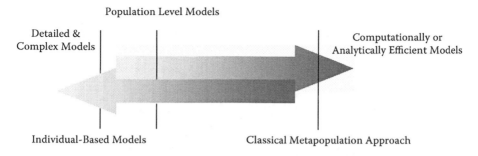

Population Level Models

Detailed & Complex Models

Computationally or Analytically Efficient Models

Individual-Based Models

Classical Metapopulation Approach

FIGURE 3.23 Tradeoff between model efficiency and level of detail.

1991; Burgman et al. 1993). At the landscape scale, black-tailed prairie dog colonies are a classic metapopulation: a spatially subdivided population whose subunits (colonies) experience periodic extinction events and are linked by migration (Levins 1970). These spatially segregated populations experience a correlation between the proportion of sites occupied and the probability of extinction (Gotelli and Kelley 1993). Most of our understanding of metapopulation dynamics has come from theoretical modeling studies rather than from empirical field studies (see reviews in Kareiva, 1990, and Fahrig and Merriam, 1994), primarily because conducting field trials requires that there be numerous, discrete, replicated habitat patches that support populations that are subject to periodic extinctions and yet are also linked by dispersal. However, the prairie dog–*Y. pestis* epidemiological system is an ideal system for linking theoretical and empirical approaches. Because plague is not native to the New World, prairie dogs have no evolutionary resistance to it; thus, colonies often suffer near-complete local extinction from plague mortality. Consequently, plague is the agent that induces metapopulation dynamics in prairie dogs.

A classical metapopulation approach (i.e., an incidence function model, or IFM) can be used to determine the relative influences of the spatial characteristics of colonies (i.e., colony extent, colony density and distribution, and distance between colonies) as well as prairie dog population density within colonies (an assay of colony "quality" as plague habitat, with higher population densities able to support higher infection rates of plague for longer durations) on plague prevalence and outbreak frequency. In this approach, individual prairie dog colonies (i.e., patches) are the unit of analysis, and the IFM uses information about patch size, isolation, and occupancy rates to determine overall risk of population extinction. It can also identify individual specific patches (or colonies) that are integral in maintaining overall metapopulation connectivity and long-term disease persistence (Hanski 1997). Commercially available software (e.g., package RAMAS) can be used to link metapopulation data with GIS-based spatial data to model spatially explicit plague population dynamics and estimate disease persistence (Akçakaya 1998).

Although it is computationally efficient, the IFM ignores many of the demographic attributes of prairie dogs and fleas that may determine potential plague threat to humans. Thus, a series of more detailed population-level epidemiological models (e.g., Susceptible-Infectious-Recovered, or SIR models) can be used to determine the influence of the spatial structure of prairie dog populations (in terms

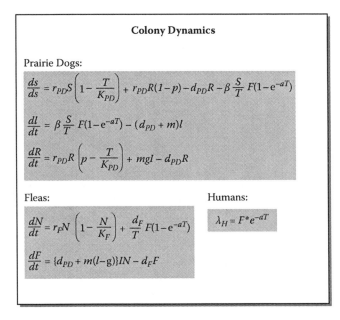

FIGURE 3.24 Starting equations for a potential SIR model of prairie dogs (susceptible, S; infectious, I; and recovered, R) and fleas (average per prairie dog, N; and free infectious fleas searching for a host, F) within a colony. The total number of prairie dogs, $T = S + I + R$. The potential for human infection can be modeled as the number of fleas that fail to find a prairie dog host, λ_H.

of both individual movements as well as population-level demographic characteristics) on the potential for plague transmission to humans. These models contain much more detail about the demographic attributes of prairie dog colonies and, as a result, are less computationally efficient. They also can be very difficult to solve analytically. Figure 3.24 illustrates the starting equations of a potential SIR model for prairie dogs within a colony (based on Keeling and Gilligan 2000a, 2000b). Empirical information about the specific details of disease dynamics within colonies is contained in the parameter estimates used within the model (Table 3.8). Because human dynamics do not affect disease behavior, they are not included in the model; however, the potential for human infection can be modeled as the number of fleas that fail to find a prairie dog host, λ_H.

SIR models typically are used to model a single population of the host species. However, metapopulation dynamics can be incorporated by linking individual colonies (each with its own SIR model) via prairie dog and flea movement parameters, μ_{PD} and μ_F. These parameters can themselves be functions of the landscape context surrounding individual colonies. Flea movement rates, in particular, may be enhanced in urban environments because of the increased density of domestic predators. Once the SIR models are developed, numerical simulations can be used to conduct a series of "experiments" that evaluate the overall risk of human infection given particular combinations of parameter values. Furthermore, by basing parameter estimates on field data, empirically based models can be developed for specific urban areas being investigated.

Finally, an individual-based model can be used to model individual prairie dogs across an entire region. This approach potentially captures the most detail about disease dynamics; however, it is very demanding computationally. One approach is to represent model dynamics via cellular automaton approximations, where fully stochastic model dynamics are approximated using cell transition probabilities, or within relatively small grids of populations with nearest-neighbor coupling (Keeling and Gilligan 2000a). Alternatively, individual-based models can incorporate elements of the above SIR model, with the spatial component being based upon a cellular automata model. This approach was used in an epidemiological study of rabies in raccoons (Mollison 1977; Sheeler 2002) and is a variation of a classical BIDE (birth + immigration − death − emigration) model. It incorporates individual-based movements (I, E) and probabilities of individual births and deaths (B, D). The model utilizes a rectangular grid of cells, and each cell would represents an area determined by the size of a typical prairie dog colony. Study area heterogeneity (habitat aggregations) can be simulated by placing initial adult prairie dogs on the grid in clumped distributions representing colonies. Values for parameters used in the model can be based on data collected from prairie dog plague surveys and data available from the literature.

3.5.2 Summary of Biological Threat Transport

Understanding the fate of biological threat agents in the environment cannot be complete unless the potential for the threat to become established in natural populations of susceptible species is considered. This is particularly important given the number of currently recognized biological threat agents that are zoonotic in origin. Statistical models can be used to predict the future occurrence of zoonosis outbreaks based on past occurrences and can be useful in identifying areas that are particularly sensitive to the introduction of biological threats. Simulation models can be used to understand the dynamics of infected populations of host species. In addition, by placing them in a spatially explicit context, they can provide insight into the possible transport of infected individuals. Nevertheless, the utility of individual-based, SIR, or metapopulation models in the context of understanding the transport of biological threats is an area that has received very little attention to date.

REFERENCES

Allen, M.D., Moss, O.R., and Briant, J.K., 1978. Dynamic shape factors for LMFBR mixed-oxide fuel aggregates, *J. Aerosol Sci.*, 10, pp. 43–48.

Allwine, K.J. and Flaherty, J.E., 2006. *Urban Dispersion Program MSG05 Field Study: Summary of Tracer and Meteorological Measurements*, PNNL–15969, Pacific Northwest National Laboratory.

Allwine, K.J., Leach, M.J., Stockham, L.W., Shinn, J.S., Hosker, R.P., Bowers, J.F., Pace, J.C., 2004. *Overview of Joint Urban 2003: An Atmospheric Dispersion Study on Oklahoma City*, preprints, Symposium on Planning, Nowcasting and Forecasting in the Urban Zone, American Meteorological Society, Seattle, WA.

Allwine, K.J., Shinn, J.H., Streit, G.E., Clawson, K.L., and Brown, M., 2002. Overview of Urban 2000: a multiscale field study of dispersion through an urban environment. *Bull. Am. Meteorol. Soc.*, 83, pp. 521–536.

Akçakaya, H.R., 1998. *RAMAS GIS: Linking Landscape Data with Population Viability Analysis* (version 3.0). Applied Biomathematics, Setauket, NY.

American Society for Testing and Materials (ASTM), 2000. *Standard Guide for Statistical Evaluation of Atmospheric Dispersion Model Performance*, ASTM, West Conshohocken, PA.

Anthes, R.A., 1983. Regional models of the atmosphere in middle latitudes, *Mon. Wea. Rev.*, 111, pp. 1306–1335.

Arya, S.P., 1999. *Air Pollution Meteorology and Dispersion*, Oxford University Press, New York.

Bacon, D.P., Ahmad, N.N., Boybeyi, Z., Dunn, T.J., Hal, M.S., Lee, P.C.S., Sarma, R.A., Turner, M.D., Waight, K. T., III, Yound, S.H., and Zack, J.W., 2000. A dynamically adapting weather and dispersion model: the Operational Multiscale Environment Model with Grid Adaptivity (OMEGA), *Mon. Wea. Rev.*, 128, pp. 2044–2076.

Baker, D.E., Ed., 1981. *Chemistry in the Soil Environment*, ASA Special Publication 40, American Society of Agronomy, Madison, WI.

Barad, M.L., 1958. Project Prairie Grass: a field program in diffusion. Geophys. Res. Paper 59, vols. 1–III, AFCRF-TR-58–235, Air Force Cambridge Research Center, Bedford, MA.

Barr, S. and Clements, W.E. 1984. Diffusion modeling: principles of application, in *Atmospheric Science and Power Production*, D. Randerson, Ed., DOE/TIC-27601, U.S. Department of Energy.

Basinger, J.M., 2003. Ricin and peanut lectin transport and influence on microbial activity in soil, Ph.D. dissertation, Texas Tech University, Lubbock, TX.

Batchelor, G.K., 1949. Diffusion in a field of homogeneous turbulence, *Aust. J. Sci. Res.*, 2, pp. 437–450.

Belcher, S.E., Jerram, N., and Hunt, J.C.R., 2003. Adjustment of a turbulent boundary layer to a canopy of roughness elements, *J. Fluid Mech.*, 488, pp. 369–398.

Benschop, H.P. and De Jong, L.P.A., 1988. Nerve agent stereoisomers: analysis, isolation, toxicology, *Accts. Chem. Res.*, 21, pp. 368–374.

Black, T.L., 1994. The new NMC mesoscale Eta model: description and forecast examples. *Wea. Forecasting* 9(2), pp. 265–278.

Blackadar, A.K., 1979. High resolution models of the planetary boundary layer, *Adv. Env. Sci. & Engr.*, 1, pp. 50–85.

Briggs, G.A., 1974. Diffusion estimation for small emissions, in *Environmental Research Laboratories: Air Resources Atmosphere Turbulence and Diffusion Laboratory 1973 Annual Report*, USAEC Report ATDL_106, National Oceanic and Atmospheric Administration.

Brown, M.J., Arya, S.P., and Snyder, W.H., 1992. Vertical dispersion from surface and elevated releases: an investigation of a non-Gaussian plume model, *J. Appl. Meteorol.*, 32, pp. 490–505.

Brown, M.J., Arya, S.P., and Snyder, W.H., 1997. Plume descriptors derived from a non-Gaussian concentration model, *Atmos. Environ.*, 31, pp. 183–189.

Brown, M.J., Lawson, R.E., Decroix, D.S., and Lee, R.L., 2001. *Comparison of Centerline Velocity Measurements Obtained around 2D and 3D Building Arrays in a Wind Tunnel*, preprint, Proceedings of the 2001 International Symposium on Environment Hydraulics.

Burgman, M.A., Ferson, S., and Akçakaya, H.R., 1993. *Risk Assessment in Conservation Biology*, Chapman and Hall, London.

Chang, C.-B. and Gill, T.E., 2005. MM5 and HPAC experiments, *Meteorol. Atmos. Phys.*, 90, pp. 127–138.

Chang, C.-B., Gill, T.E., and Conder, C., 2003. West Texas Mesonet observations and four-dimensional data assimilation for testing MM5 and HPAC, Battlespace Atmospheric and Cloud Impacts on Military Operations Conference, Naval Research Lab., Monterey, CA, September 9–12.

Chang, M., Glynn, M., and Groseclose, S., 2003. Endemic, notifiable bioterrorism-related diseases, United States, 1992–1999. *Emerg. Infect. Dis.*, 9(5), pp. 556–564.

Chesser, R.K., Bondarkov Baker, M.R.J., Wickliffe, J.K., and Rodgers, B.E., 2004. Reconstruction of radioactive plume characteristics along Chernobyl's Western Trace, *J. Environ Radioactivity*, 71, pp. 147–157.

Childs, J., Shope, R., Fish, D., Meslin, F., Peters, C., Johnson, K., Debess, E., Dennis, D., and Jenkins, S., 1998. Emerging zoonoses, *Emerg. Infect. Dis.*, 4(3), pp. 453–454.

Cionco, R.M., 1994. Overview of the Project WIND data, in *Mesoscale Modeling of the Atmosphere*, Pielke, R.A. and Pearce, R.P., Eds., American Meteorological Society, Boston, MA.

Cox, R., Bauer, B.L., and Smith T., 1998. A mesoscale model intercomparison, *Bull. Am. Meteorol. Soc.*, 79, pp. 265–283.

Cramer, H.E., 1957. *A Practical Method for Estimating the Dispersal of Atmospheric Contaminants*, Proceedings of the First National Conference on Applied Meteorology, American Meteorological Society, pp. 33–55.

Csanady, G.T., 1973. *Turbulent Diffusion in the Environment*, Reidel, Dordrecht.

Daley, R., 1991. *Atmospheric Data Analysis*, Cambridge University Press, Cambridge.

Davenport, A.G., 1960. Rationale for determining design wind velocities, *Proc. ASCE, J. Struct. Div.*, 86, pp. 39–68.

Dumais, R., Henmi, T., and Torres, M., 2003. Development of a high-resolution meteorological nowcasting system for the U.S. Army Objective Force, Proceedings of BACIMO 2003 (Battlefield Atmospheric and Cloud Impacts on Military Operations) Conference, Monterey, CA, September, 2003. U.S. Department of Defense publication.

Eastepp, E.D., 2006. The influence of individual, isolated vegetative canopies of aerosol dispersion in an urban environment, master's thesis, Texas Tech University, Lubbock, TX.

Eisen, R.J., Enscore, R.E., Biggerstaff, B.J., Reynolds, P.J., Ettestad, P., Brown, T., Pape, J., Tanda, D., Levy, C.E., Engelthaler, D.M., et al., 2007. Human plague in the southwestern United States, 1957–2004: spatial models of elevated risk of human exposure to *Yersinia pestis*, *J. Med. Entomol.*, 44(3), pp. 530–537.

Elliott, W.P., 1961. The vertical diffusion of gas from a continuous source, *Int. J. Air. Wat. Pollut.*, 4, pp. 33–46.

Ellis, J., 2003. Use of atmospheric models in response to the Chernobyl disaster, in *Tracking and Predicting the Atmospheric Dispersion of Hazardous Material Releases*, National Research Council, National Academies Press, Washington DC

Ermak, D.L. and Nasstrom, J.S., 2000. A Lagrangian stochastic diffusion method for inhomogeneous turbulence, *Atmos. Environ.*, 34, pp. 1059–1068.

Fahrig, L. and Merriam, G., 1994. Conservation of fragmented populations, *Conserv. Biol.*, 8, pp. 50–59.

Fjeld, R.A., Eisenberg, N.A., and Compton, K.L., 2007. *Quantitative Environmental Risk Analysis for Human Health*, John Wiley and Sons, Hoboken, NJ.

Galimand, M., Guiyoule, A., Gerbaud, G., Rasoamanana, B., Chanteau, S., Carniel, E., and Courvalin, P., 1997. Multidrug resistance in *Yersinia pestis* mediated by a transferable plasmid, *New Engl. J. Med.*, 337(10), pp. 677–680.

Gardner, A., Mehta, K.C., Tanner, L.J., Zhou, Z., Conder, M., Howard, R., Martinez, M.S., and Weinbeck, S., 2000. *The Tornadoes of Oklahoma City of May 3, 1999*, report by Wind Science & Engineering Research Center, Texas Tech University, under NIST (National Institute of Standards and Technology) Cooperative Agreement 70NANB8H0059.

Gifford, F.A., 1961. Use of routine meteorological observations for estimating atmospheric dispersion, *Nuclear Safety*, 2, pp. 47–51.

Gifford, F.A., 1976. Turbulent diffusion typing schemes: a review. *Nuclear Safety.* 17, pp. 68–86.

Gill, T.E., Chang, C.-B., Doggett, A.L., Conder, M.R., Midgley, C.J., Peterson, R.E., and Schroeder, J.L., 2003. *Modeling Airborne Transport of Hazards Using Advanced Atmospheric Monitoring Systems and Numerical Techniques*, phase II, final report to the U.S. Department of Defense on Contract DAAD13–02-C-0068.

Gill, T.E., Doggett, A.L., Chang, C.-B., Conder, M.R., Peterson, R.E., Burgett, W.S., Schroeder, J.L., Radell, D., and Martinez, M., 2002. *Modeling Airborne Transport of Hazards Using Advanced Atmospheric Monitoring Systems and Numerical Techniques*, final report to the U.S. Department of Defense on Contract DAAD13–01-C-0066.

Gilpin, M.E. and Hanski, I., Eds., 1991. *Metapopulation Dynamics*, Academic Press, London.

Gotelli, N.J. and Kelley, W.G., 1993. A general model of metapopulation dynamics, *Oikos* 68, pp. 36–44.

Grell, A.G., Dudhia, J., and Stauffer, D.R., 1994. *A description of the fifth-generation Penn State/NCAR mesoscale model (MM5)*, Mesoscale and Microscale Meteorology Division, NCAR, Boulder, CO.

Grenfell, B., and Harwood, J., 1997. (Meta)population dynamics of infectious diseases, *Trends Ecol. Evol.*, 12, pp. 395–399.

Gross, G., 1994. Statistical evaluation of the mesoscale model results, in *Mesoscale Modeling of the Atmosphere*, Pielke, R.A. and Pearce, R.P., Eds., American Meteorological Society, Boston, MA, pp. 137–154.

Gubler, D., 1998. Resurgent vector-borne diseases as a global health problem, *Emerg. Infect. Dis.*, 4(3), pp. 442–450.

Hanna, S.R., Brown, M.J., Camelli, F.E., Chan, S.T., Coirier, W.J., Hansen, O.R., Huber, A.H., Kim, S., and Reynolds, R.M., 2006. Detailed simulations of atmospheric flow and dispersion in downtown Manhattan: an application of five computational fluid dynamics models. *Bull. Am. Meteorol. Soc.*, 87, pp. 1713–1726.

Hanna, S.R. and Chang, J.C., 2001. Use of the Kit Fox field data to analyze dense gas dispersion modeling issues, *Atmos. Environ.*, 35, pp. 2231–2242.

Hanna, S.R., Hansen, O.R., and Dharmavaram, S., 2004. FLACS CFD air quality model performance evaluation with Kit Fox, MUST, Prairie Grass, and EMU observations, *Atmos. Environ.*, 38, pp. 4675–4687.

Hanna, S.R., Tehranian, S., Carissimo, B., Macdonald, R.W., and Lohner, R., 2002. Comparisons of model simulations with observations of mean flow and turbulence within simple obstacle arrays, *Atmos. Environ.*, 36, pp. 5067–5079.

Hanna, S.R. and Yang, R., 2000. Evaluation of mesoscale models' simulations of near-surface winds, temperature gradients, and mixing depths, *J. Appl. Meteorol.*, 39, pp. 1095–1104.

Hanski, I., 1997. Predictive and practical metapopulation models: the incidence function approach, in *Spatial Ecology: The Role of Space in Population Dynamics and Interspecific Interactions*, Tilman, D. and Kareiva, P., Eds., Princeton University Press, Princeton, NJ, pp. 21–45.

Henmi, T., 2000. Comparison and Evaluation of Operational Mesoscale Models MM5 and BFM over White Sands Missile Range (WSMR), paper presented at BACIMO (Battlefield Atmospherics and Cloud Impacts on Military Operations 2000 Conference, April 26, Fort Collins, Colorado.

Hillel, D., 1998. *Environmental Soil Physics*, Academic Press, San Diego, CA.

Hinds, W.C., 1999. *Aerosol Technology: Properties, Behavior and Measurement of Airborne Particles*, John Wiley & Sons, New York.

Hodur, R.M., 1997. The Naval Research Laboratory's coupled ocean/atmosphere mesoscale prediction system (COAMPS), *Mon. Wea. Rev.*, 125, pp. 1411–1430.

Hoke, J.E. and Anthes, R.A., 1976. The initialization of numerical models by a dynamic initialization technique, *Mon. Wea. Rev.*, 104, pp. 1551–1556.

Hong, S.-Y. and Pan, H.-L., 1996. Non-local boundary layer vertical diffusion in a medium-range forecast model, *Mon. Wea. Rev.*, 124, pp. 2322–2339.

Hosker, R.P., 1984. Flow and diffusion near obstacles, in *Atmospheric Science and Power Production*, Randerson D., Ed., DOE/TIC-27601, U.S. Department of Energy, Oak Ridge, TN, pp. 241–326.

HPAC User's Guide, 2001. Defense Threat Reduction Agency (DTRA), Alexandria, VA.

Huang, C.G., 1979. A theory of dispersion in turbulent shear flow, *Atmos. Environ.*, 12, pp. 453–463.

Inglesby, T.V, Henderson, D.A, and Bartlett, J.G., 1999. Anthrax as a biological weapon: medical and public health management, *JAMA*, 281, pp. 1735–1745.

Jacobson, A.R. and Morris, S.C., 1976. The primary air pollutants: viable particles, their occurrence, sources and effects, in *Air Pollution*, 3rd ed., Stern, A.C., Ed., Academic Press, New York.

Jury, W.A., Gardner, W.R., and Gardner, W.H., 1991. *Soil Physics*, John Wiley and Sons, New York.

Kalnay, E., 2003. *Atmospheric Modeling, Data Assimilation and Predictability*, Cambridge University Press, New York.

Kareiva, P., 1990. Population dynamics in spatially complex environments: theory and data, *Philos. Trans. R. Soc. London B*, 330, pp. 175–190.

Keeling, M. and Gilligan, C., 2000a, Bubonic plague: a metapopulation model of a zoonosis, *Proc. Biol. Sci.*, 267(1458), pp. 2219–2230.

Keeling, M. and Gilligan, C., 2000b, Metapopulation dynamics of bubonic plague, *Nature*, 407(6806), pp. 903–906.

Klemp, J.B., 2004. Next-generation mesoscale modeling: a technical overview of WRF, 20th Conference on Weather Analysis and Forecasting/16th Conference on Numerical Weather Prediction, 11.2, American Meteorological Society.

Lee, R.L., Humphreys, T., and Chan, S.T., 2004. CFD Simulations of Joint Urban Atmospheric Dispersion Study, Preprint, UCRL-CONF-204787 Department of Energy, Symposium on the Urban Environment, Vancouver, Canada.

Levins, R., 1970. Extinction, in *Some Mathematical Questions in Biology*, Gerstenhaber, M., Ed., American Mathematical Society, Providence, RI, pp. 77–107.

Loeppert, R.H., Schwab, A.P., and Goldberg, S., Eds., 1993. *Chemical Equilibrium and Reaction Models*, Special Publication 42, Soil Science Society of America, Madison, WI.

Lord, R.J., Menzel, W.P., and Pecht, L.E., 1984. ACARS wind measurements: an intercomparison with radiosonde, cloud motion, and VAS thermally derived winds, *J. Oceanic Atmos. Tech.*, 1, pp. 131–137.

MacDonald, R.W., 2000. Modelling the mean velocity profile in the urban canopy layer, *Bound. Layer Meteorol.*, 97, pp. 25–45.

Mamrosh, R.D., 1998. The use of high-frequency ACARS soundings in forecasting convective storms, Weather and Forecasting Conference, American Meteorological Society, January 12–16, 1997. Phoenix, AZ.

Mathieu, J. and Scott J., 2000. *An Introduction to Turbulent Flow*, Cambridge University Press, Cambridge.

Meoeng, C.-H. and Sullivan, P.P., 1994. A comparison of shear- and buoyancy-driven planetary boundary layer flows, *J. Atmos. Sci.*, 51, pp. 999–1022.

Mitchell, C., 1991. Vector competence of North and South American strains of *Aedes albopictus* for certain arboviruses: a review, *J. Am. Mosq. Control Assoc.*, 7(3), pp. 446–451.

Mollison, D., 1977. Spatial contact models for the ecological and epidemic spread, *J. R. Stat. Soc. Series B*, 39, pp. 283–326.

Moore, C. and Mitchell, C., 1997. *Aedes albopictus* in the United States: ten-year presence and public health implications, *Emerg. Infect. Dis.*, 3(3), pp. 329–334.

Munro, N.B., Talmage, S.S., and Griffin, G.D., 1999. The sources, fate and toxicity of chemical warfare agent degradation products, *Environ. Health Perspect.*, 107, pp. 933–974.

Naslund, E., Rodean, H.C., and Nasstrom, J.S., 1994. A comparison between two stochastic diffusion models in a complex three-dimensional flow. *Bound. Layer Meteorol.*, 67, pp. 369–384.

Nasstrom, J.S., Sugiyama, G., Leone, J.M., and Ermak, D.L., 1999. *A Real-Time Atmospheric Dispersion Modeling System*, preprint, Department of Energy UCRL-JC-135120, 11th Joint Conference on the Applications of Air Pollution Meteorology with the Air and Waste Management Association, Long Beach, CA.

National Research Council, 2003. *Tracking and Predicting the Atmospheric Dispersion of Hazardous Material Releases: Implications for Homeland Security*, National Academies Press, Washington DC.

O'Toole, T., Inglesby T.V., and Henderson D.A., 2002. Why understanding biological weapons matters to medical and public health professionals, in *Bioterrorism Guidelines for Medical and Public Health Management*, Henderson, D.A., Inglesby, T.V., and O'Toole, T., Eds., AMA Press, Chicago, pp. 1–6.

Pasquill, F., 1961. The estimation of the dispersion of windborne material, *Met. Mag. London*, 90, pp. 33–49.

Pasquill, F., 1974. *Atmospheric Diffusion: The Dispersion of Windborne Material from Industrial and Other Sources*, 2nd ed., John Wiley & Sons, New York.

Pepper, C.B., Nascarella, M.A., Marsland, E.J., Montford, J.T., Wood, L., Cox, S.B., Bradford, C.M., Burns, T.H., and Presley, S.M. 2004. Threatened or endangered? Keystone species or public health threat? The black-tailed prairie dog, the Endangered Species Act, and the imminent threat of bubonic plague, *J. Land. Resour. Environ. Law* 24, pp. 355–391.

Pielke, R.A., 2001. *Mesoscale Meteorological Modeling*, 2nd ed., Academic Press, New York.

Pielke, R.A. and Pearce, R.P., 1994. *Mesoscale Modeling of the Atmosphere*, American Meteorological Society, Boston, MA, Meteorological Monograph 47.

Rendon, J.N., 2005. Influence of tree arrangement and spacing on dispersion in finite forest canopies, master's thesis, Texas Tech University, Lubbock, TX.

Reynolds, O., 1895, On the dynamical theory of incompressible viscous fluids and the determination of the criterion, *Philos. Trans. R. Soc. London*, 186, pp. 123–164.

Roberts, O.F.T., 1924. The theoretical scattering of smoke in a turbulent atmosphere, *Proc. R. Soc. London A.*, 110, pp. 640–654

Rotz, L., Khan A., Lillibridge S., Ostroff S., and Hughes J., 2002. Public health assessment of potential biological terrorism agents, *Emerg. Infect. Dis.*, 8(2), pp. 225–30.

Sanz, C., 2003. A note on k-ε modeling of vegetation canopy air flows, *Bound. Layer Meteorol.*, 108, pp. 191–197.

Schroeder, J.L., Burgett W.S., Haynie K.B., Sonmez I., Skwira G.D., Doggett A.L., and Lipe J.W., 2005. The West Texas Mesonet: a technical overview, *J. Atmos. Oceanic Tech.*, 22, pp. 211–222.

Schroeder, J.L., Smith D.A., and Peterson R.E., 1998. Variation of turbulence intensities and integral scales during the passage of a hurricane, *J. Wind. Engr. Indust. Aerodynam.*, 77, pp. 65–72.

Schwartz, B.E. and Benjamin S.C., 1995. A comparison of temperature and wind measurements from ACARS-equipped aircraft and rawinsondes, *Wea. Forecasting*, 10, pp. 528–544.

Scott, H.D., 2000. *Soil Physics: Agricultural and Environmental Applications*, Iowa State University Press, Ames.

Selim, H.M., 1992. Modeling the transport and retention of inorganics in soils. *Adv. Agron.* 47, pp. 331–384.

Sheeler, L.L., 2002. Epidemiological model of raccoon rabies in Alabama, Ph.D. dissertation, Texas Tech University, Lubbock, TX.

Sidell, F.R, Takafuji E.T., and Franz D.R., 1997. *Medical Aspects of Chemical and Biological Warfare*, Office of the Surgeon General, TMM Publications, Washington, DC.

Skaggs, T.H., Wilson G.V., Shouse P.J., and Leij F.J., 2002. Solute transport: experimental methods, in *Methods of Soil Analysis, Part 4, Physical Methods*, Dane, J.H. and Topp, G.C., Eds., Soil Science Society of America, Madison, WI, pp. 1381–1402.

Slade, D.H., 1968. *Meteorology and Atomic Energy 1968*, U.S. Department of Energy, Oak Ridge, TN.

Snook, J.S., Cram J.M., and Schmidt J.M., 1995. LAPS/RAMS: a nonhydrostatic mesoscale numerical modeling system configured for operational use, *Tellus*, 47A, pp. 864–875.

Stull, R.B., 1989. *An Introduction to Boundary Layer Meteorology*, Kluwer Academic, Dordrecht.

Sykes, R.I., Cerasoli, C.P., and Henn, D.S., 1999. The representation of dynamic flow effects in a Lagrangian puff dispersion model, *J. Haz. Mat. A*, 64, pp. 223–247.

Sykes, R.I. and Gabruk, R.S., 1997. A second-order closure model for the effect of averaging time on turbulent plume dispersion, *J. Appl. Meteorol.*, 36, pp. 1038–1045.

Sykes, R.I. and Henn, D.S., 1995. Representation of velocity gradient effects in a Gaussian puff model. *Bull. Am. Meteorol. Soc.*, 34, pp. 2715–2723.

Sykes, R.I., Lewellen, W.S., and Parker, F.S., 1986. A Gaussian plume model of atmospheric dispersion based on second order closure, *J. Climate. Appl. Meteorol.* 25, pp. 322–331.

Sykes, R.I., Parker, S.F., Henn, D.S., Cerasoli, C.P., and Santos L.P., 1998. PC-SCIPUFF Version 1.2PD, Technical Documentation, Titan Corporation, Princeton, NJ 08543.

Taylor, G.I., 1915. Eddy motion in the atmosphere, *Philos. Trans. R. Soc. London*, 215, pp. 1–26.

Taylor, G.I., 1923. Diffusion by continuous movements, *Proc. London. Math. Soc.*, 20, pp. 196–211.

Tomilin, V. and Berezhnoĭ, R., 1985. Exposure of criminal activity of the Japanese military authorities regarding the preparation and use of bacterial warfare, *Voenno-meditsinskiĭ zhurnal*, (8), pp. 26–29.

Turner, D.B., 1970. *Workbook of Atmospheric Dispersion Estimates*, Public Health Service Publication 999-AP-26, U.S. Department of Health, Education and Welfare.

Van Genuchten, M.Th. and Wierenga, P.J., 1986. Solute dispersion coefficients and retardation factors, in *Methods of Soil Analysis, Part 1, Physical and Mineralogical Methods*, Klute, A., Ed., 2nd ed., Soil Science Society of America, Madison, WI, pp. 1025–1054.

Warner, S., Platt, N., and Heagy, J.F., 2004. Comparisons of transport and dispersion model predictions of the Urban 2000 field experiment, *J. Appl. Meteorol.*, 43, pp. 829–846.

Wieringa, J., 1980. Representativeness of wind observations at airports, *Bull. Am. Meteorol. Soc.*, 61, pp. 962–971.

Wilson, J.D. and Sawford, B.L., 1996. Review of Lagrangian stochastic models for trajectories in the turbulent atmosphere, *Bound. Layer Meteorol.*, 78, pp. 191–210.

Wu, C.C., Kuo, Y.H., Wang, W., and Yen, T.H., 1999. *Numerical Simulation of Typhoon Herb (1996) Using MM5*, preprint, 23rd Conference on Hurricanes and Tropical Meteorology, January 1999, American Meteorological Society, Dallas, TX, Vol. 2, pp. 1008–1011.

Xue, M., Droegemeier, K.K., Wong, V., Shapiro, A., and Brewster, K., 1995. *ARPS Version 4.0 User's Guide*, Center for Analysis and Prediction of Storms, University of Oklahoma, Norman.

Yamada, T., 2000. Numerical simulation of airflows and tracer transport in the southwestern United States, *J. Appl. Meteorol.*, 3, pp. 399–411.

Yamada, T. and Bunker S., 1988. Development of a nested grid, second moment turbulence-closure model and an application to the 1982 ASCOT Brush Creek data simulation, *J. Atmos. Sci.*, 27, pp. 562–578.

Yee, E. and Biltoft, C.A., 2004. Concentration fluctuations measurements in a plume dispersing through a regular array of obstacles, *Bound. Layer Meteorol.*, 111, pp. 363–415.

Zimmerman, P.D. and Loeb, C., 2004. Dirty bombs: the threat revisited, *Defense Horizons*, 38, pp. 1–11.

4 Assessment Strategies for Environmental Protection from Chemical and Biological Threats

Richard Zartman, Chai-bo Chang, George P. Cobb, Joe A. Fralick, and Steven M. Presley

CONTENTS

4.1   INTRODUCTION

Vulnerability to biological and chemical threat agents associated with terrorism is not only directed toward humans but also toward the environment upon which we rely for life. This chapter focuses on those various biological and chemical threats to our environment, including our food and fiber supply, while direct biological and chemical threats to humans are more thoroughly discussed in Chapter 8. The environment is comprised of four fundamental components—air, land, living organisms, and water. As humans, we live in and interact with each of these environmental communities and are dependent upon them for our health and well-being. Related to this chapter, modeling of biological and chemical threat agents in the environment has been previously described in Chapter 3, and individual personal protection from biological and chemical threat agents is discussed in Chapter 8.

4.1.1   AIR

The atmosphere below 10 km can be divided into the boundary layer (BL), ranging from the surface to about 3 km, and the remainder of the air known as the free atmosphere. We all spend most of our lives in the BL. Naturally, the most harmful air environmental impacts due to biological and chemical weapons occur inside the BL.

The air we breathe comprises the most ubiquitous environmental component in which we live. It is composed primarily of 78% nitrogen gas, which is inert and, in quantities, overshadows the 21% oxygen that we need to survive. The small quantities of carbon dioxide, argon, and other noble gases we inhale are benign and comprise the remaining 1%. Gaseous biological and chemical weapons generally are classified as "nerve agents" (Sidell 1997). Nerve agents are extremely toxic and were first developed secretly before World War II. Use of gas warfare in the First World War led to adoption of treaties banning such weapons. This ban, however, has not precluded the development of sarin or other gaseous biological and chemical weapons. It should be noted that blister agents such as mustards are also delivered as aerosols that, while not truly vapors, are transported to human targets via the atmosphere.

4.1.2   LAND

The land upon which we live and depend upon for our food supplies is vulnerable to biological and chemical threat agents in many ways. Natural pollution and water contamination has prevented our society from using many acres for food and fiber production. Biological and chemical threat agents that would render large areas of land uninhabitable could dramatically alter our way of life. While the Earth is large, much of the surface is covered by water and is generally uninhabitable. The soil resources that provide our livelihood are just a thin mantle of unconsolidated material at the

Earth's surface. The arable areas that we depend upon for living and food production are limited and vulnerable to biological and chemical threat agents.

Gruinard Island, for example, was used during World War II to evaluate the potential of airborne spores of *Bacillus anthracis* as a biological weapon (Manchee et al. 1994). Annual soil sampling from 1946 to 1969 indicated measurable but declining spore numbers. An intensive soil sampling survey in 1979 revealed detectable spores on portions of Gruinard Island. In 1986, treatment with formalin in seawater was successful in decontaminating the island (Manchee and Stewart 1988). This was demonstrated by grazing sheep on the island in 1987. While this is one example of soil contamination, other biological and chemical threat materials could pose other environmental problems.

While the United States has a low population density (29/km^2) compared to most industrialized countries, our arable land is needed for habitation as well as for food and fiber production. Even a limited terrorist event using radiological dispersal devices (RDDs) (colloquially known as "dirty bombs") could render areas uninhabitable for long periods of time (Helfand et al. 2002). These RDDs' materials could be from commercial sources (Ferguson et al. 2003) or waste products produced by nuclear power generation. Luckey (2003) states that the most readily available radiation for a RDD is cobalt-60, cesium-137, and strontium-90. While terrorists "anticipate more harm and deaths from fear and panic than from radiation" (Luckey 2003), the actual impact would be both economic and societal (NCRP 2001). Transuranic waste, having a long half-life, could be transported in the air and with windblown soil particles, rendering the environment uninhabitable for long periods of time (Ferguson et al. 2003).

4.1.3 WATER

While the Earth's surface is mostly water (78%), it is too saline to be potable. We depend on a small, vulnerable fraction of the Earth's water (~3%) (Manahan 2004) to sustain our lives and the crops and animals upon which we depend for food and fiber. The environmental destruction caused by biological and chemical weapons would first endanger the surface water and subsequently be transmitted to groundwater. The environmental effects of biological and chemical weapons would be catastrophic because we all depend on water. Not only would human life be compromised, but our plant and animal communities would also suffer.

Efforts to catalogue and integrate reportage from over 150,000 public water systems (i.e., water intakes, water treatment and wastewater treatment facilities) have been underway for over a decade (USEPA 1998, 2002a). Numerous facts relative to vulnerabilities associated with public water systems have been compiled and released to the public (USEPA 2002a, 2002b). One fact that demonstrates the need for rapid and sensitive monitoring programs is that 113 (26%) U.S. cities with 100,000 to 500,000 residents use surface water exclusively, as do 36 (43%) cities with populations over 500,000. It should also be noted that surface water is the primary water source for 65 (15%) and 17 (20%) cities of these respective sizes. Also important are the number of entry points for water from these surface sources. For cities of 100,000 to 500,000 and for those over 500,000, each system has 1.5±0.1 and 2.7±0.3 points of entry. These data show that a major fraction of large U.S. cities primarily utilize surface waters from multiple points of entry that are largely unprotected from intruders. The integration and continuing updating of this enormous database is of paramount importance to effectively protect our freshwater resources.

There are also issues of pathogen and pollutant occurrence in coastal regions where fisheries either exist or where the food chains originate for the fisheries. Urban, agricultural, and maritime activities have all been shown to contribute to adverse effects on these coastal ecosystems. The maritime shipping effects on ecosystems pose specific concerns due to the large volume of ships that are of foreign registry or that visit ports in areas with questionable ballast water quality (McCarthy and Khambaty 1994). For example, between 1973 and 1998, 40 *V. parahaemolyticus* outbreaks were reported to the U.S. Centers for Disease Control and Prevention (CDC), involving more than 1000 illnesses (Daniels et al. 2000). A particular *V. parahaemolyticus* serovar, 03:K6, that was hemolysin positive, apparently emerged in India in 1996 and became pandemic (Okuda et al. 1997; Barth et al. 1999; Matsumoto et al. 2000; Wong et al. 2000). Some of the clinical strains isolated in the United States during four multistate outbreaks in 1997 and 1998 belonged to the new 03:K6 clone, which had not previously been detected in U.S. coastal waters. One theory was that this strain was introduced to U.S. coastal waters in ballast water discharged from ships coming from Asia (CDC 1998, 1999; Wong et al. 2000). Another example of this would be the introduction of the zebra mussel (*Dreissena polymorpha*) to North America (Strayer et al. 1996).

While each of the environmental elements can be considered alone, it is their interrelationships that truly necessitate considering the environmental protection strategy a coherent unit. The basic hydrological cycle (Figure 4.1) relates all of these individual elements and shows their interrelationships. Air, as wind, causes the soil to move in dust storms and can transport contaminant-laden sediments. Air also transports water vapor and causes precipitation events. Water can transport soil and contaminants into and within the soil.

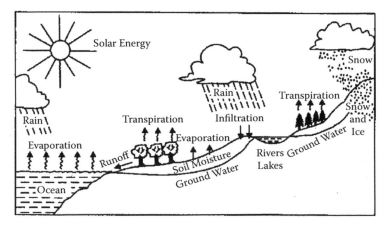

FIGURE 4.1 Hydrological cycle (Hillel, D., 1998, *Environmental Soil Physics*, Academic Press, New York.)

4.2 PRE–9/11/2001 ENVIRONMENTAL CONCERNS

As explained in Chapter 2, the probable conceptual intent of use of biological and chemical threat agents for terrorism is to create instability and paranoia within the population. A brief historic summary of biological and chemical weapon use is provided in Chapters 1 and 2, with further details available in the U.S. Department of Defense "Bluebook" (USAMRID 2005) *Medical Aspects of Chemical and Biological Warfare* (Sidell et al. 1997). The primary examples of biological terrorist materials discussed in this chapter will be toxins (aflatoxin B_1 and ricin) and *Bacillus cereus* (a surrogate for *B. anthracis*). Even before the events of September 11, 2001, the CDC had identified anthrax (*B. anthracis*) and ricin (a toxic protein produced by castor [*Ricinus communis*] plants) among the high-priority threat materials for biological terrorism. The CDC placed *B. anthracis* on the Category A list since it is a potentially deadly pathogen and may be used to contaminate food and feedstuffs. Ricin is considered a Category B biological threat agent because of its ease of dissemination, moderate morbidity rates, and low mortality rates. Another naturally occurring fungal toxin that may contaminate food and feedstuffs, and cause severe health effects in humans, is aflatoxin. Aflatoxin B_1 is a toxin and a known risk to human health that can occur on corn (*Zea mays*) (Jaynes et al. 2007).

Concern regarding the potential use of naturally occurring or intentionally produced biological pathogens and toxins as weapons has long been known. For example, in 1993 a table of biological and chemical agents was formulated by the U.S. Department of Defense (see Table 4.1). The physical properties of these materials (gas, liquid, solid), as well as their chemical properties, determine their transport, movement, and ultimate fate in the environment. More specific information about the physical and chemical properties of biological and chemical threat agents can be found in Chapters 2, 6, and 8, as well as information from the more extensive treatise that is available online from the National Library of Medicine (toxnet.nim.nih.gov/index.html).

4.3 POST–9/11/2001 ENVIRONMENTAL CONCERNS

After the events of 9/11, security issues with respect to the potential use of both biological and chemical threat agents have been an increasing concern. These security concerns led to the passage of the Public Health Security and Bioterrorism Preparedness and Response Act of 2002 (Bioterrorism Act). The Bioterrorism Act takes steps to protect the public from a threatened or actual terrorist attack on the U.S. food supply. There were arrests in Great Britain in 2003 for the potential contamination of the military food supply with ricin (Risen and van Natta 2003). A somewhat similar story is the polonium poisoning of ex-KGB agent Litvinenko (Vergano 2006). ^{210}Polonium (^{210}Po) is too expensive to be used in a large-scale attack; however, it can be used to contaminate a salad to target an individual, with deadly effect.

The primary concern about contamination of our food supply is that it is a very "low-technology" threat. This threat, however, continues to be largely ignored by the agricultural and environmental community. Brandenberger and McGlynn (2003) in the December 2003 Council for Agricultural Science and Technology (CAST)

TABLE 4.1
Toxic Chemicals and Their Precursors

Chemical Agents
Schedule 1
1. O-Alkyl ($\leq C_{10}$, incl. cycloalkyl) alkyl (Me, Et, n-Pr or i-Pr)-phosphonofluoridates
2. O-Alkyl ($\leq C_{10}$, incl. cycloalkyl) N,N-dialkyl (Me, Et, n-Pr or i-Pr) phosphoramidocyanidates
3. O-Alkyl (H or $\leq C_{10}$, incl. cycloalkyl) S-2-dialkyl (Me, Et, n-Pr or i-Pr)-aminoethyl alkyl (Me, Et, n-Pr or i-Pr) phosphonothiolates and corresponding alkylated or protonated salts
4. Sulfur mustards
5. Lewisites
6. Nitrogen mustards
7. Saxitoxin
8. Ricin (a biologic agent)

Schedule 2
9. Amiton: O,O-Diethyl S-[2-(diethylamino)ethyl] phosphorothiolate and corresponding salts
10. PFIB: 1,1,3,3,3-Pentafluoro-2-(trifluoromethyl)-1-propene
11. BZ: 3-Quinuclidinyl benzilate (*)

Schedule 3
12. Phosgene: Carbonyl dichloride
13. Cyanogen chloride
14. Hydrogen cyanide
15. Chloropicrin: Trichloronitromethane

Precursors
Schedule 1
1. Alkyl (Me, Et, n-Pr or i-Pr) phosphonyldifluorides
2. O-Alkyl(H or $\leq C_{10}$, incl. cycloalkyl) O-2-Dialkyl (Me, Et, n-Pr or i-Pr)-aminoethyl alkyl (Me, Et, n-Pr or i-Pr) phosphonites and corresponding alkylated or protonated salts
3. Alkylphosphonochloridates

Schedule 2
4. Chemicals, except those listed in Schedule 1, containing a phosphorus atom to which is bonded one methyl, ethyl or propyl (normal or iso) group but no further carbon atoms
5. Arsenic trichloride
6. 2,2-Diphenyl-2-hydroxyacetic acid
7. Quinuclidine-3-ol
8. *N,N-Dialkyl (Me, Et, n-Pr or i-Pr) aminoethanol deriviatives*
9. Thiodiglycol: Bis(2-hydroxyethyl)sulfide
10. Pinacolyl alcohol: 3,3-Dimethylbutane-2-ol

Schedule 3
11. Phosphorus chloride compounds
12. Poly-alkylphosphites
13. Sulfur chlorides and sulfur oxychlorides
14. Poly-ethanolamines

Source: U.S. Department of Defense (DoD), 1993, Defense treaty readiness program, Chemical weapons convention, Annex on chemicals, available at http://www.dtirp.dtra.mil/TIC/treatyinfo/cwc.cfm.

Commentary stated, "Therefore, *any* disease-causing microbes present on fruits or vegetables are there because of *inadvertent contamination*" [emphasis added]. This is *not* true in the ricin example previously presented. The agricultural and environmental communities need to be aware of intentional contamination and available remediation methods. Although there is abundant public geospatial information that could aid terrorists, it is not the greatest environmental vulnerability. "Human expertise" is a much more critical problem in the "more demanding attack planning part of the targeting problem" (Baker et al. 2004).

Pathogenic microorganisms such as anthrax or various toxins need not be "weaponized" to contaminate food. While the anthrax letters in the U.S. Postal Service after September 11, 2001, had weaponized *B. anthracis* spores (Matsumoto 2003), ricin in the letter (October 15, 2003) was determined not to be a weaponized form (CDC 2003). The sophistication of these bioterrorists is in their subtle use of low-technology methods. Contamination of fresh fruits and vegetables by *B. anthracis* spores and toxins would be an example of a sophisticated, low-technology terrorist attack. The food and feedstuffs supply chain is highly vulnerable to this type of attack throughout the complete on-farm production through consumer sales cycle. It is a standard practice to keep vegetables fresh in the supermarket with water rinses. If the water sources used to rinse fresh produce were contaminated by *B. anthracis* spores or ricin, the consumer would be harmed without immediate knowledge. Our vulnerability to this form of covert biological attack against our food supply may be enhanced by the fact that the persons washing the fresh fruits and vegetables are generally low-paid employees who might easily be bribed for their participation. It is also important to note that *B. anthracis* can cause gastrointestinal tract infections and cutaneous infections in addition to respiratory infections. Thus, consumption of food contaminated with *B. anthracis* spores, or water contaminated with such spores, can lead to infections. Bacteria thrive in warm, moist environments such as fresh foods.

4.3.1 AIR CONCERNS

The key controlling factor of boundary-layer (BL) depth and processes in the air is heating. Daytime heating results in a deep unstable BL with rapid turbulent mixing in the vertical, while nighttime cooling causes a shallow stable BL with very little vertical mixing. The stability of the BL affects the vertical movement of air. For example, particulate plumes released above the stable BL are rarely dispersed onto the ground because of limited turbulence. On the other hand, if particulate materials are released inside the stable BL, higher concentrations are expected nearer to the ground. To a lesser degree, the surface roughness (e.g., flat vs. rugged terrain) also plays a role in the BL evolution (Stull 1989).

The horizontal airflow in the BL often consists of synoptic-scale (> 1000 km) and mesoscale (10–1000 km) circulations, and microscale (~ 1 km) turbulence. Their time scales are days, 1 to 24 h, and less than 1 h, respectively. The impact of toxic dispersion on a synoptic scale or large mesoscale is generally not threatening to health, while for small mesoscale (< 100 km) and microscale, the impact can be life threatening because of accumulated doses over a relatively small area. Small-scale weather and BL dispersion are strongly influenced by the fast-changing local conditions such as clouds and surface heating (Ray 1986). These facts, in essence, make the small-scale problems crucial yet challenging for evaluating the air-environment impacts. Specific examples of this complexity are discussed in Chapter 3.

4.3.2 AIR–LAND CONCERNS

Concerns regarding the air–land interface are of great significance. If biological and chemical threat agent materials are to be employed against "open-air" targets, the meteorological conditions must be known. For the maximum effectiveness of an aerosolized biological or chemical weapon agent, the material must remain 1 to 5 m above the ground (Sidell et al. 1998). Biological and chemical threat agent attacks are most likely to occur at night, daybreak, or sundown. At these times, there is little mixing of air, and the materials generally remain at ground level. Wind speeds between 8 and 40 km/h are best for agent dispersal (Sidell et al. 1998). Wind speeds below this range limit effective dispersion of biological and chemical threat agents, while wind speeds exceeding this range impose loss of physical integrity. In contrast, liquid and dry materials may be applied over a wide range of conditions. In general, liquids perform best in humid environments while dry materials do best in arid environments.

The Gruinard Island example from War II indicates the potential effectiveness and efficacy that may result from proper dispersion of airborne biological and chemical weapon agents. Procedures and equipment for producing liquid materials are simple, but the dissemination of them is relatively difficult (Patrick 1996). Producing dried biological and chemical materials is more difficult and requires more sophisticated equipment. Even after the release of these materials into the atmosphere, their ultimate destination depends not only upon air current, relative humidity, and sunlight but also the method of dispersal (Calder 1957). Between 1941 and 1969, the U.S. government conducted multiple large-scale aerosol vulnerability tests using several bacteria (Franz et al. 1997). These tests showed the vulnerability and vagary of open air-testing. [For an interesting account of open-air testing from a plaintiff's perspective, the reader is urged to read *Clouds of Secrecy* by Cole (1988)].

4.3.3 SOIL–LAND CONCERNS

While the general population considers all soil to be the same, site-specific differences in the soil may greatly affect the environmental impacts of an attack with biological and chemical weapons. The dynamics of the three-phase soil system determine the potential fate of biological and chemical materials in the soil. Would the soils mitigate or exacerbate the biological and chemical weapon problem? The answer depends upon the agent used, the texture of the soil, the clay mineral in the soil, the climatic conditions, and the depth to groundwater. Biological and chemical threat agents that readily sorb to the soil solids would be rapidly detoxified with soils having large quantities of 2:1 clay content. Soils having high sand content, however, will be more likely to transmit both biological and chemical weapon materials to the groundwater.

Landscape position and slope will influence the harm posed by biological and chemical weapons. Low-lying areas will most likely be regions of biological and chemical weapon material gain, while upper slope regions will more likely be areas of increased agent effectiveness loss. Just as low-lying areas are typically areas of high organic matter due to increased localized water content, these areas will be sinks for biological and chemical threat agents. The steeper the slope, the more likely the biological and chemical weapons will be to run off (Sidell et al. 1998).

Soil contaminated by biological and chemical weapons would be a different problem than waters contaminated by biological and chemical weapons. Soils contaminated by transuranic waste from RDDs would be a long-lasting problem. The radioactive particles would travel through the air with the soil particles moved by wind. The radioactive particles would travel within the soil if the particles are moved by water. Biological and chemical materials that are sorbed to the soil would move within the soil by convection-dispersion equation (see Chapter 3) or become retransported with the soil in the air or water.

4.3.4 Water Concerns

The introduction of biological and chemical threat agents into bodies of water poses several different problems. The nature of the problems posed will depend on the biological and physical properties of the threat agent used and the nature of the water body. To a large extent, the chemical characteristics of the biological and chemical agents (see Table 4.1) and the water quality parameters of a water body control solubility of the material as well as the sorption of the material to soil particles (suspended or bed sediments). Many of the transport phenomena are discussed in Chapter 3.

Most chemical threat materials will have high water solubility but may also have high hydrolytic degradations rates. In most incidences, solubility of biological threat materials will not be high, but this is largely irrelevant because biological materials require that target organisms experience exposure to only a few microbes or toxin molecules.

Flowing waters tend to fall into one of two categories: well aerated with low total organic carbon (TOC) or poorly aerated with high TOC. Lakes can span the same range of dissolved organic carbon (DOC) when considering those situated in alpine regions as compared to standing water bodies in lowland areas (impoundments, oxbow lakes, marshes). As rivers empty into estuaries, flow slows and suspended sediments settle to form sediments with high nutrient content. This process will also deposit toxins and toxicants into sediments in estuarine regions. It would seem that this would remove the contaminants. Organisms that encounter these toxins or toxicants, however, may experience adverse effects, and those that encounter biological materials could serve as reservoirs for incubation of hazardous microbes. This is well documented with the contamination of shellfish by ballast water from ships entering U.S. harbors. For example, ballast waters have been shown to contain microbial pathogens (McCarthy and Khambaty 1994), and such pathogens have adversely impacted significant fisheries and estuarine ecosystems.

There is an interesting difference in the effect that organic matter will have on chemical versus biological materials. TOC will bind chemical materials and toxins, thereby decreasing availability to organisms. However, TOC, especially suspended particles, can provide growth substrates for microbes that may be introduced as a

TABLE 4.2

Accidental Releases of Chemicals or Microbes into the Environment

Area	Type	Toxin/Toxicant	Extent	Duration
Bhopal	Industrial	MeICN	40 mi^2	Hours
Sevesto	Industrial	Dioxin	6 km^2	20 min.
SRS	Defense/Energy	Radionuclides	1100 ha	Decades
Valdez	Maritime	Oil	11,000 mi^2	> 10 days
CA	Railway	Pesticide??		
Upstate SC	Pipeline	Heating Oil	23 miles of river bank	11+ min.
NC Flooding	Agricultural Hog Farm	Coliforms	60+ km^2	72-96 hours
Times Beach	Residential	Dioxin	2 km^2	Months-1 year

biological material. There is a wealth of data emerging that describes the different types of organic matter and the ways that chemical contaminants interact with these different types of organic matter (Pignatello 2006; Schaumann 2006; Simpson and Johnson 2006). Accidental releases of chemicals or pathogens to the environment provide sound examples of the extent and duration of certain types of contaminants (Table 4.2).

The nature of human exposure to biological and chemical threat agents largely depends upon the uses for the affected water body. For example, irrigation waters in agricultural areas would be particularly susceptible to microbial materials, whereas municipal water intakes would be less susceptible given the extensive treatments that are used to purify drinking water. It should also be noted that poisoning remote areas could cause significant damage to natural resources such as fisheries or areas for water sports.

4.3.5 Agroterrorism Concerns

Agricultural terrorism, or agroterrorism, particularly as previously discussed regarding the contamination of food and feedstuffs, may be a low-tech threat yielding high-impact results. Current agricultural production, commodity storage, transport, and marketing practices offer many vulnerabilities that have been largely ignored in assessments of the actual threats we face. U.S. agriculture involves more than 2 million farms that extend over 1 billion acres that account for a $230 billion per year industry. One area of concern rests in food production and distribution. This is especially true in the case of food products received from developing countries (e.g., highly biologically active contaminants in Chinese feedstuffs, South American grapes, and others).

Data are emerging that demonstrate the need to carefully evaluate food production, processing, and transportation within the United States and its North American Free Trade Agreement (NAFTA) partners. Highly concentrated populations of livestock animals for breeding, feeding, and rearing, such as cattle dairies and feedlots, poultry and swine farms, and other animal production operations, offer a vulnerability through which an outbreak of a contagious disease would be very difficult to contain, especially if it is airborne. Operations at which large numbers of livestock are concentrated in closed spaces and are maintained for fattening and ultimate human consumption are termed *confined animal feeding operations* (CAFOs). Successful attacks could trigger quarantines, food shipment cutoffs, and destruction of exposed animals, and cause large economic losses to the American agriculture industry. Agroterrorists have a large menu of biological threat agents from which to choose, including bacteria, viruses, fungi, and insects that target crops or livestock. These materials are usually environmentally hardy, are not the focus of livestock vaccination or seed development programs, are readily available in countries where they naturally infect crops or livestock, and can be easily smuggled across state, national, and international borders. Many of these biological threats do not have to be "weaponized," and in some cases, such as foot-and-mouth disease (FMD), the causative virus can be effectively transmitted by simply walking onto a farm with contaminated boots. Furthermore, while most biological threat agents that may be used for agroterrorism infect only plants or livestock, some of these agents may also cause disease in man such as *Brucella*, anthrax, *Salmonella*, and various zoonotic diseases caused by viruses (see Chapter 2 for more information on zoonoses). There are several studies that document the detection and occurrence of chemical contaminants, microbes, and fungal spores in dusts emanating from concentrated animal feeding operations. See Table 2.2 from Chapter 2. This fact is more important given the high airborne emission factors for these facilities (NRC 2002). It is also interesting to note that dusts serve as vehicles for transboundary atmospheric transport of contaminants, often on transcontinental scales. The distribution of microbes, spores, or toxicants could arise from agricultural areas in neighboring countries where little or no monitoring and control is possible by U.S. officials or allies. This mode of environmental transport is covered in more detail in Chapter 3.

4.4   COUNTERING THREAT AGENTS IN THE ENVIRONMENT

The primary focus of this chapter is upon the significant research findings regarding countering biological and chemical threat agents in various environments that war fighters and homeland security personnel must work. This research was conducted by scientists associated with the Zumwalt Program and other academic and scientific agencies and groups. The chapter is divided into sections addressing both terrestrial and aquatic environments, and the influences and effects biological and chemical threat agents may have on each. The atmospheric compartment of the environment will not be discussed in detail since it serves as a transport medium for biological and chemical threat agents, and not a sorbent material. Detailed discussion on the movement of biological and chemical threat agents is provided in Chapter 3. The impetus for this text is the concept that terrorism seeks to create instability and paranoia. Events of September 11, 2001, documented that principal at the national level. With current terrorists' threats and activities, this project has taken on added relevancy. This chapter focuses upon ricin as a biological threat agent that could plausibly be used to attack human populations. Although there are many other possible toxins and toxicants that pose threats, many of the same concepts and factors controlling threats from ricin could also be considered in evaluating threats from other biological and chemical threat agents.

Ricin is a toxic protein naturally produced by castor plants (*Ricinus communis*) and considered a biological threat agent, or toxin. The toxin ricin is derived from the endosperm of the castor seed and is composed of two polypeptides that together are highly toxic to humans. These polypeptides are known as the ricin-A chain and ricin-B chain. The A chain and B chain are held together by a disulfide bond. The B chain has the ability to bind to complex galactosides on the cell membranes of eukaryotic cells. After being engulfed, both chains are transported through the cell and eventually to the ribosomes. In the ribosomes, the A chain depurinates the adenine located in the 28S ribosomal RNA subunits (Lord et al. 1994). A single ricin molecule can inactivate over 1,500 ribosomes (Olsnes and Saltvedt 1975). This toxin has long been known, and more than 750 cases of ricin poisoning have been reported (Rauber and Heard 1985). The public's perception of ricin as poison most likely stems from the Georgi Markov incident. An assassin in London used a modified umbrella as a weapon to insert a hollow, ricin-filled, stainless steel pellet into Markov's leg (Crompton and Gall 1980). Markov subsequently died from ricin poisoning. This event has been repopularized in Tom Clancy's book *Red Rabbit*. A somewhat similar story is the ^{210}Po poisoning of Litvinenko discussed previously.

Ricin clearly has the potential to be used by terrorists because of the ease of cultivation of the castor plant and its global use and availability as an ornamental plant (Atlas 1999). Measurable concentrations of ricin in soil indicate either an intentional contamination or the cultivation of a large castor crop (Hennessy 2000). Ricin has been contemplated as a biological weapon agent in the United States without much notoriety or success (Tucker and Pate 2000). In one event, four antigovernment activists in Minnesota allegedly conspired to assassinate local and federal law enforcement officials using ricin. "The four had acquired ricin ... by ordering materials and instructions through the mail" (Tucker and Pate 2000). They considered using a ricin-DMSO-aloe vera mixture as a cream to be delivered by smearing it on a doorknob or inside the victim's shoes. They additionally considered putting dried ricin powder in a syringe and injecting it into a car. This method might have been lethal. When ricin is inhaled, it produces severe tissue damage, leading to hemorrhagic pneumonia and death (Wannemacher et al. 1990).

Ricin as a potential biological threat agent has received much popular press. In 2003, suspects were arrested in London for "making" ricin from castor seed in their apartment (Risen and van Natta 2003). The popular press has speculated that if ricin were made, it could be used to contaminate food in military mess halls. All of these instances indicate that biological and chemical materials may be a potential terrorist weapon to compromise the safety of the military and civilian food supply or other vulnerable areas. Given the extreme toxicity of ricin, the relative ease with which it can be obtained, and the fact that references to its use have been discovered in terrorist haunts, the ability to accurately and precisely detect ricin is a critical need.

4.4.1 SOIL FACTORS INFLUENCING PREDICTION, DETECTION, PROTECTION, AND RESPONSE

Many factors influence the sorption to and fate of biological and chemical weapons in the soil, including the predominant factors such as soil texture, cation exchange capacity, organic matter content, and pH. While soil forms the land surface, it is very site specific when it comes to the fate of biological and chemical weapons in the environment. Soil, along with the air and water, determine the fate of biological and chemical weapons.

4.4.1.1 Soil Texture

Soil texture is the percentage of sand, silt, and clay content of soil. Each of these textural content components has different effects on the fate of biological and chemical threat agents in the environment. The sand fraction with large size (2- to 0.05-mm effective diameter) has low surface area and lacks the ability to sorb chemicals. At the opposite end of the soil fraction spectrum is the clay-sized particle (less than 0.002 mm in effective diameter). The clay fraction is most likely to sorb biological and chemical threat agents due to its very large surface area per unit per unit mass. Cationic chemicals readily sorb to the anionic clay particles. Although this surface area sorption is responsible for our ability to fertilize the soil and grow plants, it is also capable of holding transuranic components of RDDs. The sand fraction would allow the cationic biological and chemical threat agents to pass rapidly through the soil and contaminate shallow ground water. Silt size fractions (0.05 to 0.002 mm in effective diameter) neither sorb cations nor readily transmit chemicals.

Therefore, the soil texture may determine the environmental fate of biological and chemical threat agents. To sequester, sorb, and remove such materials, a clay-textured soil is advantageous. Sand-textured soils facilitate biological and chemical materials within the environment. In studies presented below, the quantity and type of clay will be discussed with respect to the sorption and movement of toxin and microbial spores.

4.4.1.2 Cation Exchange Capacity

The high cation exchange capacity (CEC) of clays is due not only to the size of the material but also to how the charge was derived. There are hundreds of types of clay minerals. For the purpose of this discussion, however, we shall only identify silicate clays (1:1 and 2:1) and nonphyllosilicate materials. The silicate clays derive their names from the numbers of layers and the types of layers they contain. A 1:1 clay is a silicate clay having one silica tetrahedral layer and one alumina octahedral layer (Schulze 1989). The typical common name for the dominate clay in this family is kaolinite. This clay generally occurs in "old" soils and is typical of the clay mineralogy that is found in the southeastern portion of the United States. This soil has a low CEC (2 to 10 centimoles per kilogram of material) and is not capable of sorbing large quantities of cationic biological and chemical weapons. While this type of clay has low CEC, at low soil pHs (less than 5), these clays possess anion exchange capacity (AEC). The AEC allows these clay minerals to sorb anions within the soil.

A 2:1 clay is a silicate clay that has two silica tetrahedral layers that "sandwich" an alumina octahedral layer between them. There are two general groups of 2:1 clays, and they differ by where the isomorphic substitution occurs. Soils that have isomorphic substitution in the alumina octahedral layer are known to have high CEC and high shrink-swell capabilities. These soils expand when wet and shrink when dry. Sorption of cationic biological and chemical agents is significant for these clays. The other group of 2:1 clays has isomorphic substitution in the silica tetrahedral layer. Since the charge arises closer to the clay mineral surface, these clays generally do not expand or shrink. In general the silica tetrahedral layer-substituted clays do not sorb as much cationic chemicals.

4.4.1.3   Organic Matter

Organic matter in the soil also sorbs many ionic and nonionic chemicals. This is because the macromolecules that comprise organic matter contain both polar and nonpolar regions. These polar regions contain nitrogen, oxygen, and sulfur functional groups. On a weight-per-weight basis, organic matter sorbs many more cations than do the inorganic matter clays. The presence of organic matter within the soil is depth and geographic region specific. Because organic matter is derived from plant material, it is generally greater near the soil surface than at lower depths. Soils in the hotter, drier areas (e.g., southwest United States) generally have very low soil-organic matter contents. It is not unusual for soils in the southwestern United States to have less than 1% organic matter in the soil surface horizons. The soil organic matter percentages decrease significantly with depth. In contrast with these low organic matter levels, soils in the northeastern United States have higher soil organic matter levels. Soils in this region have increased organic matter production since they have increased precipitation. Additionally, the soils in this region have lower mean annual temperatures, resulting in less organic matter decay. The high organic matter production and the lower organic matter decay result in a greater net soil organic matter content.

4.4.1.4   pH

Soil pH is another major factor that determines the fate of biological and chemical weapon agents in the environment. The normal pH of soils ranges from 1.5 to 10.5, with most soils expressing pH within the range of 4.5 to 8.5. Soils outside this pH range are usually not vegetated to any great extent. A neutral pH of 7 is good for crop production, with slightly acid pH most favorable for a majority of row crops. Acid soils are often located in the more humid regions of the country due to leaching of cations with the water.

Soil pH controls the sorption of contaminants by the natural attraction of the cations and anions. Soils and water do not, however, remain the same pH throughout time. Natural processes of acidification take place from chemical reactions in the air, land, and water. The acidification is generally caused by sulfur and nitrogen reactions. Nitrogenous fertilizers often contribute to the soil acidification through the process of nitrification. Soils can also become more alkaline if there are increased divalent cations present. Plants often contribute to this phenomenon by base recycling through leaf litter accumulation. Soil CEC is also dependent upon pH changes. Many of the recognized biological and chemical threat agents also have pH dependant reactions. Proteins have isoelectric points that allow them to sorb or desorb as a function of pH.

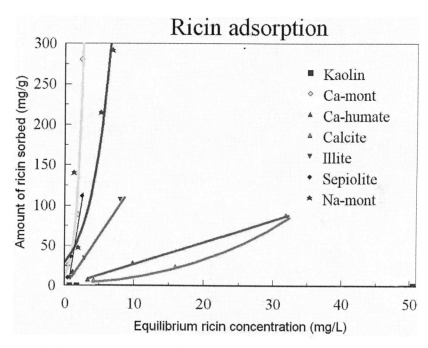

FIGURE 4.2 (See color insert following page 46.) Ricin sorption isotherms for several different clay minerals. (Zartman, R E., Green, C.J., San Francisco, M.J., Zak, J.C., Jaynes, W.F., and Boroda, E., 2002, Mitigation of ricin contamination in soils: Sorption and degradation, Joint Services Scientific Conference on Chemical and Biological Defense Research, Hunt Valley, MD, November.)

4.4.1.5 Soil Mineralogy Effects

The mineralogy that comprises the clay component of the soil is also critical to contamination of soils by biological and chemical threat agents. For example, ricin was sorbed to four different clay minerals as described above (Figure 4.2). Kaolinite, a 1:1 clay mineral, sorbed very little ricin. Sepiolite, a fibrous clay mineral, sorbed much more ricin. Illite, a tetrahedrally substituted 2:1 clay mineral, sorbed similar amounts of ricin as the sepiolite. The octahedrally substituted clay mineral, montmorillonite, sorbed much greater quantities than all of the other clay minerals. Even with this clay mineral, the cation that was dominantly sorbed to the clay made a difference. The Na-saturated montmorillonite sorbed more than the Ca-saturated montmorillonite. This is thought to be due to the swelling of the interlayers of the clays. Na-saturated clays swell more so than do Ca-saturated ones. Similar results are shown with aflatoxin B_1 (Jaynes et al. 2007).

4.4.1.6 Soil pH–Mineralogy Interactions

The soil pH–mineralogy interactions are shown in Figure 4.3. At pH values above the isoelectric point of 7.1 for ricin, there is very little ricin sorbed within the clay lattice (part a). As the soil pH decreases, more ricin is trapped within the clay interlayer.

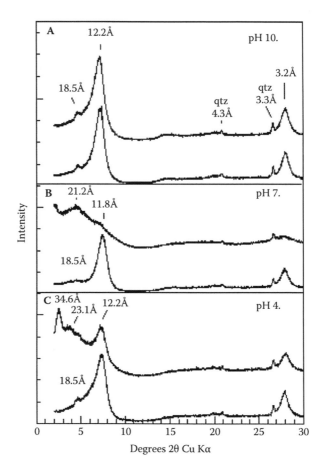

FIGURE 4.3 Influence of pH on the sorption of ricin within the interlayers of SWy-2 clay mineral. (Jaynes, W.F., Zartman, R.E., Green, C.J., San Francisco, M.J., and Zak, J.C., 2005, Castor toxin adsorbed to clay minerals, *Clays and Clay Miner.*, 53(3), 267–276.)

This is shown by the greater distance between the two lines in parts b and c. Note that at a pH of 4, there are large quantities of ricin trapped within the clay lattice.

4.4.2 MOVEMENT OF BIOLOGICAL AND CHEMICAL THREAT AGENTS

Transport of chemical threat agents with dust results from cationic materials adsorbing to clay particles and semivolatiles sorbed to clays or organic matter (Roth et al. 2005; Cincinelli et al. 2007; Karanasiou et al. 2007; Stracquadanio and Trombini 2006; Clymo et al. 2005; Lohmann and Lammel 2004; Goss 2004). Some biological threat agents can also sorb to dusts, thereby undergoing atmospheric transport. There are current estimates of the amount of particulate matter emanating from CAFOs, and there are also data that demonstrate the presence of sorbed chemical and biological contaminants. To date, there has been little if any effort devoted to integrating these datasets to determine the significance of off-site transport for these particulate-adsorbed contaminants.

FIGURE 4.4 Dust storm on the Texas Tech University campus, Lubbock, TX. (Courtesy of Dr. Tom Gill from April 2006.)

Examples of cationic material transport on dust include the extremely toxic castor protein (ricin) and the reactor-produced radioisotopes from RDDs. Ricin as purified crystals or powder forms may be disseminated as aerosols (Audi et al. 2005). The localized dispersal of powdered ricin could potentially affect large areas due to transport in dust. Note that the dust in Figure 4.4 is severe enough to limit visibility to less than one quarter of a mile and the street lights are illuminated. Figure 4.5 is

FIGURE 4.5   (See color insert following page 46.) Dust storm along a U.S. highway. (Courtesy of Dr. Tom Gill, UTEP.)

a dust storm along a U.S. highway. If this dust were contaminated with biological or chemical threat agents or materials, vehicles could transport these materials long distances and cause massive environmental contamination.

Protection from exposure to biological and chemical threat agent-contaminated particulate matter may come from breathing masks. Endotoxin exposure that causes "farmer's lung" was decreased by using breathing masks (Muller-Wening and Repp 1989). Masking will not be of benefit for gaseous nerve agents such as sarin, soman, tabun, or VX, or for vesicants (sulfur mustard), cyanide, pulmonary agents optoids, or cholinergics (National Research Council 1999). For further discussion of protection from aerosols, see Chapter 8 on personal protection devices.

Movement of the biological material ricin with soil-size fractions are shown in Figures 4.6 and 4.7. Studies on peanut (*Arachis hypogaea*) seed lectin show similar results (Zartman et al. 2005). These lectin data are similar in distribution to published values (Ravi et al. 2004). The inhalation of dust generated from ricin-contaminated soils could pose a serious hazard to war fighters. Work of A. H. Corwin cited in Lamanna (1961) stated that ricin particles with a median diameter of 2 μm are 2.75 times as toxic as particles with a maximal particle size of 4.2 μm.

As the above data demonstrates, even a coarse-textured soil contains enough fine particles to yield significant dust. An unpaved road usually provokes more annoyance than health concerns. A powdered toxin spread along an unpaved roadway, however, could be dispersed with the dust generated by vehicular traffic. A powdered biological material or toxin, such as ricin, is less dense than soil particles, is more readily moved by wind, and consequently has a tendency to be concentrated in dust. A water-soluble toxin, such as ricin, might continue to pose a threat until it is dissolved by rain and washed into the soil. A less soluble toxin or toxic material (e.g., bacterial spores) might pose a threat long after dispersal.

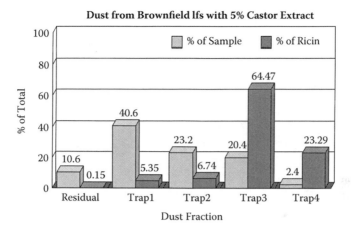

FIGURE 4.6 Dust fractions and ricin from Brownfield soil. (Zartman, R.E., Jaynes, W.F., Green, C.J., San Francisco, M.J., and Zak, J.C., 2005, Dust transport of castor toxin, Salt Lake City, UT.)

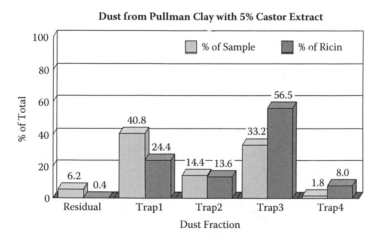

FIGURE 4.7 Dust fractions and ricin from Pullman soil. (Zartman, R.E., Jaynes, W.F., Green, C.J., San Francisco, M.J., and Zak, J.C., 2005, Dust transport of castor toxin, Salt Lake City, UT.)

4.4.2.1 Ricin Sorption and Desorption: Fruits and Vegetables

Food security issues have been an increasing homeland security and environmental concern. This concern is due to the simplicity of food supply contamination and that it is a very "low-technology" threat. To better understand food contamination by a biological threat agent, ricin sorption experiments were conducted. Ricin sorption was evaluated for the following fruits and vegetables (F/V): broccoli (*Brassica oleracea*), lettuce (*Lactuca* spp.), celery (*Apium graveolens*), grape (*Vitus* spp.), carrot

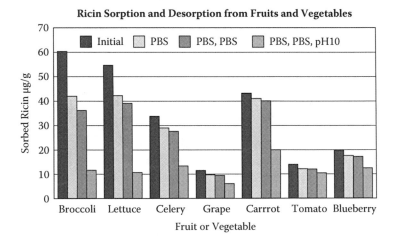

Ricin Sorption and Desorption from Fruits and Vegetables

FIGURE 4.8 **(See color insert following page 46.)** Amount of ricin sorbed by various fruits and vegetables before and after rinsing. (Zartman, R.E., Green, C.J., San Francisco, M.J., Zak, J.C., and Jaynes, W.F., 2003, Food sorption of ricin and anthrax simulants, Joint Services Scientific Conference on Chemical and Biological Defense Research, Towson, MD, November.)

(Daucus carota), tomato *(Lycopersicon esculentum)*, and blueberry *(Vaccinium* spp.). Additionally, four wash treatments were evaluated, resulting in four solutions per experiment: a solution containing the ricin that was not retained by the fruits and vegetables (initial), a solution containing ricin desorbed after one washing with phosphate-buffered saline (PBS), a solution containing ricin desorbed after a second PBS wash, and a solution containing ricin desorbed after a final pH 10 buffer wash (Zartman et al. 2003).

The results of the ricin sorption experiments (Figure 4.8) indicate high (initial) sorption ~60 μg ricin/gram fresh weight to broccoli. Lesser amounts of ricin were sorbed onto lettuce (~55 μg ricin/gram fresh weight), carrots (~40 μg ricin/gram fresh weight), and celery (~33 μg ricin/gram fresh weight). Those fruits and vegetables with a smooth waxy surface retained the least ricin. Grapes, tomatoes, and blueberries sorbed less than 20 μg ricin/gram fresh weight (Zartman et al. 2003).

For ricin desorption, the fruits and vegetables were washed with PBS, washed a second time with PBS, and finally with a pH 10 buffer solution. After these desorption washes, approximately 10 μg ricin/gram fresh weight was still retained by most of the fruits and vegetables. While the amount of sorbed ricin decreased after each wash (Figure 4.8), grapes retained 6 μg ricin/gram fresh weight, whereas carrots retained 20 μg ricin/gram fresh weight. Greater ricin sorption by fruits and vegetables might be achieved using a more concentrated ricin solution. For all the fruits and vegetables, the ricin concentrations in the pH 10 buffer wash were much higher than in the previous PBS wash. This indicated that ricin was not simply retained as an aqueous film adhering to the F/V surface but actually was sorbed to the fruits and vegetables. These amounts of sorbed ricin by the F/V are sufficient to cause fatality. A 1-oz (28 g) serving of fruits or vegetables with 20 μg/g of sorbed ricin contains more than the 400-μg minimum lethal dose (MLD) of ricin for a 150-pound person based on the ricin intraperitoneal MLD in mice of 0.001 μg ricin N/g body weight (Merck 2001).

From studies on ricin sorption to fruits and vegetables, it is clear that significant amounts of ricin were retained. Broccoli, lettuce, celery, and carrots sorbed 30–60 μg ricin/gram fresh weight. Grapes, tomatoes, and blueberries, which have smooth, waxy surfaces, only sorbed 10–20 μg ricin/gram fresh weight. Much of the ricin sorbed to the fruits and vegetables persisted through multiple aqueous rinses. After fruits and vegetables have been contaminated with ricin, it is unlikely that aqueous washing will render them safe for human consumption (Zartman et al. 2003).

All of the fruits and vegetables (Figure 4.8) adsorbed ricin. Broccoli, lettuce, and carrots adsorbed 40–60 μg ricin g^{-1}, whereas tomatoes and grapes adsorbed only 11–14 μg ricin g^{-1}. The smooth, waxy surface of grapes and tomatoes might have inhibited ricin adsorption. Sliced or peeled tomatoes and grapes would likely adsorb more ricin. The greater surface area of smaller pieces of the other fruits and vegetables would probably also increase ricin adsorption due to the increased surface area. A reduced particle diameter to one half doubles the surface area. The first and second PBS washes desorbed some adsorbed ricin, but the pH 10 extract was most effective. The effectiveness of the pH 10 buffer in removing adsorbed ricin from fruits and vegetables suggests that the alkalinity of detergents and soaps might also effectively remove adsorbed ricin.

Ricin adsorption by fruits and vegetables was about 1000 times less than adsorption by clays (i.e., μg g^{-1} versus mg g^{-1} for the clays). The lower ricin adsorption is in part due to the very small specific surface area of fruits and vegetables relative to clays. Ricin-contaminated fruits and vegetables, however, may still pose a health risk. Ricin adsorption to clays might decrease or eliminate the toxicity. Yet, without evidence to the contrary, it must be assumed that ricin adsorbed to fruits and vegetables will remain toxic. A 1-oz (28.35 g) serving of broccoli or lettuce would contain 1.7 mg or more of adsorbed ricin. Adsorbed ricin is resistant to removal by washing, and this resistance to removal indicates a contamination risk of ricin-contaminated fruits and vegetables even after washing. Yet the aqueous film adhering to fruits and vegetables after contamination with a concentrated ricin solution might easily contain ten times as much ricin. A small dinner salad contains about 4 oz of vegetables. A single 1-oz serving of fruits and vegetables contaminated with a concentrated ricin solution might contain 17 mg of ricin if it is not washed after contamination. Three castor beans (~0.3 g bean^{-1}) contain ~17 mg of ricin (Zartman et al. 2003). Orally ingested ricin is much less toxic than inhaled or intravenous ricin (i.e., intraperitoneal MLD is about 0.45 mg for a 77-kg body weight). However, 1.7 mg of ricin would likely cause toxic effects and 17 mg might produce death. Washing might reduce the amount of ricin in the surface, but it would not eliminate the toxicity of ricin-contaminated fruits and vegetables.

4.4.2.2 *Bacillus Cereus* Adsorption and Desorption to Fruits and Vegetables

Another biological threat agent that could potentially be used in a terrorist attack is bacterial spores. This is similar to the anthrax incidents of 2001. In our studies, *Bacillus cereus* was used as a surrogate for *B. anthracis* to evaluate the sorption of

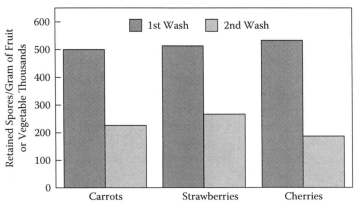

FIGURE 4.9 Sorption of *B. cereus* spores on various fruit and vegetables after rinsing. (Zartman, R.E., Green, C.J., San Francisco, M.J., Zak, J.C., and Jaynes, W.F., 2003, Food sorption of ricin and anthrax simulants, Joint Services Scientific Conference on Chemical and Biological Defense Research, Towson, MD, November.)

bacterial spores on food. Work by Helgason et al. (2000) indicated these bacteria were taxonomically very similar. Foods evaluated for sorption of *B. cereus* were carrots (*Daucus carota*), strawberries (*Fragaria* spp.), and cherries (*Prunis* spp.). The produce was sterilized before the experiment began using a dilute chlorine bleach solution. The produce was rinsed in sterile distilled water and incubated at room temperature in aqueous *B. cereus* solution containing 10^7 spores/100 mL for 10 min. The produce was washed in sterile-water for 10 min. (first wash) and washed a second time in sterile-water containing glass beads for 10 min. (second wash). *B. cereus* sorbed to the produce was determined by dilution plating of the wash solutions.

The results are presented in Figure 4.9. The figure presents the sorption of *B. cereus* spores on carrots, strawberries, and cherries. After the first wash, cherries sorbed a greater proportion (~94%) of the spores than did either the carrots or strawberries (~80% each). After a second, more vigorous wash that included glass beads, the cherries retained the least proportion of *B. cereus* spores. We do not know why the cherries retained the greatest number of spores after the gentle wash. The amount of *B. cereus* spores retained after the final wash, however, can be attributed to the rough surfaces of the carrots and strawberries. Spores retained on these rough surfaces would not be dislodged as easily as those on smooth surfaces (Zartman et al. 2003).

From our studies (Table 4.3), *Bacillus cereus* spore retention by fruit (strawberries and cherries) and vegetables (carrots) was 40–50% of 10^7 applied spores that adhered to the produce surfaces (~30 gram fresh weight) after light rinsing. This is a surprisingly high amount, ~500,000 spores/gram fresh weight and ~200,000 spores/gram after vigorous rinsing. Such large quantities of *B. anthracis* spores would very likely be hazardous if ingested.

TABLE 4.3

Estimate of Spores Adhering to Vegetable or Fruit Surfaces after Gentle Washing (First Wash) Followed by Agitation with Glass Beads (Second Wash)

Vegetable/Fruit	Infection Dose	Spores Remaining First Wash	Spores Remaining Second Wash
Baby carrots	1.7×10^7	1.5×10^7	6.8×10^6
Strawberries	1.7×10^7	1.54×10^7	8×10^6
Cherries	1.7×10^7	1.6×10^7	5.6×10^6

Source: Zartman, R.E., Green, C.J., San Francisco, M.J., Zak, J.C., and Jaynes, W.F., 2003, Food sorption of ricin and anthrax simulants, Joint Services Scientific Conference on Chemical and Biological Defense Research, Towson, MD, November.

4.4.3 Summary of Biological and Chemical Threat Agents on Soil and Food

The soil is a dynamic three-phase system of solids, liquids, and gas. All of these three phases influence the sorption of biological and chemical threat agents or weapons. While the solid phase most governs the sorption and degradation of biological and chemical threat agents, there are several factors that govern these reactions. One factor is the soil texture (percentage sand, silt, and clay fractions), while the other is the type of clay mineral. The textural effect can be generalized using Figure 4.10 for the United States. The more clayey areas, the more likely the biological and chemical weapons are to be sorbed. Texture, however, is not the sole determination for biological and chemical weapon sorption; the type of clay mineralogy must also be considered. The more the 2:1 clays, the more sorption of biological and chemical weapons.

Degradation of biological and chemical threat agents is dependent on moisture and temperature. Moisture and temperature regions in the United States are presented in Figures 4.10 and 4.11. Moisture dramatically influences biological and chemical agent movement and fate in two different ways. The greater the moisture, the more likely the biological and chemical agents are to move deeply into the soil. Additionally, the more the moisture, the more likely the microbial community is to degrade the biological and chemical weapons. Temperature regime also influences the environmental fate of biological and chemical weapons. The greater the mean annual temperature, the greater the probability the biological and chemical threat agents will degrade. In cold areas, the indigenous microflora are less likely to degrade the biological and chemical weapon material.

The interaction of soil texture and moisture also poses a threat of environmental contamination when there is high precipitation and high sand content. Areas having these conditions are likely to have groundwater contamination by biological and chemical threat agents if they are introduced into the environment.

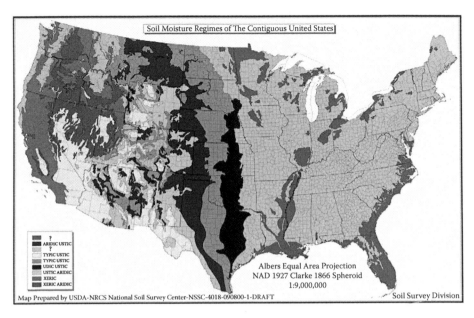

FIGURE 4.10 **(See color insert following page 46.)** Soil moisture regimes for the contiguous United States (http://soils.usda.gov/use/thematic/moist_regimes.html).

The interaction of environmental surface and temperature influences the persistence of a liquid biological and chemical threat agent. As with other liquids, evaporation increases as the temperature rises. Sarin is reported to evaporate in half the time at 43°C compared to a temperature of 10°C from a sandy soil (Sidell et al. 1998). For evaporation of sarin from a "chemical-resistant surface," the times are 2 min. and 15 min., respectively (Sidell et al. 1998).

4.4.4 STRATEGY FOR COUNTERING AND MITIGATING ENVIRONMENTAL TERRORISM

While it would be nice to be able to state a cogent, definitive countermeasure to biological and chemical terrorism, this cannot be achieved. A critical countermeasure to such events is vigilance. Care must be taken to become aware of possible terror scenarios. As the National Research Council states, "terrorist incidents involving biological agents ... are likely to be very different from those involving chemical agents" (National Research Council 1999). The National Research Council also states that military responses would be difficult or impossible to implement in a civilian population. For countering environmental terrorism, the National Research Council has five specific research and development needs. They are as follows: (1) better computer software, (2) field-testing of current and prospective atmospheric dispersion models, (3) models of other possible vectors, (4) customizable simulation software, and (5) better information on biological and chemical materials (National Research Council 1999).

Mitigation of biological and chemical terrorism is site- and threat agent specific. Personal protection equipment (PPE) must be threat agent specific for protection of emergency responders and populations at large. First responders must not exacerbate

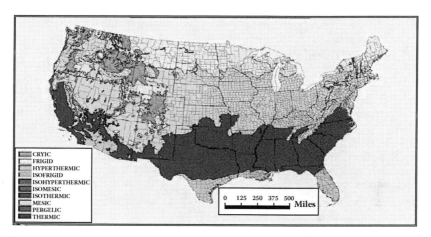

FIGURE 4.11 (See color insert following page 46.) Soil temperature regimes of the contiguous United States (http://soils.usda.gov/use/thematic/temp_regimes.html).

the problem by increasing the contaminated area. When decontaminating an area, low-kinetic-energy methods should be used so as not to spread the problem. As mentioned earlier, soils containing high quantities of 2:1 clay would sorb more biological and chemical threat agents than other soils. Depending on the biological and chemical threat agent used, the area should be cordoned off and decontaminated as appropriate. While the decontamination equipment and materials are improving, decontamination "remains problematic" (Fitch et al. 2003).

REFERENCES

Atlas, R.M., 1999. Combating the threat of biowarfare and bioterrorism, *BioScience*, 49, pp. 465–477.

Audi, J., Belson, M., Patel, M., Schier, J., and Osterloh, J., 2005. Ricin poisoning: A comprehensive review, *JAMA* 294(18), pp. 2342–2351.

Baker, J.C. et al., 2004. *Mapping the Risks: Assessing the Homeland Security Implications of Publicly Available Geospatial Information*, RAND Corp., Santa Monica, CA, p 40.

Barth, S.S., Del Rosario, L.S., Baldwin, T., Kingsley, M., Headley, V., Ray, B., Wiles, K., DePaola, A., Cook, D., Kaysner, C., Puhr, N., Daniels, N., Kornstein, L., and Nishibuchi, M., 1999. Analysis by PFGE of a *Vibrio parahaemolyticus gastroenteritis* outbreak in Texas, Abstract C-57, p. 116, in *Abstracts of the 99th General Meeting of the American Society for Microbiology, May, 1999*, American Society for Microbiology, Washington, DC.

Brandenberger, L. and McGlynn, W., 2003. Food safety and fresh produce, *CAST Commentary*, December, available at http://www.cast-science.org/websiteUploads/publicationPDFs/QTA2003–1.pdf.

Calder, K.L., 1957. Mathematical models for dosage and casualty coverage resulting from single point and line source releases of aerosol near ground level, Report #BWL-TS-3, Army Biological Laboratories, Fort Detrick, MD.

Centers for Disease Control and Prevention (CDC), 1998. Outbreak of *Vibrio parahaemolyticus* infections associated with eating raw oysters: Pacific Northwest, 1997, *Morbidity Mortality Wkly. Rep.*, 47(22), pp. 457–462.

Centers for Disease Control and Prevention (CDC), 1999. Outbreak of *Vibrio parahaemolyticus* infection associate with eating raw oysters and clams harvested from Long Island Sound: Connecticut, New Jersey, and New York, 1998, *JAMA* 281, pp. 603–604.

Centers for Disease Control and Prevention (CDC), 2003. Investigation of a ricin-containing envelope at a postal facility: South Carolina, 2003. *Morbidity Mortality Wkly. Rep.*, 52(46), pp. 1129–1131.

Cincinelli, A., Del Bubba, M., Martellini, T., Gambaro, A., and Lepri, L., 2007. Gas-particle concentration and distribution of *n*-alkanes and polycyclic aromatic hydrocarbons in the atmosphere of Prato (Italy), *Chemosphere*, 68, pp. 472–478.

Clymo, A.S., Shin, J.Y., and Holmen, B.A., 2005. Herbicide sorption to fine particulate matter suspended downwind of agricultural operations: field and laboratory investigations, *Environ. Sci. Technol.*, 39, pp. 421–430.

Cole, L.A., 1988. *Clouds of Secrecy: The Army's Germ Warfare Tests over Populated Areas*, Rowman and Littlefield, Totowa, NJ.

Crompton, R. and Gall, D., 1980. Georgi Markov: Death in a pellet, *Med. Leg. J.*, 48, pp. 51–62.

Daniels et al., 2000. *Vibrio parahaemolyticus* infections in the United States, 1973–1998. *J. Infectious Disease* 181(5), pp. 1661–1666.

Ferguson, C.D., Kazi, T., and Perera, J., 2003. Commercial radioactive sources: surveying the security risks, Occasional Paper 11, Monterey Institute of International Studies, Monterey, CA.

Fitch, J.P., Raber, E., and Imbro, D.R., 2003. Technology challenges in responding to the biological or chemical attacks in the civilian sector, *Science*, 302, pp. 1350–1354.

Franz, D.R, Parrott, C.D., and Takafuji, E.T., 1997. The U.S. biological warfare and biological defense programs, in *Medical Aspects of Chemical and Biological Warfare*, Sidell, F.R., Takafuji, E.T., and Franz, D.R., Eds., Office of the Surgeon General U.S. Army, Falls Church, VA, pp. 425–436.

Goss, K.U., 2004. The air/surface adsorption equilibrium of organic compounds under ambient conditions, *Crit. Rev. Environ. Sci. Technol.*, 34, pp. 339–389.

Helgason, E., Okstad, O.A., Caugant, D.A., Johansen, H.A., Fouet, A., Mock, M., Hegna, I., and Kolsto, A.B., 2000. *Bacillus anthracis, Bacillus cereus,* and *Bacillus thuringiensis:* one species on the basis of genetic evidence, *Appl. Environ. Micro.*, 66(6), pp. 2627–2630.

Helfand, I., Forrow, L., and Tiwa, J., 2002. Nuclear terrorism, *Brit. Med. J.*, 324, pp. 356–359.

Hennessy, T., 2000. Safety moves to the soil, *Progressive Grocer*, 79, pp. 83–88.

Hillel, D., 1998. *Environmental Soil Physics*, Academic Press, New York.

Jaynes, W.F., Zartman, R.E., Green, C.J., San Francisco, M.J., and Zak, J.C., 2005. Castor toxin adsorbed to clay minerals, *Clays and Clay Miner.*, 53(3), pp. 267–276.

Jaynes, W.F., Zartman, R.E., and Hudnall, W.H., 2007. Aflatoxin B_1 adsorption by clays from water and corn meal, *Appl. Clay Sci.*, special issue, 36(1–3), pp. 197–205.

Karanasiou, A.A., Sitaras, I.E., Siskos, P.A., and Eleftheriadis, K., 2007. Size distribution and sources of trace metals and *n*-alkanes in the Athens urban aerosol during summer, *Atmos. Environ.*, 41, pp. 2368–2381.

Lamanna, C., 1961. Immunological aspects of airborne infection: Some general considerations of response to inhalation of toxins, *Bacteriol. Rev.*, 25(3), pp. 323–330.

Lohmann, R. and Lammel, G., 2004. Adsorptive and absorptive contributions to the gas-particle partitioning of polycyclic aromatic hydrocarbons: State of knowledge and recommended parametrization for modeling, *Environ. Sci. Technol.*, 38, pp. 3793–3803.

Lord, M.J., Roberts, L.M., and Robertus, J.D., 1994. Ricin: Structure, mode of action, and some current applications, FASEB J., 8, pp. 201–208.

Luckey, T.D., 2003. Nuclear triage and the dirty bomb, *Rad. Protect. Manage.*, 20(1), pp. 11–17.

Manahan, S.E., 2004. *Environmental Chemistry*, 8th ed., CRC Press, Boca Raton, FL.

Manchee, R.J., Broster, M.G., Stagg, A.J., and Hibbs, S.E., 1994. Formaldehyde solution effectively inactivates spores of *Bacillus anthracis* on the Scottish Island of Gruinard, *Appl. Environ. Microbiol.*, 60(11), pp. 4167–4171.

Manchee, R.J. and Stewart, W.D.P., 1988. The decontamination of Gruinard Island, *Chem. Britain*, 24, pp. 690–691.

Matsumoto, G., 2003. Anthrax powder: state of the art? *Science*, 302, pp. 1492–1497.

Matsumoto, C., Okuda, J., Ishibashi, M., Iwanaga, M., Garg, P., Rammamurthy, T., Wong, H.C., DePaola, A., Kim, Y.B., Albert, M.J., and Nishibuchi, M., 2000. Pandemic spread of an 03:K6 clone of *Vibrio parahaemolyticus* and emergence of related strains evidenced by arbitrarily primed PCR and *tox* RS sequence analyses, *J. Clin. Microbiol.*, 38, pp. 578–585.

McCarthy, S.A. and Khambaty, F.H., 1994. International dissemination of epidemic *Vibro cholerae* by cargo ship ballast and other nonpotable waters, *Appl. Environ. Microbiol.*, 60(7), pp. 2597–2601.

Merck, 2001. *The Merck Index: An Encyclopedia of Chemicals, Drugs, and Biologicals*, 13th ed., Merck, Whitehouse Station, NJ.

Muller-Wening, D. and Repp, H., 1989. Investigation on the protective value of breathing masks in farmer's lung using an inhalation provocation test, *Chest*, 95, pp. 100–105.

National Council on Radiation Protection and Measurements (NCRP), 2001. Management of terrorist events involving radioactive materials, NCRP report 138, Bethesda, MD.

National Research Council (NRC), 1999. *Chemical and Biological Terrorism: Research and Development to Improve Civilian Medical Response*, National Academy Press, Washington, DC.

National Research Council (NRC), 2002. *Air Emissions from Animal Feeding Operations: Current Knowledge, Future Needs*, National Academy Press, Washington, DC.

Okuda, J., Ishibashi, M., Hayakawa, E., Nishino, T., Takeda, Y., Mukopadhyay, A.K., Garg, S., Bhattacharya, S.K., Nair, G.B., and Nishibuchi, M., 1997. Emergence of a unique 03:K6 clone of *Vibrio parahaemolyticus* in Calcutta, India, and isolation of strains from the same clonal group from Southeastern Asian travelers arriving in Japan, *J. Clin. Microbiol.*, 35, pp. 3150–3155.

Olsnes, S. and Saltvedt, E., 1975. Conformation-dependent antigenic determinants in the toxic lectin ricin, *J. Immunol.*, 114, pp. 1743–1748.

Organization for the Prohibition of Weapons, 1994. Annex on chemicals in convention on the prohibition of the development, production, stockpiling and use of chemical weapons and on their destruction, Technical Secretariat of the Organization for the Prohibition of Weapons, pp. 47–54. Available at htpp://dtirp.dtra.mil/IIC/treaty.info/cwc.cfm.

Patrick, W.C., III, 1996. Biological terrorism and aerosol dissemination, *Politics Life Sci.*, 15, pp. 208–210.

Pignatello, J.J., 2006. Fundamental issues in sorption related to physical and biological remediation of soils, in *Viable Methods of Soil and Water Pollution Monitoring, Protection and Remediation*, 69, pp. 41–68. NATO Science Series. Springer, Netherlands.

Rauber, A. and Heard, J., 1985. Castor bean toxicity re-examined: a new perspective, *Vet. Hum. Toxico.*, 27, pp. 498–502.

Ravi, S., D'Odorico, P., Over, T.M., and Zobeck, T.M., 2004. On the effect of air humidity on soil susceptibility to wind erosion: the case of air-dry soils, *Geophys. Res. Lett.*, 31L09501, pp. 1–4.

Ray, P.S., Ed., 1986. *Mesoscale Meteorology and Forecasting*, American Meteorological Society, Boston, MA.

Risen, J. and van Natta, D., 2003. Bases the target in ricin poison plot theory. *Sydney Morning Herald*, January 25, available at http://www.smh.com.au/articles/2003/01/24/1042911548292.html.

Roth, C.M., Goss, K.U., and Schwarzenbach, R.P., 2005. Sorption of a diverse set of organic vapors to urban aerosols, *Environ. Sci. Technol.*, 39, pp. 6638–6643.

Schaumann, G.E., 2006. Soil organic matter beyond molecular structure, Part II: Amorphous nature and physical aging, *J. Plant Nutr. Soil Sci.* [Zeitschrift Fur Pflanzenernahrung Und Bodenkunde], 169, pp. 157–167.

Schulze, D.G., 1989. An introduction to soil mineralogy, in *Minerals in Soil Environments*, Dixon, J.B. and Weed, S.B., Eds., 2nd ed., Soil Science Society of America, Madison, WI, pp. 1–34.

Sidell, F.R., 1997. Nerve agents, in *Medical Aspects of Chemical and Biological Warfare*, Sidell, F.R., Takafuji, E.T., and Franz, D.R., Eds., Office of the Surgeon General U.S. Army, Falls Church, VA, pp. 129–179.

Sidell, F.R, Patrick, W.C., III, and Dashiell, T.R., 1998. *Jane's Biological and Chemical Handbook*, Jane's Information Group, Alexandria, VA.

Sidell, F.R., Takafuji, E.T., and Franz, D.R., Eds., 1997. *Medical Aspects of Chemical and Biological Warfare*, Office of the Surgeon General U.S. Army, Falls Church, VA.

Simpson, M.J. and Johnson, P.C.E., 2006. Identification of mobile aliphatic sorptive domains in soil humin by solid-state C-13 nuclear magnetic resonance, *Environ. Toxicol. Chem.*, 25, pp. 52–57.

Stracquadanio, M. and Trombini, C., 2006. Gas to particle (PM10) partitioning of polycyclic aromatic hydrocarbons (PAHs) in a typical urban environment of the Po Valley (Bologna, Italy), *Fresenius Environ. Bull.*, 15, pp. 1276–1286.

Strayer, D.L., Powell, J., Ambrose, P., Smith, L.C., and Pace, M.L., 1996. Arrival, spread, and early dynamics of a zebra mussel (*Dreissena polymorpha*) population in the Hudson River Estuary, *Can. J. Fisheries Aquatic Sci.*, 53, pp. 1143–1149.

Stull, R.B., 1989. *An Introduction to Boundary Layer Meteorology*, Kluwer Academic, Dordrecht, p. 666.

Tucker, J.B. and Pate, J., 2000. The Minnesota Patriots Council (1991), in *Assessing Terrorist Use of Chemical and Biological Weapons*, Tucker J.B., Ed., Harvard University Press, Cambridge, MA, pp. 159–183.

U.S. Department of Defense (DoD), 1993. Defense treaty readiness program, Chemical weapons convention, Annex on chemicals, available at http://www.dtirp.dtra.mil/TIC/treatyinfo/cwc.cfm.

U.S. Environmental Protection Agency (USEPA), 1998. Information Available from the Safe Drinking Water Information System, Office of Water, U.S. Environmental Protection Agency, EPA 816-F-98-006.

U.S. Environmental Protection Agency (USEPA), 2002a, Community Water System Survey: 2000. Vol. 1, Office of Water, U.S. Environmental Protection Agency, EPA 815-R-02-005A.

U.S. Environmental Protection Agency (USEPA), 2002b, Community Water System Survey: 2000. Vol. 2, Office of Water,. U.S. Environmental Protection Agency, EPA 815-R-02-005A.

USAMRIID's Medical Management of Biological Casualties Handbook, 2005. 6th ed., available at http://www.usamriid.army.mil/education/instruct.htm.

Vergano, D., 2006. Litvinenko autopsy will require extra precautions, *USA Today*, December 1.

Wannemacher, R.W., Creasia, D.A., Hines, H.B., Thompson, W.L., and Dinterman, R.E., 1990. Toxicity, stability and inactivation of ricin, *Toxicologist*, 10, p. 166.

Wong, H.C., Liu, S.H., Wang, T.K., Lee, C.L., Chiou, C.S., Liu, D.P., Nishibuchi, M., and Lee, B.K., 2000. Characteristics of *Vibrio parahaemolyticus* 03:K6 from Asia, *Appl. Environ. Microbiol.*, 66, pp. 3981–3986.

Zartman, R.E., Green, C.J., San Francisco, M.J., Zak, J.C., and Jaynes, W.F., 2003. Food sorption of ricin and anthrax simulants, Joint Services Scientific Conference on Chemical and Biological Defense Research, Towson, MD, November.

Zartman, R E., Green, C.J., San Francisco, M.J., Zak, J.C., Jaynes, W.F., and Boroda, E., 2002. Mitigation of ricin contamination in soils: Sorption and degradation, Joint Services Scientific Conference on Chemical and Biological Defense Research, Hunt Valley, MD, November.

Zartman, R.E., Jaynes, W.F., Green, C.J., San Francisco, M.J., and Zak, J.C., 2005. Dust transport of castor toxin, American Society of Agronomy Meetings, Salt Lake City (UT).

5 Chemical Threat Agent Induced Latent (Delayed) Neurodegeneration

Jean Strahlendorf and Howard Strahlendorf

CONTENTS

5.1 INTRODUCTION

5.1.1 ANTICHOLINESTERASES AS CHEMICAL WARFARE AGENTS

Anticholinesterases are agents that inhibit the enzyme acetylcholinesterase (AChE) and other cholinesterases that degrade acetylcholine (ACh). The first known anticholinesterase agent was Calabar bean, known as "ordeal poison" or "ordeal bean"; its active ingredient is the carbamate compound physostigmine (Koelle 1975). It is thought that the Calabar bean was used by native tribesmen of Western Africa who used it in witchcraft rituals. The principle behind its use was that if a tribal member was accused of a capital crime, he or she would be subjected to a trial by ordeal in which they were forced to eat Calabar beans. If a person was innocent, he or she would quickly eat the beans with little apprehension, which would quickly induce vomiting and prevent death. On the other hand, if the person felt guilty and apprehensive, he or she might ingest the beans more slowly and obtain a lethal dose in the absence of vomiting. In 1864, an extract and elixir of Calabar bean was used medicinally, when it was named eserine; the structure was determined in 1925 and synthesized in 1935 (Koelle 1975). Although they are used as insecticides, carbamates also are utilized medicinally in the treatment of glaucoma and myasthenia gravis and include pyridostigmine bromide, ambenonium, neostigmine, and physostigmine (Koelle 1975).

In addition to carbamates, another class of anticholinesterase agents are the organophosphate (OP) chemical threat agents (nerve agents) consisting of the G agents (named for Germany, the country that synthesized them). Organophosphates are compounds that contain a phosphorus oxygen double bond in their chemical structure. The G-series includes tabun (GA), sarin (GB), soman (GD), and GF (cyclosarin). Sarin, tabun, and soman are some of the most potent synthetic agents known with percutaneous LD_{50} of 1700 mg, 1000 mg, and 350 mg, respectively (LD_{50} is the amount of a material, given all at once, that causes the death of 50% of a group of test animals).

As a rule, chemical threat agents are readily aerosolized or dermally active and are routinely manufactured in industrial chemical laboratories. The first use of a nerve gas in war occurred in 1915 during the second battle of Ypres, when Germany used chlorine against the French and Algerian troops (Duffy 2002). Phosgene and mustard gas (or Yperite) succeeded the use of chlorine and proved to be more deadly, causing violent coughing and choking. At the end of World War I, Germany, France, and Britain were using all three forms of gas (chlorine, phosgene, and mustard) in artillery shells (Duffy 2002). Although chlorine was the first chemical threat agent used for military purposes, its use was surpassed by the first generation of OP chemical threat agents known as the G-series (Tucker 2006). The V-series (apparently standing for venomous) was developed by Porton Down in the United Kingdom and are considered the second generation of OP chemical threat agents, and include VE, VG, VM, and VX, with VX being ten times more potent than the G-series agent sarin. The V-agents are persistent agents, do not degrade or wash easily, and can remain on clothes for long periods. The chemical structures of representative OP chemical threat agents from the G- and V-series are shown in Figure 5.1.

A new and extremely lethal "third generation" of nerve agents was developed in 1973–1976 as part of a Soviet program code-named "Foliant" and referred to as Novichok (translated as "newcomer"). This group of agents includes Substance 33, Novichok-7 (its volatility is similar to soman and ten times more effective), A-230, A-232 (aka Novichok-5), and A-234 (the ethyl analog of A-232). These agents are binary chemical weapons that consist of two relatively nontoxic ingredients that, when mixed together, yield a lethal chemical agent. Specifically, the two agents are benign industrial and agricultural chemicals that can be made quickly in large quantities, and thus do not need to be stockpiled. The intent of the program was aimed at developing new chemical threat agents that would be covert and unrelated to chemical agents and thus not detectable by inspectors. In general, Novichok agents are considered five to ten times more effective than soman or VX. The currently

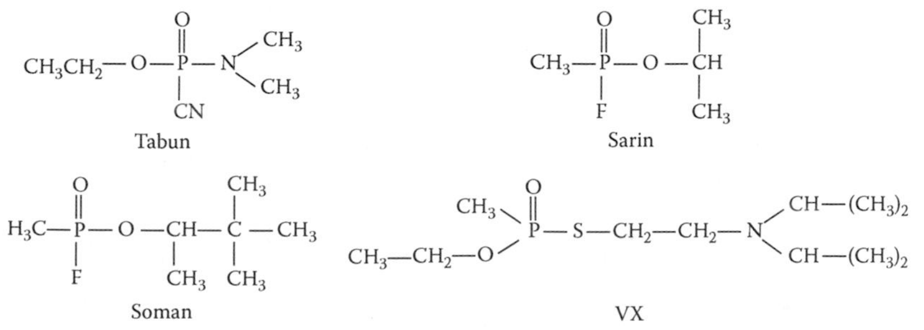

FIGURE 5.1 Chemical structures of some representative organophosphate-chemical threat agents.

available agents are 15-fold to a 100-fold more potent than the agents used in World War I (Sokolski 2000; Tucker 1996).

5.1.2 PHYSIOLOGY OF ACETYLCHOLINE

Acetylcholine is released in response to an action potential at the axon terminal and binds to either the nicotinic or muscarinic classes of cholinergic postjunctional receptors. Receptor activation is terminated rapidly by the cleavage of ACh by AChE. Acetylcholine is one of the few transmitters that is solely enzymatically inactivated/degraded followed by the active reuptake of its precursor, choline, back into the presynaptic terminal for resynthesis of ACh. In the presence of carbamate or organophosphate inhibitors of AChE, ACh is not inactivated because AChE is inhibited, allowing the accumulation of endogenous ACh in the vicinity of the cholinergic receptors. Continuous opening of the cholinergic receptor-linked ion channels, secondary to the inhibition of the enzyme that inactivates ACh, leads to the toxicity associated with cholinesterase inhibitory compounds.

Acetylcholine is a major transmitter in the autonomic nervous system, the somatic nervous system, and the central nervous system (CNS; brain). Figure 5.2 illustrates the physiology of the cholinergic system, including its location within the body. It is released from all preganglionic neurons in both the sympathetic and parasympathetic nervous system, and activates nicotinic type ionotropic, cholinergic receptors. It also is released from postganglionic parasympathetic neurons and activates the muscarinic type metabotropic, cholinergic receptors located on cardiac muscle, smooth muscles (in particular the pulmonary and gastrointestinal systems), and glands (respiratory and gastrointestinal systems). Acetylcholine released from the parasympathetic nervous system leads to glandular secretion and smooth muscle contraction. In the somatic nervous system, ACh is released from motor neurons that reside in the ventral horn of the spinal cord and cranial nerve nuclei to directly stimulate nicotinic receptors located within skeletal muscle fibers that promote muscle contraction. This rather ubiquitous distribution of ACh underscores the broad-based effects of AChE inhibition in the brain and body (Koelle 1975).

(A)

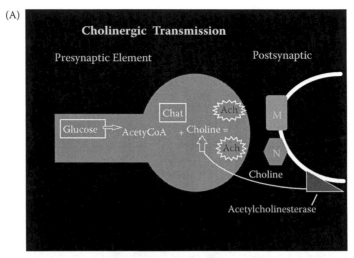

(B)

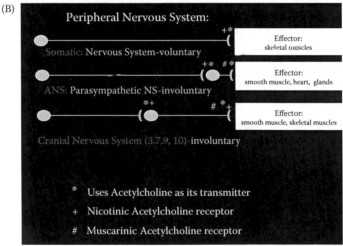

FIGURE 5.2   (See color insert following page 46.) Presynaptic and postsynaptic regions of the acetylcholine neuron, emphasizing the synthesis and degradation of acetylcholine and the cholinergic receptor subtypes (Panel [A]); summary of the peripheral nervous system that utilizes acetylcholine as the transmitter (Panel [B]).

5.1.3   CLASSIFICATION OF ACETYLCHOLINESTERASES

There are three primary cholinesterases (ChE) in the body. Acetylcholinesterase is located in the vicinity of ACh receptors at neuronal and neuromuscular junctions. Acetylcholinesterase terminates ACh activity via hydrolysis into choline and acetic acid. The positively charged choline portion of the ACh molecule attaches to the anionic site and the acetyl region attaches to the esteratic site on the AChE molecule. Following the attachment of the two regions, choline is rapidly released to be recycled back into the presynaptic nerve terminal, and the acetyl group reacts with water to form acetate, which allows the enzyme to be reactivated. The hydrolysis of ACh is considered one of the fastest enzymatic reactions in the body; AChE possesses the capacity to hydrolyze 6×10^5 ACh molecules per molecule of enzyme per minute. In addition to AChE, there are two additional cholinesterases that catalyze the hydrolysis of acetylcholine-linked esters located either in or on erythrocytes (red blood cells, RBC), referred to as erythrocyte, RBC-cholinesterase or true ChE. Cholinesterase located in the blood is referred to as butyrocholinesterase or pseudocholinesterase. The preferred substrate for the ChE in blood is butyrylcholine in serum and plasma, whereas ACh is the preferred substrate for both RBC-cholinesterase and AChE. Functionally, the blood enzymes appear to serve as buffers for AChE in that AChE is not inhibited until RBC-cholinesterase and butyrocholinesterase are inhibited (Koelle 1975; Sidell and Franz 1997).

Dissimilar to ACh, OPs in general bind only to the esteratic site of the AChE molecule, and the binding stability of organophosphates for this site is structure dependent. Alkyl groups larger than the ethyl moiety bind to the site with a very high stability, and the hydrolysis from the site may not occur for hours. A phosphorylated form of the enzyme may require *de novo* synthesis of AChE because the binding is permanent in duration, resulting in an irreversible inhibition (Koelle 1975).

5.1.4 Acute Toxicity of Organophosphate Chemical Threat Agents

The eyes, nose, lung, and gastrointestinal tract are affected by OP chemical threat agents, and the biological effects observed typify exacerbated cholinergic tone. To summarize the increase in *muscarinic* activity in the presence of these agents, the acronym SLUD—which stands for increased salivation, lacrimation, urination, and defecation—has been used (Koelle 1975). With vapor exposure, signs and symptoms that correspond to an overaccumulation and muscarinic receptor stimulation by ACh are observed within seconds to minutes. In the case of the eyes, miosis (constricted pupils due to the contraction of the sphincter muscle) accompanied by eye pain, dim vision (ascribed to miosis), and blurred vision is often experienced (Rengstorff 1985; Sidell 1974, 1990). The nose and mouth are sensitive to OP toxicity and elicit copious secretions and rhinorrhea. The individual also may experience "tightness in the chest" and dyspnea that is attributable to bronchiolar contractions and increased secretions from bronchial goblet cells, potentially leading to cyanosis and respiratory collapse. Organophosphate chemical threat agents also suppress the central drive from the CNS on the respiratory system, as well as induce weakness and flaccid paralysis of the diaphragm and intercostal musculature (Adams et al. 1976; de Candole et al. 1953; Karczmar 1985; Krop and Kunkel 1954; Meeter and Wolthuis 1968; Rickett et al. 1986; Wright 1954). In the case of the gastrointestinal system, excessive cholinergic tone is manifested as nausea, diarrhea, fecal and urinary incontinence, and vomiting. With the exception of an occasional arrhythmia and bradycardia, the cardiovascular system is maintained until the individual reaches the terminal stage (Koelle 1975).

Excessive nicotinic cholinergic tone in the somatic muscle groups following OP chemical threat agent exposure causes stimulation of muscle fibers and muscle groups that can lead to fasciculations, twitches, fatigue, and paralysis, depending upon the concentration. Generalized fasciculations are often a signature of severe exposure to these agents and represent activity in muscle fibers innervated by a single motor neuron. Recruitment of large muscle groups can induce convulsive jerks, rigidity, hyperextension, and convulsions that can terminate in flaccid paralysis (Koelle 1975). In total, signs and symptoms of an OP chemical threat agent exposure for both muscarinic and nicotinic activation can be remembered by the acronym BUBBLES, which refers to bradycardia, urination, bronchial constriction, bronchospasm, lacrimation, emesis, and seizures (Langerman and Bachi 2007).

Concerning CNS effects of OP chemical threat agents, numerous behavioral changes have been reported. These effects usually start within a few hours and can last several weeks, if death is prevented. The symptoms that appear may vary in their occurrence, but include fatigue, withdrawal, depression, insomnia (often accompanied by unusual or frightening dreams), irritability, anxiety, mental confusion, and slowness. These behavioral effects have been reported to persist for days, weeks, months, or years, depending partly on the severity of exposure (Grob and Harvey 1953, 1958; Hoskins et al. 1986; Landauer and Romano 1984; Lynch et al. 1986; Rylands 1982; Sidell and Groff 1974). These actions need to be studied because they may be related to delayed neurotoxicity, possibly associated with neurodegeneration.

In addition to an inhibitory action directed at AchE, chemical threat agents also inhibit serine esterases, an enzyme that degrades various transmitters that include substance P, enkephalins, and endorphins (O'Neil 1981). These noncholinesterase effects in the brain appear to be associated with the long-lasting analgesia following sarin and soman exposure, and can be reversed by the opiate antagonist naloxone (O'Neil 1981). Furthermore, it has been reported that OP chemical threat agents are inhibitors of proteases that would increase endogenous opiates levels (Clement and Copeman 1984).

Organophosphate chemical threat agents have been reported to induce a delayed neuropathy by inhibiting a neuropathy-targeted esterase (Abou-Donia 1981; Johnson 1975, 1992). Specifically, subneurotoxic doses of sarin given to animals pretreated with atropine and the oxime P2S, to block increased cholinergic tone, induced a delayed neuropathy secondary to the inhibition of neuropathy-targeted esterases (Davies and Holland 1972). Prior to the Gulf War, the military opinion was that OP-induced delayed neuropathy required very high levels, far exceeding lethal doses. However, mice that received daily inhalations of sarin at doses that failed to elicit cholinergically related illnesses exhibited typical delayed neuropathy after a 10-day exposure (Husain et al. 1993). It could be speculated that some of the Gulf War illnesses may be related to OP-induced delayed neuropathies.

5.1.5 ACUTE TREATMENT FOLLOWING ORGANOPHOSPHATE CHEMICAL THREAT AGENT EXPOSURE

The goal for the medical treatment of individuals exposed to an OP chemical threat agents is to antagonize the excessive increase in cholinergic tone at the parasympathetic end organs (smooth muscles, cardiac muscle, and exocrine glands). Due to its limited side effects, atropine is the drug of choice for blocking the muscarinic cholinergic receptor; specifically decreasing gastrointestinal motility, hypersecretions in the nose, mouth, and airways, and relaxation of the bronchioles (Koelle 1975; Sidell 2003; McDonough and Shih 2007). Since atropine does not address the increased cholinergic nicotinic activation of the skeletal muscles, oximes are administered with atropine to block muscular fasciculations and twitching. Oxime 2-PAM Cl [2-(pyridine aldoxime methyl chloride) competes with the OP for its binding site on the AChE site, and will displace and promote the hydrolysis of a reversible "unaged" organophosphate (Eyer and Worek 2007). The so-called aging phenomenon of the OP involves a change in the kinetics of the enzyme-poison complex. Initially, the binding of OPs to the AChE molecule is reversible, which implies that the enzyme is restored to its reactive state by the undocking of the organophosphate. However, with time, the OP can alter its kinetics, secondary to a loss of the alkyl group. This process leads to the permanent inactivation of the enzyme and is referred to as "aging." In the case of soman exposure, aging can occur within 2 min., by which time 50% of the molecule has formed an irreversible bond with the AChE molecule and is refractory to oxime reactivation. Pyridostigmine, a carbamate, circumvents the rapid aging of the AChE-soman complex by transiently (reversibly) binding to the active site of AChE, and thus preventing the binding of soman. Pyridostigmine can only be given prior to OP chemical threat agent exposure, because if given after exposure, it will actually increase the efficacy of the OP (Sidell 2003).

Accordingly, war fighters routinely are issued three MARK 1 kits (Meridian Medical, Columbia, MD) in the case of exposure to OP chemical threat agents, with each kit equipped with autoinjectors containing the anticholinergic, atropine (2 mg), and the cholinesterase reactivator oxime 2-PAM Cl (600 mg). For more severe exposures to OP chemical threat agents, exposed individuals can inject more than three applications of atropine, if necessary, but not more than three applications of oxime (Sidell 2003). In more severe cases in which the individual cannot self-administer the MARK 1 kit injections, it is advised that diazepam 10 mg be injected, with or without the presence of convulsions, at 10-min. intervals (Sidell 2003; Marrs and Sellstroulm 2007). Accordingly, military personnel are supplied with three autoinjectors containing diazepam. Diazepam is a benzodiazepine that acts as an anxiolytic, anticonvulsant, sedative, and skeletal muscle relaxant, by enhancing GABA (gamma-aminobutyric acid)-mediated inhibition by serving as a positive allosteric modulator of the GABA receptor (Twyman et al. 1989).

5.1.6 DELAYED NEUROPATHOLOGY OF ORGANOPHOSPHATE CHEMICAL THREAT AGENTS

Chemical threat agents pose significant, well-characterized, acute, and lethal threats to military and civilian personnel. Less recognized, however, are delayed long-term degenerative actions on specific body organ systems such as the cardiac, musculoskeletal, endocrine, and central and peripheral nervous systems. Delayed toxicity and degenerative effects in the CNS are particularly important and potentially debilitating because the CNS is essential for the maintenance of life, and it is selectively vulnerable to these agents (Sidell 2003). As previously discussed, OPs are highly toxic chemicals that target and inhibit all nervous system functions. In recent years, the devastating effects of acute poisoning by OPs have become evident in several military conflicts and two terrorist attacks (see Chapter 1). Chemical threat agents act to produce acute lethality by inhibiting AChE, thereby triggering a chain of neurochemical events that rapidly culminates in severe seizures and death, if untreated. Current prophylactic and antidotal strategies are directed at reducing excessive cholinergic neurotransmission with pyridostigmine, atropine (or analogues), and oxime 2-PAM Cl administration, and reducing seizures with benzodiazepines, which enhance GABA transmission (Eyer and Worek 2007; Marrs and Sellstroulm 2007; McDonough and Shih 2007; Shih and McDonough 1997). However, excessive cholinergic and diminished GABA neurotransmission, secondary to OP threat agent exposure, disrupts homeostatic activity throughout the brain, leading to excitotoxicity characterized as excessive and abusive glutamate-mediated damaging events (see Section 5.2). Underscoring the importance of glutamate neurotransmission in OP intoxication, recent experimental data indicate a greater level of protection to CNS neurons and glia is achieved by combinations of classical agents (MARK 1 therapy) and compounds directed at blocking some of the glutamate receptor subtypes, yet protection is still incomplete (Solberg and Belkin 1997). Even though these treatments offer protection from terminal convulsions and acute lethality, little is known regarding the neuronal intracellular processes initiated by exposure to these chemical threat agents and the long-term consequences to brain function.

While their potential use as military or terrorist threat agents is more limited by their physiochemical properties, chemical threat agents that target the CNS pose additional threats to humans. A large number of neurotoxins act by inhibiting or activating ionic channels in membranes of excitable cells, including neurons of the central and peripheral nervous systems. Inhibition of potassium channels or activation of sodium or calcium channels leads to excessive membrane depolarization, excessive release of neurotransmitters (including acetylcholine and glutamate), abusive receptor stimulation, seizures, and neuropathology (Choi 1987, 1988; Mattson 2003; Olney et al. 1983). Excess glutamate neurotransmission also can generate excessive intracellular free radicals and disrupt cellular energy stores, thus potentially activating a self-perpetuating, self-amplifying cascade of deleterious cellular processes (Choi 1987, 1988; Mattson 2003). Abusive glutamate neurotransmission causes excitotoxicity characterized by ill-defined intracellular cell-death cascades, which suggest mechanisms, common to those utilized by OPs, may be recruited in cases of poisoning by chemical threat agents that cause CNS hyperexcitability (Figure 5.3). Data regarding long-term nervous system effects following exposure to low-level, sublethal chemical and biological threat agent insults are virtually nonexistent. Thus, excitotoxicity is a commonly accepted process responsible for the delayed neurotoxicity and degeneration seen following chemical threat agent exposures (Olney et al. 1983; Choi 1987; Braitman and Sparenborg 1989; Lallement et al. 1991; Sparenborg et al. 1992; Solberg and Belkin 1997; Shih and McDonough 1997; McDonough and Shih 1997).

Delayed neuropathology leading to permanent disability potentially occurs in survivors of episodic or long-term, low-level exposures that do not elicit the full

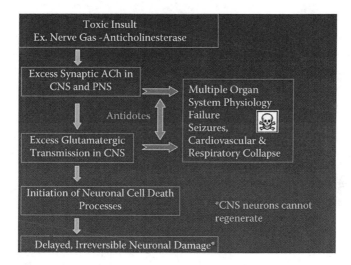

FIGURE 5.3   (See color insert following page 46.) Role of excitotoxicity in mediating neuronal cell death induced by nerve gas exposure.

spectrum of overt poisoning symptoms (Rothestein et al. 1993), or when immediate life-threatening symptoms from lethal exposures to chemical threat agents, are ameliorated by symptomatic therapies directed at quelling seizures and preventing respiratory collapse and cardiovascular failure (Collingridge and Lester 1989). Even if the immediate life-threatening insult is suppressed or aborted, an occult, continuous intracellular cascade of molecular events, including genetic reprogramming, can be set into motion and can ultimately lead to self-initiated elimination of cells via programmed cell death (PCD) (Monaghan et al. 1989). Analogous cell death processes also can be initiated following a nonlethal exposure to crude, non-weapons-grade chemical threat agents. This is of paramount importance considering that these crude, non-weapons-grade materials could be used in a terrorist attack. Such an exposure potentially leads to delayed nervous system degeneration and disease in a large population of civilian or military personnel years later. Programmed cell death has a major role in many, if not all, progressive neurodegenerative conditions, including Alzheimer's disease, Parkinson's disease, amyotropic lateral sclerosis (ALS or Lou Gehrig's disease), spinocerebellar ataxia, and various behavioral and cognitive disorders. Accumulating evidence suggests that exposure to sublethal chemical threat agents can also, in fact, have long-term effects leading to debilitating neurodegenerative disorders such as ALS, Parkinson's disease, Alzheimer's disease, neuropsychiatric disturbances, as well as many other lesser or unrecognized disorders (Beal 1992; Churchill et al. 1985; Dugan and Choi 1994; Meldrum and Garthwaite 1990; Plaitakis et al. 1982; Rothestein et al. 1993; Solberg and Belkin 1997). Numerous reports of health problems in personnel from the Vietnam conflict (Agent Orange, dioxin exposure) and Persian Gulf War (potential exposure to sarin nerve gas and other unknown agents) have been grouped collectively under the monikers of the Vietnam War or Gulf War syndromes. Significantly, a recent report suggests an increased incidence of ALS in Gulf War personnel following a potential low-level chemical threat agent exposure (Solberg and Belkin 1997). Thus, there is an urgent need to define and understand neuronal molecular signal-transduction processes initiated by chemical treat agent exposure in order to identify personnel at future risk, and to design therapeutic interventions targeted at the known pathologic processes to prevent delayed neurodegenerative and other diseases in these individuals.

5.2 MULTIPLE PATHWAYS LEADING TO NEURONAL DEGENERATION

Inhibition of AChE by OP chemical threat agents leads to excessive synaptic cholinergic (acetylcholine) tone, which induces excessive glutamatergic neurotransmission (Solberg and Belkin 1997; Figure 5.3). It has been estimated that 50% of the synapses in the CNS utilize glutamate as their transmitter and that it is responsible for sustaining soman-induced seizures and promoting the development of status epilepticus (Olney et al. 1983; Wade et al. 1987; Braitman and Sparenborg 1989; Sparenborg et al. 1992; Fosbraey et al. 1990; Solberg and Belkin 1997). Excessive glutamatergic tone culminates in neuronal death in selectively vulnerable brain regions, a condition termed excitotoxicity (Dugan and Choi 1994), and occurs when glutamate abusively activates various glutamate receptor subtypes, including AMPA (α-amino-3-hydroxy-5-methyl-4-isoxazolepropionic acid, a glutamate analogue), NMDA (N-methyl-D-aspartate, a glutamate analogue), and metabotropic glutamate subtypes (Beal 1992). Glutamate toxicity encompasses a host of neuronal and cellular responses, including excessive generation of free radicals, excessive intracellular calcium concentrations, disruption of cellular energy homeostasis, and induction of a multitude of enzymatic signal-transduction pathways in neurons and glia (Green and Reed 1998; Neary 1997; Solberg and Belkin 1997; Xia et al. 1995). Parallel and converging intracellular cascades combine to disrupt the normal ionic, metabolic, and biochemical homeostasis of the neuron and elicit neurodegeneration in selectively vulnerable areas in the CNS. These actions can be either directly caused within the neuron by the neurotoxicant or indirectly elicited via factors released or activated in glia (Ashkenazi and Dixit 1998; Green and Reed 1998; Neary 1997). Therefore, currently employed therapeutic interventions directed only at cell surface receptors may be consistently inadequate to afford full protection against acute and delayed neurodegeneration. Investigations examining novel approaches directed at convergence points of many of these cell death pathways could yield valuable results.

Excitotoxicity underlies both acute neuronal necrotic death and chronic neurodegenerative processes of PCD, including apoptosis, dark cell degeneration, and autophagy (Beal 1992; Rothestein et al. 1993). A signature response of neurons exposed to excitotoxic challenge is an elevation of intracellular calcium ($[Ca^{2+}]_i$), that in acute and severe instances causes swelling, lysis, and necrosis of neurons, and in less severe subacute challenges, triggers a myriad of enzymatic and genetic events, leading to PCD of the neuron. Necrosis and PCD represent two distinct modes of neuronal death in the CNS (Rothestein et al. 1993; Collingridge and Lester 1989). Classic necrosis is considered a passive response of a cell induced by sudden changes in the neuronal environment and is characterized by rapid rupture of the plasma membrane and extravasation of the internal milieu, followed by a secondary inflammatory response (Rothestein et al. 1993; Monaghan et al. 1989). Programmed cell death is an active, energy-dependent, and gene-directed type of cell death that is characterized by several stereotyped morphological changes that fall on a spectrum of cellular transformations with features of classic apoptosis at one end and necrosis at the other (Leist and Jaattela 2001). Recently, the notion that necrosis is purely a passive response of the neuron to external stressors has undergone revision with the elaboration of intracellular enzymatic cascades, leading to active necrosis characterized by cellular morphological changes and death similar to classic necrosis (Leist and Jaattela 2001). Forms of PCD, particularly apoptosis, represent more physiological forms of cell disposal observed during development and pathologically following physical and chemical stress (Rothestein et al. 1993; Collingridge and Lester 1989; Monaghan et al. 1989). Many early-response genes have Ca^{2+}-responsive elements that are expressed secondary to elevations in $[Ca^{2+}]_i$. Programmed cell death is associated with activation of selected prodeath genes and several principal enzymatic cascades, including caspases, a family of cysteine proteases that resemble the ced-3 gene from *Caenorhabditis elegans*, and calpains, cysteine proteases that are ubiquitously expressed as two isoforms. The two isoforms are calpain 1 and 2, also known as μ- and m-calpain, which are activated by micromolar and millimolar concentrations of $[Ca^{2+}]_i$, respectively.

In order to refine therapeutic regimes and thus prevent chemical warfare agent-induced excitatory neurodegeneration in exposed individuals (Figure 5.4), a more comprehensive characterization of the destructive signal-transduction mechanisms and intracellular enzymatic cascades associated with glutamate receptor mediated delayed excitotoxicity (PCD) is needed. Accordingly, the refinement of therapeutic

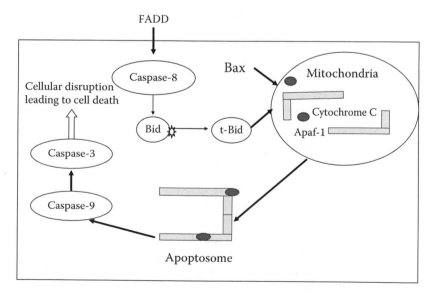

FIGURE 5.4 (See color insert following page 46.) Cellular processes underlying the activation of caspases and their role in cellular disruption associated with apoptosis.

regimes could be expedited by discovery of various proteins that mediate delayed excitotoxicity, a process that appears to be a common final process responsible for neurodegeneration, and may explain some of the long-term health effects experienced by personnel during the Vietnam conflict and Gulf War. Furthermore, neuroprotection against neurodegenerative effects elicited by chemical threat agents can be afforded by therapeutic interventions directed at specific common intracellular cell death mediators.

5.2.1 ACTIVATION OF CASPASES (CYSTEINE REQUIRING ASPARTATE PROTEASES)

The consensus in the literature is that nerve-agent-induced seizures lead to the development of glutamate-mediated excitotoxicity (Olney et al. 1983; Choi 1987; Braitman and Sparenborg 1989; Lallement et al. 1991; Sparenborg et al. 1992; Solberg and Belkin 1997; Shih and McDonough 1997; McDonough and Shih 1997). It is widely appreciated that the majority of soman-induced, seizure-related brain damage results from glutamate excitotoxicity and corresponding delayed calcium overload (Olney et al. 1983; Braitman and Sparenborg1989; Sparenborg et al. 1992; Solberg and Belkin 1997). The increase in calcium occurs through ligand-gated glutamate channels (Jayakar and Dikshit 2004), voltage-gated calcium channels, intracellular release of calcium from endoplasmic reticulum (via inositol triphosphate and ryanodine receptor activation; Verkhratsky and Toescu 2003), and mitochondria (secondary to loss of mitochondrial transmembrane potential and release of proapoptotic factors; Halestrap et al. 2002). The delayed calcium overload that ensues after exposure to soman-induced, seizure-related brain damage could promote a pathological sequence that is characterized by the activation of several potentially damaging enzymes, which include the mitochondrial release of cytochrome C and the activation of caspases (Figure 5.4). Caspases are known to be major mediators of PCD, and are induced by either an extrinsic or intrinsic pathway. The extrinsic pathway involves the activation of a death receptor such as FADD, leading to caspase-8 activation and Bid truncation (tBid) and its activation (Figure 5.4). The intrinsic mitochondrial-related pathway involves the release of proapoptotic factors from the mitochondria that include cytochrome C, apoptosis-inducing factor (AIF), and a protein named Smac, following a transition in the outer mitochondrial membrane. Release of mitochondria-derived proapoptotic molecules occurs secondary to translocation of one or more members of the proapoptotic Bcl-2 protein family (Bax, Bid, Bad) or opening of the permeability transition pore, which is often accompanied by depolarization of the mitochondrial potential ($\Delta\Psi_m$) (Mayer and Oberbauer 2003; Zamzami and Kroemer 2001). Smac promotes apoptosis by binding to, and neutralizing, an inhibitor of apoptosis proteins (IAPs) that subsequently allows the release of caspases and their activation. Released cytochrome C is known to form an "apoptosome" consisting of cytochrome C, d-ATP (or ATP), procaspase-9, and Apaf-1, resulting in the activation of caspase-9 and subsequently, caspases-3 and -7, ultimately culminating in cell death (Figure 5.4; Ashkenazi and Dixit 1998; Fadeel et al. 1999; Green and Reed 1998; Zamzami and Kroemer 2001).

Caspase activation has been shown to be a critical step in several models of excitotoxic-induced neuronal apoptosis. Du et al. (1997) reported that glutamate induced an increase in caspase-3 activity in cultured cerebellar granule cells, and this increase correlated with apoptosis. Using cultured rat cerebral cortical neurons, numerous studies have reported a correlation between enhanced caspase-3 activity and excitotoxic insults, with agents that include glutamate, AMPA, kainate (Hirashima et al. 1999; Nath et al. 1998), and NMDA (Hirashima et al. 1999; Nath et al. 1998; Tenneti and Lipton 2000). Nath et al. (1998) observed that AMPA-induced excitotoxicity of cultured cortical neurons elicited a caspase-specific α-spectrin breakdown product that was attenuated by caspase inhibitors. Furthermore, in these studies, caspase inhibitors effectively attenuated the apoptotic-like neuronal death.

A previous study (Strahlendorf et al. 2003) conducted by the author also has shown that AMPA-induced PCD in both the cerebellum and the hippocampus is mediated in part by the activation of caspases. Immunohistochemical analyses revealed the active (cleaved) form of caspase-3 was markedly and significantly increased in both the Purkinje neurons and pyramidal neurons in the CA1 and CA3 regions of the hippocampus. Figure 5.5 (Panels A–D) depicts temporal changes in mean active caspase-3 immunofluorescence in both regions of the brain. These findings indicate that one of the processes involved in an excitotoxic-mediated neuronal cell death, possibly triggered by chemical threat agents in both cerebellar and hippocampal neurons, is the activation of caspase-3, one of the major executor caspases in the brain for the induction of PCD processes. These findings would suggest that the use of a caspase-3 inhibitor would offer a potential therapeutic approach for the treatment of excitotoxic-mediated PCD.

As further assessment of caspase involvement in mediating AMPA-induced excitotoxicity in cerebellar and hippocampal slices and as potential therapeutic targets in delayed neurodegenerative processes, the effectiveness of a pan caspase inhibitor FK011 [BAF, Boc-Asp(me)FMK, Enzyme System Products] to attenuate AMPA-induced toxicity in the hippocampus was ascertained. As shown in Figure 5.5, panel E, the caspase inhibitor effectively attenuated the number of pyramidal neurons that exhibited AMPA-induced dark cell degeneration (DCD; a form of excitotoxicity that resembles apoptosis; Strahlendorf et al. 1996, 1999). However, in the case of the cerebellum, a marked increase in the occurrence of edematous necrosis was evident in the presence of AMPA and the caspase antagonist. These findings strongly suggest that DCD involves activation of caspases as observed using immunohistochemistry, and suggests the need to exercise caution when advocating for the use of caspase inhibitors in light of their potential to induce necrosis.

5.2.2 ACTIVATION OF CALPAIN (CALCIUM-DEPENDENT NEUTRAL PROTEASE)

Cysteine proteases, called calpains, are known to be activated by sustained elevations in intracellular free calcium. Once activated, calpains degrade the cytoskeleton, transmitter and membrane channels, and metabolic enzymes (Hou and MacManus 2002; Mattson 2003; Nicholls 2004). Functionally, calpains have been characterized as pivotal mediators of both active necrotic cell death and PCD (Wang 2000) following cell-damaging stressors and insults such as soman exposure, excitotoxic challenges, toxins, free radicals, UV radiation, acute hypoxia, traumatic brain injury, cytokines, heat, and in chronic neurodegenerative conditions (Fischer et al. 1991; Caner et al.

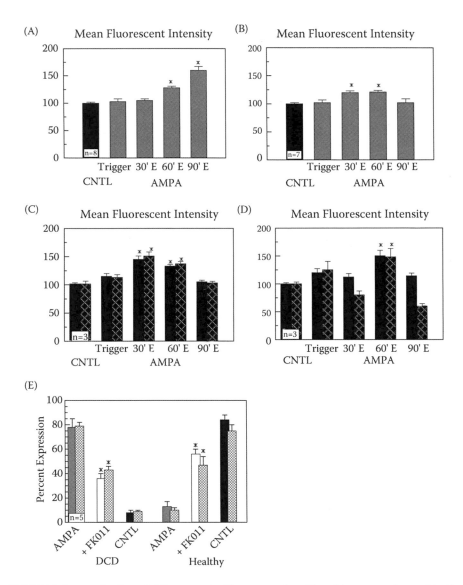

FIGURE 5.5 Panels (A) and (B) summarize the temporal profiles of increases observed in the active form of caspase-3 and active caspase-9 immunoreactivities in cerebellar Purkinje neurons following exposure to 30 μM AMPA for 30 min., respectively. Each bar represents the mean percent increase ($\pm$ S.E.) in fluorescence intensity. (Strahlendorf, J. et al.). AMPA-induced dark cell degeneration of cerebellar Purkinje neurons involves activation of caspases and apparent mitochondrial dysfunction. *Brain Res.*, vol. 994, p. 150, figure 1 (Panel [A]), p. 151, figure 2 (Panel [B]), graph II. Elsevier, 2003.) Panels (C) and (D) summarize the temporal profiles of increases in the active form of caspase-3 in the cytosol (filled bars) and nucleus (open bars) and of the CA1 and CA3 regions of the hippocampus, respectively, following exposure to 100 μM AMPA for 30 min. followed by a 90-min. AMPA-free expression period. Panel (E) is a summary of experiments demonstrating the effectiveness of the pan caspase inhibitor FK011 (50 μM) to attenuate AMPA-induced DCD and increase healthy pyramidal neurons in the CA1 (solid bars) and CA3 (stripped bars) regions of the hippocampus. Statistical significance ($p \leq 0.05$) is denoted by an asterisk.

1993; Xia et al. 1995; Neary 1997; Vanderklish and Bahr 2000; Nixon 2003). Intracellular Ca^{2+}-induced activation of calpain promotes pathophysiological cytoskeletal reorganization, specifically targeting the cytoskeletal proteins α-spectrin, neurofilaments, actin, and microtubule associated proteins (Potter et al. 1998; Fischer et al. 1991; Banik et al. 1997). Thus, both calpains and caspases can be activated by similar cell-death triggers, and potentially could act in concert to dismantle intracellular structures and organelles, to ultimately destroy a cell. Moreover, recent findings suggest that caspase and calpain pathways reciprocally interact at the intracellular level such that caspase activity is dependent on calpain activation (McCollum et al. 2002; Gil-Parrado et al. 2002). Calpains therefore represent a potential parallel pathway to caspases leading to PCD but with some common convergence points.

Previous studies have indicated that calpains are involved in excitotoxic-mediated cell death (Bi et al. 1998; Brorson et al. 1994; Choi, 1988; Lopez-Picon et al. 2006; Markgraf et al. 1998; Rami 2003; Siman et al. 1985). Cell death associated with ischemia, which is known to involve excitotoxicity, can be attenuated with calpain antagonists (Choi, 1988; Markgraf et al. 1998). Brorson et al. (1994) have reported that kainate-induced cell death in cultured Purkinje neurons was reduced by calpain antagonists. Interestingly, in the hippocampus, an α-spectrin cleavage product was evident in the CA1 region within 20 min. of an ischemic insult (Rami 2003), supporting an early involvement of calpains in mediating excitotoxic-induced cell death in this CNS region, presumably through the disassembly of cytoskeletal elements. The marked cytoskeletal changes in response to abusive AMPA receptor activation, coupled with increased intracellular Ca^{2+} concentration, suggests activation of various destructive enzymes such as calpains, in addition to caspases.

A recent study using immunohistochemical analyses revealed that calpain-mediated α-spectrin breakdown product was markedly and significantly increased in both the dendritic and somatic regions of Purkinje neurons, whereas control Purkinje neurons failed to exhibit increases in α-spectrin breakdown immunoreactivity (Mansouri et al. 2007). AMPA-treated, fragmented α-spectrin immunopositive Purkinje neurons exhibited the typical shrunken appearance of DCD, implying that the induction of DCD is intimately related to calpain activation. Furthermore, a series of studies was conducted to determine whether AMPA-induced DCD could be attenuated by PD150606, a noncompetitive calpain antagonist. This antagonist effectively attenuated both the morphological changes and the increases in calpain-cleaved α-spectrin immunoreactivity associated with AMPA-induced DCD (Figure 5.6), further establishing a causal relationship between the two events (Mansouri et al. 2007).

A previous report has implicated a role of calpain in mediating the tissue injury caused by the chemical threat agent sulfur mustard. Specifically, tissue homogenates from mouse ear skin exposed to sulfur mustard displayed a marked increase in calpain activity (170% increase 24 h after treatment; Powers et al. 2000). These findings underscore the need to identify effective antiproteases with therapeutic use in reducing, or eliminating, tissue injuries. Since excitotoxicity is related directly to calpain activation, it can be surmised that sulfur mustard exposures may be linked to excitotoxicity.

These findings provide a direct link between AMPA-induced excitotoxicity in the cerebellum and the activation of the Ca^{2+}-dependent enzyme calpain. Based upon

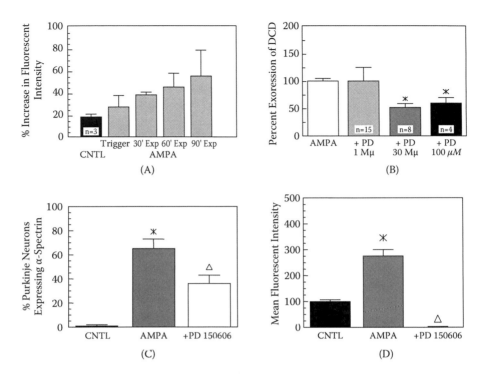

FIGURE 5.6 Panel (A) represents temporal elevations of α-spectrin immunoreactivity in Purkinje neurons following AMPA treatment. Statistical significance ($p \leq 0.05$) is denoted by an asterisk. Panel (B) summarizes the effectiveness of calpain antagonist PD150606 to attenuate AMPA-elicited increases in α-spectrin immunoreactivity in cerebellar Purkinje neurons at the 90-min. expression period. Panel (C) illustrates a significant reduction in the percent of Purkinje neurons expressing in α-spectrin immunoreactivity at the 90-min. expression in the presence of 100 μM PD150606. Panel (D) illustrates a significant reduction in the mean fluorescent intensity of AMPA-induced increases in α-spectrin immunoreactivity at the 90-min. expression in the presence of 100_ μM PD150606. Statistical significance ($p \leq 0.05$) is denoted by an asterisk. (Mansouri, B., Henne, W.M., Oomman, S.K., Bliss, R., Attridge, J., Finckbone, V., Zeitouni, T., Hoffman, T., Bahr, B.A., Strahlendorf, H.K., and Strahlendorf, J.C., 2007, Involvement of calpain in AMPA-induced toxicity to rat cerebellar Purkinje neurons, *Eur. J. Pharmacol.*, 557, 106–114.)

these studies, it could be concluded that calpain inhibitors would present a rational approach to the pharmacological treatment of AMPA-induced excitotoxicity. By extrapolation, exposure to OP chemical threat agents would best be therapeutically approached by the use of calpain pharmacological antagonists, in order to protect the neurons from functional losses.

The utility and attractiveness of novel general neuroprotection afforded by common and otherwise innocuous agents is becoming increasingly recognized. Such agents would not narrowly target specific cell surface receptors or intracellular transduction events but rather would induce some general change in the physiology of a cell to render it less susceptible to toxic insults. In theory, this would circumvent the need to develop many target-selective protective agents for multiple toxic compounds that differentially activate, or trigger, common cell-death pathways via various cell membrane receptors or ion channels.

5.3   CONCLUSIONS

Caspase activation has been shown to be a critical step in several models of excitotoxic-induced neuronal apoptosis. However, the usefulness of caspase inhibitors in the treatment of delayed excitotoxicity will need to be carefully evaluated since these agents appear to increase the expression of necrotic toxicity. These findings suggest that any therapeutic approach addressing toxicity associated with excessive activation of AMPA receptors would need to incorporate caspase inhibitors, as well as inhibitors that would protect against edema and necrosis. This latter type of toxicity, if allowed to be expressed, would be more damaging because it would promote inflammation and more widespread cell death. Thus, a single therapeutic intervention to address excitotoxicity may promote more damaging phenomena rather than the desired protective action. Based upon these findings, it would appear necessary to ascertain the intracellular mechanisms associated with excitotoxic-induced necrosis by continuing to explore the involvement of other types of proteases, such as calpains and cathepsins. Thus, inactivation of caspase activity, while effectively reducing delayed DCD, is not a promising potential therapeutic strategy by itself in the cerebellum.

The marked effectiveness of AMPA to generate calpain-derived cleavage products in cells that express DCD and the effectiveness of the calpain antagonist PD150606 to attenuate both AMPA-induced DCD morphology and increases in calpain cleavage products need to be carefully evaluated for their potential therapeutic interventions in neurodegenerative conditions. There appears to be a direct link between abusive AMPA receptor activity and activation of calpains leading to production of an α-spectrin fragment resulting in DCD in the cerebellum. Unlike caspase inhibition, calpain inhibition did not unmask an edematous state, which would suggest that calpain antagonists represent a better therapeutic choice to address AMPA-elicited excitotoxicity (Strahlendorf et al. 2003; Mansouri et al. 2007). Figure 5.7 illustrates the proposed involvement of both caspases and calpains in mediating AMPA-induced excitotoxicity.

5.4   RESEARCH RECOMMENDATIONS

One of the major premises of this chapter is that exposure to chemical threat agents, in scenarios that do not cause acute and immediate lethality, initiates a cascade of intraneuronal events that lead to delayed or latent neurotoxicity and neurodegeneration. In point of fact, the first magnetic resonance imaging study on individuals exposed to low concentrations of sarin nerve gas in the 1991 Persian Gulf War has shown the presence of significant loss of cortical and subcortical regions of the brain (Heaton et al. 2007). Central nervous system damage in these individuals was correlated to behavioral deficits in fine motor skills (coordination) and cognitive functions (memory). Coordination and cognition are functions of the cerebellum and hippocampus, respectively, and since both regions display selective vulnerability to

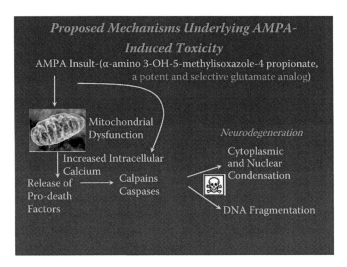

FIGURE 5.7   (See color insert following page 46.) Proposed involvement of caspases and calpains in mediating AMPA-induced excitotoxicity.

glutamate toxicity via AMPA receptors, loss of neuronal activity in these regions could be responsible for the reported deficits. Therefore, continued investigation in this realm should uncover important, and significant, common intracellular transduction pathways responsible for delayed neurodegenerative changes following low-level nonlethal exposure to chemical threat agents.

Future approaches to address, in a more comprehensive and qualitative manner, the complexity of the multiple pathways triggered by chemical threat agents, could utilize protein microarray technology to directly identify complexes of proteins that work in molecular ensembles to carry out the pathophysiologic events. Not only does this approach translate to more efficient discovery of cellular processes, but it also fosters a rapid progression and rational development of therapeutic interventions that prevent catastrophic, lifelong neurological problems following exposure to nonlethal amounts of chemical threat agents, by targeting proteins identified as major contributors to neuronal PCD in selected brain regions. Because cells are endowed with multiple, complex, and redundant pathways that are each independently capable of causing PCD, and each of these pathways may operate simultaneously, in parallel, or through numerous branching interconnections that initiate or amplify the actions of others, it is highly unlikely that a single cellular entity will be found that could serve as a sole target for therapeutic intervention. To devise safe and effective therapeutic agents and strategies, it will be necessary to more completely define and elaborate components of the cell-death programs; hopefully, a select number of common mediators will emerge as important targets for pharmacologic interventions. Using this knowledge, a blend or "cocktail" of agents will more than likely constitute a vital and important advancement in treating individuals exposed to low- and nonlethal levels of chemical threat agents to abort and prevent any insidious neurotoxic cascades leading to latent nervous system damage, degeneration, and diminished normal motor and cognitive functions.

ACKNOWLEDGMENT

The P. I. (primary investigator) is grateful to RDECOM for their financial support of this project, especially for the great help from Drs. William Lagna, John White, Ronald Kendall, Steve Presley, and Galen Austin.

REFERENCES

Abou-Donia, M.B., 1981. Organophosphorus ester-induced delayed neurotoxicity, *Ann. Rev. Pharmacol. Toxicol.*, 21, pp. 511–548.

Adams, G.K., III, Yamamura, H.I., and O'Leary, J.F., 1976. Recovery of central respiratory function following anticholinesterase intoxication, *Eur. J. Pharmacol.*, 38, pp. 101–112.

Ashkenazi, A. and Dixit, V.M., 1998. Death receptors: signaling and modulation, *Science*, 281, pp. 1305–1308.

Banik, N. L., Matzelle, D. C., Gantt-Wilford, G., Osborne, A., and Hogan, E. L., 1997. Increased calpain content and progressive degradation of neurofilament protein in spinal cord injury, *Brain Res.*, 752, pp. 301–306.

Beal, M.F., 1992. Mechanisms of excitotoxicity in neurological diseases, *FASEB J.*, 6, pp. 3338–3344.

Bi, X., Chen, J., and Baudry, M., 1998. Calpain-mediated proteolysis of GluR1 subunits in organotypic hippocampal cultures following kainic acid treatment, *Brain Res.*, 781, pp. 355–357.

Braitman, D.J. and Sparenborg, S., 1989. MK-801 protects against seizures induced by cholinesterase inhibitor soman, *Brain Res. Bull.*, 23, pp. 145–148.

Brorson, J.R., Manzolillo, P.A., and Miller, R.J., 1994. Ca^{2+} entry via AMPA/KA receptors and excitotoxicity in cultured cerebellar Purkinje cells, *J. Neurosci.*, 14, pp. 187–197.

Caner, H., Collins, J.L., Harris, S.M., Kasdell, N.J., and Lee, K.S., 1993. Attenuation of AMPA-induced neurotoxicity by a calpain inhibitor, *Brain Res.*, 607, pp. 354–356.

Choi, D.W., 1987. Ionic dependence of glutamate neurotoxicity, *J. Neurosci.*, 7, pp. 369–379.

Choi, D.W., 1988. Glutamate neurotoxicity and diseases of the nervous system, *Neuron*, 1, pp. 623–634.

Churchill, L., Pazdernik, T.L., Jackson, J.L., Nelson, S.R., Samson, F.E., McDonough, J.H., and McLeod, C.G., 1985. Soman induced brain lesions demonstrated by muscarinic receptor autoradiography, *Neurotoxicology*, 6, pp. 81–90.

Clement, J.G. and Copeman, H.T., 1984. Soman and sarin induce a long-lasting naloxone-reversible analgesia in mice, *Life Sci.*, 34, pp. 1415–1422.

Collingridge, G.L. and Lester, R.A.J., 1989. Excitatory amino acid receptors in the vertebra central nervous system, *Pharmacol. Rev.*, 40, pp. 143–210.

Davies, D.R. and Holland, P., 1972. Effect of oximes and atropine upon the developme of delayed neuropathy in chickens following poisoning by DFP and sarin, *Biochen Pharmacol.*, 21(3), pp. 145–51.

de Candole, C.A., Douglas, W.W, Lovatt, E.C., Holmes, R., Spencer, K.E.V., Torrance, R.W and Wilson, K.M., 1953. The failure of respiration in death by anticholinesterase po soning, *Brit. J. Pharmacol.*, 8, pp. 466–475.

Du, Y., Bales K.R., Dodel, R.C., Hamilton-Byrd, E., Horn, J.W., Cyzilli, D.L., Simmons L.K., Ni, B., and Paul, S.M., 1997. Activation of caspase-3 related cysteine protease in required for glutamate-mediated apoptosis in cultured granule neurons, *Proc. Natl Acad. Sci. USA*, 94, pp. 11657–11662.

Duffy, M., 2002. Weapons of war: poison gas, available at http://www.firstworldwar.com/weaponry/gas.htm.

Dugan, L.L. and Choi, D.W., 1994. Excitotoxicity, free radicals, and cell membrane changes, *Ann. Neurol.*, 35, S17–S21.

Eyer, P.Q. and Worek, F., 2007. Oximes, in *Chemical Warfare Agents: Toxicology and Treatment*, 2nd ed., Marrs, T.C., Maynard, R.L., and Sidell, F., Eds., John Wiley & Sons, West Sussex, England, chap. 15.

Fadeel, B., Zhivotovsky, B., and Orrenius, S., 1999. All along the watchtower: on the regulation of apoptosis regulators, *FASEB J.*, 13, pp. 1647–1657.

Fischer, I., Romano-Clarke, G., and Grynspan, F., 1991. Calpain-mediated proteolysis of microtubule associated proteins MAP1B and MAP2 in developing brain, *Neurochem. Res.*, 16, pp. 891–898.

Fosbraey, P., Wetherell, J.R., and French, M.C., 1990. Neurotransmitter changes in guinea-pig brain regions following soman intoxication, *J. Neurochem.*, 54, pp. 72–79.

Gil-Parrado, S., Fernandez-Montalva A., Assfalg-Machleidt, I., Popp, O., Bestvater, F, Holloschi, A., Tobias, A., Knoch, T.A., Auerswald, E.A., Welsh, K., Reed, J.C., Fritz, H., Fuentes-Prior, P., Spiess, E., Salvesen, G.S., and Machleidt, W., 2002. Ionomycin-activated calpain triggers apoptosis, *J. Biol. Chem.*, 277, pp. 27217–27226.

Green, D.R. and Reed, J.C., 1998. Mitochondria and apoptosis, *Science*, 281, pp. 1309–1312.

Grob, D. and Harvey, A.M., 1953. The effects and treatment of nerve gas poisoning, *Am. J. Med.*, 14, pp. 52–63.

Grob, D. and Harvey, J.C., 1958. Effects in man of the anticholinesterase compound sarin (isopropyl methyl phosphonofluoridate), *Johns Hopkins Med. J.*, 37, pp. 350–368.

Halestrap, A.P., McStay, G.P., and Clarke, S.J., 2002. The permeability transition pore complex: another view, *Biochimie*, 84, pp. 153–166.

Heaton, K.J., Palumbo, C.L, Proctor, S.P., Killiany, R.J., Yurgelun-Todd, DA., and White, R.F., 2007. Quantitative magnetic resonance brain imaging in U.S. Army veterans of the 1991 Gulf War potentially exposed to sarin and cyclosarin, *NeuroToxicology*, 28, pp. 761–769.

Hirashima, Y., Kurimoto, M., Nogami, K., Endo, S., Saitoh M., Ohtani, O., Nagata, T., Muraguchi, A., and Takaku, A., 1999. Correlation of glutamate-induced apoptosis with caspase activities in cultured rat cerebral cortical neurons, *Brain Res.*, 849, pp. 109–118.

Hoskins, B., Fernando, J.C.R., Dulaney, M.D., Lim, D.K., Liu, D.D., Watanabe, H.K., and Ho, I.K., 1986. Relationship between the neurotoxicities of soman, sarin and tabun, and acetylcholinesterase inhibition, *Toxicol. Lett.*, 30, pp. 121–129.

Hou, S.T., and MacManus, J.J.P., 2002. Molecular mechanisms of cerebral ischemic-induced neuronal death, *Intern. Rev. Cytol.*, 221, pp. 93–149.

Husain, K., Vijayaraghavan, R., Pant, S.C., Raza, S.K., and Pandey, K.S., 1993. Delayed neurotoxic effect of sarin in mice after repeated inhalation exposure, *J. Appl. Toxicol.*, 13, pp. 143–145.

Jayakar, S.S. and Dikshit, M., 2004. AMPA receptor regulation mechanisms: future target for safe neuroprotective drugs, *Int. J. Neurosci.*, 114, pp. 695–734.

Johnson, M.K., 1975. The delayed neuropathy caused by some organophosphorous esters: mechanisms and challenge, *CRC Crit. Rev. Toxicol.*, 3, pp. 289–316.

Johnson, M.K., 1992. Molecular events in delayed neuropathy: experimental aspects of neuropathy target esterase, in *Clinical and Experimental Toxicology of Organophosphates and Carbamates*, Ballantyne, B. and Marrs, T.C., Eds., Butterworth and Heinemann, London.

Karczmar, A.G., 1985. Present and future of the development of anti-OP drugs, *Fundam. Appl. Toxicol.*, 5, pp. S270–S279.

Koelle, G.B., 1975. Anticholinesterase agents, in *The Pharmacological Basis of Therapeutics*, 5th ed., Goodman, L.S. and Gilman, A., Eds., Macmillan, New York, p. 445.

Krop, S. and Kunkel, A.M., 1954. Observations on pharmacology of the anticholinesterases sarin and tabun, *Proc. Soc. Exp. Biol. Med.*, 86, pp. 530–533.

Lallement, G., Carpentier, P., Collet, A., Pernot-Marino, I., Baubichon, D., and Blanchet, G., 1991. Effects of soman-induced seizures on different extracellular amino acid levels and on glutamate uptake in rat hippocampus, *Brain Res.*, 563, pp. 234–240.

Landauer, M.R. and Romano, J.A., 1984. Acute behavioral toxicity of the organophosphate sarin in rats, *Neurobehav. Toxicol. Teratol.*, 6, pp. 239–243.

Langerman, B. and Bachi, K., 2007. State of Maine nerve agent antidote kit training module, available at http://www. maine.gov/dps/ems/docs/.

Leist, M. and Jaattela, M., 2001. Four deaths and a funeral: from caspases to alternative mechanisms, *Nat. Rev.*, 2, pp. 1–10.

Lopez-Picon, F.R., Kukko-Lukjanov, T., and Holopainen, I.E., 2006. The calpain inhibitor MDL-28170 and the AMPA/KA receptor antagonist CNQX inhibit neurofilament degradation and enhance neuronal survival in kainic acid-treated hippocampal slice cultures, *Eur. J. Neurosci.*, 23, pp. 2686–2694.

Lynch, M.R., Rice, M.A., and Robinson, S.W., 1986. Dissociation of locomotor depression and ChE activity after DFP, soman and sarin, *Pharmacol. Biochem. Behav.*, 24, pp. 941–947.

Mansouri, B., Henne, W.M., Oomman, S.K., Bliss, R., Attridge, J., Finckbone, V., Zeitouni, T., Hoffman, T., Bahr, B.A., Strahlendorf, H.K., and Strahlendorf, J.C., 2007. Involvement of calpain in AMPA-induced toxicity to rat cerebellar Purkinje neurons, *Eur. J. Pharmacol.*, 557, pp. 106–114.

Markgraf, C.G., Velayo, N.L., Johnson, M.P., McCarty, D.R., Medhi, S., Koehl, J.R., Chmielewski, P.A., and Linnik, M.D., 1998. Six-hour window of opportunity for calpain inhibition in focal cerebral ischemia in rats, *Stroke*, 29, pp. 152–158.

Marrs, T.C. and Sellstroulm, A., 2007. The use of benzodiazepines in organophosphorus nerve agent intoxication, in *Chemical Warfare Agents: Toxicology and Treatment*, 2nd ed., Marrs, T.C., Maynard, R.L., and Sidell, F., Eds., John Wiley & Sons, West Sussex, England, chap. 16.

Mattson, M.P., 2003. Excitotoxic and excitoprotective mechanisms: review article, *Neuromol. Med.*, 3, pp. 65–94.

Mayer, B. and Oberbauer, R., 2003. Mitochondrial regulation of apoptosis, *News Physiol. Sci.*, 18, pp. 147–157.

McCollum, A.T., Nasr, P., and Estus, S., 2002. Calpain activates caspase-3 during UV-induced neuronal death but only calpain is necessary for death, *J. Neurochem.*, 82, pp. 1208–1220.

McDonough, J.H. and Shih, T.-M., 1997. Neuropharmacological mechanism of nerve agent-induced seizure and neuropathology, *Neurosci. Biobehav. Rev.*, 21, pp. 559–579.

McDonough, J.H. and Shih, T.-M., 2007. The use of benzodiazepines in organophosphorus nerve agent intoxication, in *Chemical Warfare Agents: Toxicology and Treatment*, 2nd ed., Marrs, T.C., Maynard, R.L., and Sidell, F., Eds., John Wiley & Sons, West Sussex, England, chap. 14.

Meeter, E. and Wolthuis, O.L., 1968. The spontaneous recovery of respiration and neuromuscular transmission in the rat after anticholinesterase poisoning, *Eur. J. Pharmacol.*, 2, pp. 377–386.

Meldrum, B. and Garthwaite, J., 1990. Excitatory amino acid neurotoxicity and neurodegenerative disease, *TIPS*, 11, pp. 379–387.

Monaghan, D.T., Bridges, R.J. and Cotman, C.W., 1989. The excitatory amino acid receptors: their classes, pharmacology and distinct properties in the function of the central nervous system, *Ann. Rev. Pharmacol. Toxicol.*, 1, pp. 365–402.

Nath, R., Probert, J.R., McGinnis, K.M., and Wang, K.K., 1998. Evidence for activation of caspase-3 like protease in excitotoxin- and hypoxia/hypoglycemia-injured neurons. *J. Neurochem.*, 71, pp. 186–195.

Neary, J.T., 1997. MAPK cascades in cell growth and death, *News Physiol. Sci.*, 12, pp. 286–293.

Nicholls, D.G., 2004. Molecular switches deciding the death of injured neurons, *Toxicol. Sci.*, 74, pp. 4–9.

Nixon, R.A., 2003. A "protease activation cascade" in the pathogenesis of Alzeheimer's disease, *Ann. NY Acad. Sci.*, 924, pp. 117–131.

Olney, J.W., DeGubareff, T., and Labruyere, J., 1983. Seizure-related brain damage induced by cholinergic agents, *Nature*, 301, pp. 520–522.

O'Neill, J.J., 1981. Non-cholinesterase effects of anticholinesterases, *Fundam. Appl. Toxicol.* 1, pp. 154–160.

Plaitakis, A., Berl, S., and Yahr, M.D., 1982. Abnormal glutamate metabolism in adult-onset degenerative neurological disorder, *Science*, 216, pp. 193–196.

Potter, D.A., Tirnauer, J.S., Janssen, R., Croall, D.E., Hughes, C.N., Fiacco, K. A., Mier, J. W., Maki, M., and Herman, I. M., 1998. Calpain regulates actin remodeling during cell spreading, *J. Cell Biol.*, 141, pp. 647–662.

Powers, J.C., Chih-Min, D., Ricketts, K.M., and Casillas, R.P., 2000. Cutaneous protease activity in the mouse ear vesicant model, *J. Appl. Toxicol.*, 20, pp. S177–S182.

Rami, A., 2003. Ischemic neuronal death in the rat hippocampus: the calpain-calpastatin-caspase hypothesis, *Neurobiol. Dis.*, 13, pp. 75–88.

Rengstorff, R.H., 1985. Accidental exposure to sarin: vision effects, *Arch. Toxicol.*, 56, pp. 201–203.

Rickett, D.L., Glenn, J.F., and Beers, E.T., 1986. Central respiratory effects vs. neuromuscular actions of nerve agents, *Neurotoxicology*, 7, pp. 225–236.

Rothestein, J.C., Jin, L., Dykes-Hoberg, M., and Kuncl, R.W., 1993. Chronic inhibition of glutamate uptake produces a model of slow neurotoxicity, *Proc. Natl. Acad. Sci. USA*, 90, pp. 6591–6595.

Rylands, J.M., 1982. A swimming test for assessing effects of drugs upon motor performance in the guinea-pig (*Cavia porcellus*), *Neuropharmacology*, 21, pp. 1181–1185.

Shih, T.-M., and McDonough, J.H., Jr., 1997. Neurochemical mechanisms in soman-induced seizures, *J. Appl. Toxicol.*, 17, pp. 255–264.

Sidell, F.R., 1974. Soman and sarin: clinical manifestations and treatment of accidental poisoning by organophosphates, *Clin. Toxicol.*, 7, pp. 1–17.

Sidell, F.R., 1990. Clinical notes on chemical casualty care, in USAMRICD Technical Memorandum 90–1, USAMRICD, Ed., U. S. Army Medical Research Institute of Chemical Defense, Aberdeen Proving Ground, MD.

Sidell, F.R., 2003. Chemical warfare agents, in *Preventive Medicine: Mobilization and Deployment*, Vol. 1, Borden Institute, Department of Defense, Office of the Surgeon General, U.S. Army, Washington, DC, pp. 611–625.

Sidell, F.R. and Franz, D.R., 1997. Overview: defense against the effects of chemical and biological warfare agents, in *Medical Aspects of Chemical and Biological Warfare*, Borden Institute, Walter Reed Army Medical Center, Office of the Surgeon General, U.S. Army, Washington, DC, pp. 1–7.

Sidell, F.R. and Groff, W.A., 1974. The reactivatibility of cholinesterase inhibited by VX and sarin in man, *Toxicol. Appl. Pharmacol.*, 27, pp. 241–252.

Siman, R., Gall, C., Perlmutter, L.A., Christian, C., Baudry, M., and Lynch, G., 1985. Distribution of calpain I, an enzyme associated with degenerative activity in rat brain, *Brain Res.*, 347, pp. 399–403.

Sokolski, H., 2000. Rethinking bio-chemical dangers, *Orbis*, pp. 183–195, excerpted from Foreign Policy Research Institute program, "America the Vulnerable: Three Threats and What to Do about Them," Philadelphia, PA, October 7–8, 1999.

Solberg, Y. and Belkin, M., 1997. The role of excitotoxicity in organophosphorous nerve agents central poisoning, *TIPS*, 18, pp. 183–185.

Sparenborg, S., Bennecke, L.H., Jaax, N.K., and Braitman, D.J., 1992. Dizocilprin (MK-81) arrests status epilepticus and prevents brain damage induced by soman, *Neuropharm.*, 31, pp. 357–368.

Strahlendorf, J.C., Box, C., Attridge, J., Diertien, J., Finckbone, V., Henne, W.M., Medina, M., Miles, R., Oomman, S., Schneider, M., Singh, H., Veliyaparambil, M., and Strahlendorf, H., AMPA-induced dark cell degeneration of cerebellar Purkinje neurons involves activation of caspases and apparent mitochondrial dysfunction. *Brain Res.*, vol. 994, p. 150, figure 1 (Panel A), p. 151, figure 2 (Panel B), graph II. Elsevier, 2003.

Strahlendorf, J.C., Acosta, S., and Strahlendorf, H.K., 1996. Diazoxide and cyclothiazide convert AMPA-induced dark cell degeneration of Purkinje cells to edematous damage in the cerebellar slice, *Brain Res.*, 729, pp. 197–204.

Strahlendorf, J.C., Box, C., Attridge, J., Diertien, J., Finckbone, V., Henne, W.M., Medina, M., Miles, R., Oomman, S., Schneider, M., Singh, H., Veliyaparambil, M., and Strahlendorf, H., 2003. AMPA-induced dark cell of cerebellar Purkinje neurons involves activation of caspases and apparent mitochondrial dysfunction, *Brain Res.*, 994, pp. 146–159.

Strahlendorf, J.C., McMahon, K., Border, B., Barenberg, P., Miles, R., and Strahlendorf, H., 1999. AMPA-induced dark cell degeneration of cerebellar Purkinje neurons has characteristics of apoptosis, *Neurosci. Res. Comm.*, 25, pp. 149–161.

Tenneti, L. and Lipton, S.A., 2000. Involvement of activated caspase-3 like proteases in N-Methyl-D-Asparate-induced apoptosis in cerebrocortical neurons, *J. Neurochem.*, 74, pp. 134–142.

Twyman, R.E., Rogers, C.J., and MacDonald, R.L., 1989. Differential regulation of gamma aminobutyric acid receptor channels by diazepam and phenobarbital, *Ann. Neurol*, 25, pp. 213–220.

Tucker, J.B., 1996. Viewpoint: converting former Soviet CW-plants, *Nonproliferation Rev.*, pp. 78–89.

Tucker, J.B., 2006. *War of Nerves: Chemical Warfare from World War I to Al-Qaeda*, Pantheon Books, New York.

Vanderklish, P.W. and Bahr, B.A., 2000. The pathogenic activation of calpain: a marker and mediator of cellular toxicity and disease states, *Int. J. Exp. Pathol.*, 81, pp. 323–339.

Verkhratsky, A. and Toescu, E.C., 2003. Endoplasmic reticulum calcium homeostasis and neuronal death, Neuroscience Review Series, *J. Cell Mol. Med.*, 7, pp. 351–361.

Wade, J.V., Samson, F.E., Nelson, S.R., and Pazdernik, T.L., 1987. Changes in extracellular amino acids during soman- and kainic acid-induced seizures, *J. Neurochem.*, 49, pp. 645–650.

Wang, K.W., 2000. Calpain and caspase: can you tell the difference? *Trends Neurosci.*, 23, pp. 20–26.

Wright, P.G., 1954. An analysis of the central and peripheral components of respiratory failure produced by anticholinesterase poisoning in the rabbit, *J. Physiol.*, 126, pp. 52–70.

Xia, A., Dickens, M., Raingeaud, J., Davis R.J., and Greenberg, M.E., 1995. Opposing effects of ERK and JNK-p38 Map kinases on apoptosis, *Science*, 270, pp. 1326–1331.

Zamzami, N. and Kroemer, G., 2001. The mitochondrion in apoptosis: how Pandora's box opens, *Nat. Rev. Mol. Cell Biol.*, 2, pp. 67–71.

6 Sensing Biological and Chemical Threat Agents

Gopal Coimbatore, Steven M. Presley, Jonathan Boyd, Eric J. Marsland, and George P. Cobb

CONTENTS

6.1   INTRODUCTION

Recent history has heightened our awareness of the dangerous nature of the current world situation resulting from the global war on terror. The tragic events of September 11, 2001, took more than 3,000 lives and caused roughly $100 billion in direct and indirect economic losses (Freedonia Group 2005). Perhaps more important, that day represented an awakening of the world to the potential for mass destruction by any means necessary at the hands of terrorists. Acts of terrorism, however, had occurred before by more conventional means. The September 11, 2001, tragedy was by no means the first act of terrorism in a major city, nor will it be the last. One incident in Japan sets itself apart because of the use of unconventional chemical threat agents during a terroristic act. In 1995, a religious cult calling itself Aum Shinrikyo released a chemical threat agent known as sarin within a portion of the subway system in Tokyo, Japan. The attack caused 12 deaths, injured thousands, and terrorized at least as many. While there was personal tragedy, the event exposed the underdeveloped management and response capabilities of the Japanese infrastructure to handle such events (Institute of Medicine 1999). This use of the chemical threat agent sarin is considered by many to be the first significant terrorist attack using biological and chemical threat agents in modern times (Institute of Medicine 1999).

The Aum Shinrikyo cult in Japan proved that subnational groups can obtain the expertise and ingredients to threaten a society with chemical threat agents. Their use of this chemical threat agent as a terror weapon has been shown to be the modern original event that spawned the progression from potential terrorist threats utilizing chemical threat agents to actual aggression against civilians (Institute of Medicine 1999). The potential use of chemical threat agents in attacks on civilian populations has intensified the need for the development of sensors to detect chemical threat agents and their precursors. Sensors should demonstrate low detection limits and high selectivity when operated in complex environments, and deliver analysis results in near-real time (Zemel 1990).

6.1.1   EARLY DETECTION DEVICES DEPLOYED FOR THE PROTECTION OF CIVILIAN POPULATIONS

Most extensive risk assessments for chemical threat agent release scenarios have been conducted with military personnel in mind; however, within the current global environment, terrorist threats pose a genuine problem for civilians as well. It is critical to note key differences in strategies for protecting military populations and civilian populations against chemical threat agents. An excerpt from *Chemical Exposure Guidelines for Deployed Military Personnel* (USCHPPM 2003) elucidates this from the outset: "The military population, for which these guidelines are developed, is assumed to be 'healthy and fit' and often believed to be less susceptible to the adverse health effects caused by chemical exposures than the general (civilian) population."

Military personnel have extensive training in decontamination procedures following exposure to biological or chemical threat agents to mitigate the physiological effects of the agent. Unfortunately, an event using a chemical threat agent in a civilian population would in all likelihood be the result of a terrorist act. The potentially exposed population and the affected area would be undefined until symptoms of exposure to a chemical threat agent became evident in the victims. Preventive treatment is typically not feasible, as the exposed population would not be aware of an impending covert terrorist act, and hence a large number of citizens would be exposed during such an incident. Also, personal protective equipment (PPE) to minimize exposure before or immediately following the release of a chemical threat agent is typically not available to civilians and is limited to first responders. Decontamination would be extremely difficult as evidenced in the decontamination of the Hart Building (Washington, D.C.) following exposure to anthrax spores (*Bacillus anthracis*) after the September 11, 2001, attacks on New York City and Washington, DC. Medical therapy would also be limited by the availability of local emergency response resources, number of local hospitals, and properly trained physicians on staff.

Early detection and effective evacuation procedures from an area where a biological or chemical threat agent has been released has long been considered the best means to prevent massive exposure of people for both military and civilian populations (USAMRICD 1995). By utilizing early warning detection systems, an entire population of untrained individuals may be able to circumvent the need for PPE, decontamination, and medical treatment. The cost of supplying the public at large with the necessary equipment and training for responding to a chemical threat agent release justifies the need for advanced early detection systems. Early detection (assessment) will also help identify risks for individuals within the exposed area, assist with defining the appropriate decontamination strategy, and may greatly enhance emergency medical response, if the situation warrants.

Taking early assessment one step further, the ability to detect the presence of a chemical threat agent and associated precursors will allow protection of large populations in the event of a chemical threat agent's release. The ability to detect agents during the production process or even when precursors are in transit to production facilities would greatly enhance the government's ability to enforce the international ban on chemical threat agents.

6.1.2 Chemical Warfare Agent Detection Technology

Methods for detection of toxic chemicals originated thousands of years ago in the form of a "royal taster." Prior to eating a meal, a person of ruling or royal status would have a servant taste or sample the food to ensure that the meal was safe for consumption. Significant advances in detection technology have been made since those early days of using human sampler technology. More sophisticated detection methods were developed and employed during and after World War I in response to the need for rapid detection of chemical threat agents on the battlefield (Zemel 1990).

Methods and devices continue to emerge that promise to improve the ability to isolate, identify, and quantify biological and chemical threat agents in a wide array of environments, and the field of chemical threat agent sensor development is a rapidly growing area of research. The Freedonia Group (Cleveland, OH), an international business research company providing market research data analysis and prediction, shows that the business aspect of the sensor development field is expected to grow at a rate of approximately 7.5% per annum until 2009, covering an estimated market of approximately $3 billion (Freedonia Group 2005).

The current research focus within the sensor development field seems to be concentrated on miniaturization while incorporating multiple quantitative analytical capabilities. Other high-demand characteristics are shorter response time, minimal hardware requirements, multiple analyte and media capabilities, and improved sensitivity, selectivity, and specificity (Zemel 1990). Advancements and improvements for both biological and chemical threat agent sensors will have numerous other benefits to diverse applications, such as quality and process control, biomedical analysis, medical diagnostics, fragrance analysis, environmental pollution monitoring, and control forensics.

Sensors for detection of either biological or chemical threat agents can be generally divided into two major categories: (1) physical sensors, which measure physical quantities (e.g., mass, distance, time, temperature, pressure); and (2) chemical sensors, which measure the quantity or quality of a substance (e.g., hormones, enzymes, pH, pesticides, glucose) by eliciting a physical or chemical response. Biological sensors, a subcategory of chemical sensors, are used to detect the presence of a living microorganism or a by-product produced or derived from a living organism. Detection of biological materials may utilize immunologic, genomic, proteomic, or other components associated with a specific organism or biological toxin. Often the focus of biological sensor research is the development of a detector for specific microorganisms or biological toxins that can be weaponized (see Chapter 2).

An excellent review of technologies for and approaches to detecting and sensing biological and chemical threat agents, both clinically and in the environment, is provided in a joint report by the Institute of Medicine and the National Research Council (Institute of Medicine 1999). The target area of research has shifted focus from primarily chemical sensing to mixed sensing technologies, where both chemical and biological agents can be detected simultaneously. Recently, nucleic acid fragments have been used as a DNAzyme (target-specific DNA fragment) for detection of metals such as mercury and obtaining sensitivities similar to those from state-of-the-art atomic absorption spectroscopic methods, but at a chip-device scale (an area of 1 cm^2 with dramatically reduced power requirements; Wernette et al. 2006). Molecular-level recognition of other heavy metals, such as arsenic and manganese, has also been accomplished with similar single-stranded DNA sensors (Wang 2002).

6.1.3 THE NEED FOR A FIELD-DEPLOYABLE SENSOR

As discussed in Chapters 1 and 2, threats from the use of biological and chemical threat agents have become issues of national and international importance. The ease in obtaining chemical precursors and the availability of literature for producing chemical threat agents have made terrorism with these agents a reality. Precedence of their use globally has converted the threat from a possibility to an actuality.

Nerve agents are among the most toxic biological and chemical threat agents known. They are chemically similar to organophosphate pesticides and exert their physiological effects by inhibiting acetylcholinesterase enzymes. More detailed descriptions of mechanisms of action are described in Chapter 5. Nerve agents have been classified according to their physical properties as G, GB, GA, GD, and VX. The G-type agents are clear, colorless, and tasteless liquids that are miscible in water and most organic solvents. The GB-type chemical is odorless and is the most volatile nerve agent; however, it evaporates at about the same rate as water. The class GA agents have a slightly fruity odor, and GD agents have a slight camphorlike odor. The group of agents belonging to the VX class are clear, amber-colored, odorless, and oily liquids. They are miscible with water and soluble in all organic solvents and are the least volatile nerve agents known (USAMRICD 1995).

Vapors of nerve agents are heavier than air, and their odor does not provide adequate warning for detection. Nerve agents are readily absorbed in the respiratory tract. Rhinorrhea and tightness in the throat or chest begin seconds to minutes after exposure. The estimated LC_{t50} ranges from 10 mg-min./m^3 for VX-type to 400 mg-min./m^3 for GA-type agents (USAMRICD 1995). LC_{t50} is the cumulative exposure, expressed as the concentration of a chemical or biological material integrated over the time period of exposure (e.g., integrated air concentraiton [(gm-sec)/m^3]) that produces lethal effect.

Detection devices for automated near real-time detection and monitoring of biological and chemical threat agent materials are currently not available. Most nerve agents typically decompose in the environment soon after release. Hence, current detection techniques rely on the analysis of the decomposition products, primarily the hydrolysis products. These techniques have been limited to gas chromatography (GC), liquid chromatography (LC), and allied mass spectroscopic methods. Although these methods are sensitive and highly selective, they cannot be performed readily in the field due to the size, weight, power, and environmental requirements of instrumentation. Furthermore, these methods are not fully automated and often require substantial operator intervention. Despite advances in technology and miniaturization of instrument components, sensor development for rapidly monitoring chemical threat agents has been limited to the measurement of hydrolysis products in water (Zemel 1990)

This chapter describes two approaches to develop sensors for biological and chemical threat agents, in particular nerve agent surrogates for sarin (GB) and soman (GD). These can be released into the atmosphere in vapor or aerosol form. Promising results have been achieved using an inexpensive liquid-crystal-based diagnostic sensor. The primary aim of the liquid crystal sensor is not to quantify target analytes but to provide a diagnostic detection of the nerve agents at sufficiently low concentrations to warn potentially exposed personnel before they experience irreversible damage. Furthermore, the design allows easy assembly and operation, is extremely lightweight, and requires minimal user input for detection of chemical threat agents.

6.2   GENERAL SENSOR CHARACTERISTICS

Nature has devised some of the most refined and effective chemical sensors that are in existence. The olfactory system is an excellent example of what nature can do. Much work has been done in the field of sensors to create an artificial nose that can detect and differentiate virtually unlimited varieties of chemical mixtures at extremely low concentrations (often at parts per trillion levels; Di Natale 1998).

A chemical sensor consists of a detector and transducer. The detector is the chemically responsive part that interacts with the environment and generates a chemical response to the analyte's presence. The transducer converts the response generated by the detector into a machine-readable form, usually an electrical signal. Properties that determine the selection of a specific interactive material for use in a sensor are the sensitivity, selectivity, reproducibility, and reversibility of the corresponding sensing mechanisms (Di Natale 1998). The selectivity of the sensor depends upon the chemical response of the detector. Often, chemical selectivity is achieved by molecular recognition through reactivity, sorption, or chemical bonding. A plethora of sensor technologies exists commercially, and newer technologies are in development or in research stages. The following sections provide an overview of currently available sensor technologies.

6.3   CHEMICAL THREAT AGENT SENSOR OVERVIEW

Sensor technologies are developed using principles of physical and chemical sciences and their interrelationships. Some examples are fiber-optic-based, semiconductor-based, conducting polymer-based, mechanical, electrochemical, calorimetric, and colorimetric sensors. In this section these sensor technologies are discussed with respect to their functions, advantages, capabilities, and requirements.

6.3.1   SENSORS BASED ON FIBER OPTIC TECHNOLOGY

Walt (1998) described the principle of optical sensing using fibers. An optical fiber consists of two concentrically arranged optically transparent media: an inner ring, called the *core*, carries the optical signal, and a thin outer ring, called the *clad* (made of a lower refractive index material). The refractive index mismatch at the interface of the two media acts as a mirror to help the transmission of light from one end of the fiber to the other end. The phenomena in play here is that of total internal reflection (Walt 1998).

An indicator material immobilized on the tip, at one end of the fiber optic cable, acts as the sensing tip. A change in optical property brought about by the chemical interaction of the indicator material with the analyte of interest is transmitted to the detection system by the optical fiber where the signal is analyzed (Wolfbeis 2000). The change can be in the form of change in fluorescent intensity, wavelength, or a spectral shift. Sensors that use fiber optic technology rely either on a direct (measures intrinsic property of material) or an indirect sensing (measures the signal generated by a material's response to chemical interaction with the analyte) methodology.

The phenomenon of fiber luminescence is one where the optical fiber is coated with a chemical to produce light upon a chemical reaction. This light is transmitted by the fiber to the detector. Fiber luminescence has been extended to fiber imaging to detect and study analytes. For example, quenching of luminescence is observed in some dyes, such as porphyrins, upon exposure to very minute concentrations of oxygen. Fiber optic imaging can be used to evaluate these low-oxygen concentrations (Yeh et al. 2006).

Suitably modified fiber optic sensors can also be used for detecting gas vapors, humidity, ions, and organic compounds. Fiber inclusions that show length variation were used to develop humidity sensors, whereas ion-responsive lipid bilayers formed the basis for the detection of inorganic ions. Immobilized neutral and ionic crown ethers in polymeric membranes were designed as sensors for determination of barium and copper (Wolfbeis 2000).

Modified fiber-optic-based sensors can be used for sensing pollutants, explosives, drugs, pharmaceuticals, and miscellaneous organics (Yeh et al. 2006). Optical fibers coated with porous silica can be used to detect the presence of chlorinated hydrocarbons. Alternatively, these compounds can also be detected using fiber-optic-coupled surface plasmon resonance methods. Aromatic compounds were detected by evanescent wave absorption spectroscopy. Suitably modified fiber-optic array tips can be used to detect presence of explosive materials (Wolfbeis 2000).

Neural networks coupled to fiber optic sensors are best suited to discriminate between organic volatiles using pattern recognition (Wolfbeis 2000). Walt (1998) reported that essential features of a sensor such as small size, array format, and cross reactivity can be incorporated into optical sensor arrays. The information-rich optical signals can be collected and the signals deconvoluted for signal intensities, phase changes, and polarization information at different wavelengths. This generates a significant amount of data in a very short time. Detailed computer-assisted analyses of these data permit unambiguous identification of individual organic compounds in a mixture. Fiber optic sensors have become the backbone of modern-day sensor technology (Walt 1998).

6.3.2 Sensors Based on Conducting Properties

6.3.2.1 Semiconductive Gas Sensors

Surface conductance of semiconducting oxides such as tin oxide (SnO_2) and zinc oxide (ZnO) are influenced by the composition of gas present in its immediate environment. Chemically reducible gases like carbon monoxide, hydrogen, and so forth can be easily detected using SnO_2-based semiconductors (Roy et al. 2000). Some advantages of semiconductor-based sensors include high sensitivity, low cost, fast response, and low power consumption. Disadvantages of these sensors, however, are factors such as long-term drift, relatively low selectivity, interference of humidity, and susceptibility to temperature variations (Mehregany et al. 2000).

6.3.2.2 Silicon Carbide (SiC) Based Gas Sensors

Silicon carbide–based sensors can operate at higher temperatures (above 600°C) because of a wide band gap and low intrinsic carrier concentration availability of SiC. Taking advantage of this trait, silicon carbide semiconductors have been used as Schottky diode gas sensors (Hunter et al. 1995). Schottky diodes are semiconductor diodes with a low forward voltage drop and a very fast switching action, which usually operates at higher temperatures. The gases, which are not detectable at low temperatures using traditional semiconductor-based sensors, such as hydrogen and hydrocarbons, are detectable by SiC-based sensors. Fergus et al. (2007) investigated the metal-insulated semiconductor capacitor or metal semiconductor construction with silicon carbide as the semiconductor. Using this approach, Fergus et al. (2007) have been successful in detecting hydrogen and hydrocarbons at concentrations as low as 2.5 ppm and temperature as high as 8000°C. Fergus et al. (2007) also reported the sensing capacities for hydrocarbons such as methane, ethane, and propane with a response time in the millisecond range. Hunter et al. (2000) used silicon carbide as a semiconductor that responded to an analyte producing large signals for low concentrations, thereby increasing the sensitivity of the sensor. They developed the silicon carbide–based sensor by depositing palladium on a silicon carbide diode. Palladium was used because of its superior ability to absorb hydrogen (Hunter et al. 1999). The surface conductivity changes due to the dipole layer formation by hydrogen dissociation, thereby altering the electrical properties of the device in proportion to the hydrogen present in its immediate environment. This was exploited by Mehregany and coworkers to detect hydrogen in very low concentrations (Mehregany et al. 2000).

6.3.2.3 Conducting Polymer-Based Sensors

These sensors have an array of elements with polymer coatings that have conductive properties. The measurement of change in electrical resistance upon exposure to different kinds of analytes aids in identification and quantification of gaseous vapors (Freund and Lewis 1995). The conducting polymer-based sensor approach aims mainly at controlling the chemical properties of a polymer layer to be coated on the sensor. This control is achieved during the polymer deposition, thus controlling the binding properties of the chemi-resistor elements. Exposure of thin conducting polymer films immobilized on nonconducting surfaces to different analytes produces different electrical resistances. An array of such engineered polymer films produces a signal pattern for specific analytes, enabling unambiguous identification (Freund and Lewis 1995). Lang and coworkers (1999) exposed a similar sensor array to eight different solvent vapors simultaneously and reported their individual identification. The polymer films were functional and stable over the test period. Signal analysis and compound identification was carried out by either a software- or a hardware-based neural network analysis (Lang et al. 1999).

6.3.2.4 Molecularly Imprinted Polymers (MIPs)

Molecularly imprinted polymer recognition units are based upon template polymerization techniques (Haupt and Mosbach 2000). The MIP recognition units are formed in the presence of a template molecule that is later leached out or extracted, thus leaving complementary cavities embedded in the final structure of the polymer. These polymers display high chemical-binding affinity for molecules with structural similarities to the template molecule. Hence, MIPs can be used to fabricate sensors with very high selectivity, displaying a binding mechanism that is similar to those of antibodies or enzymes. A MIP-incorporated sensor for the detection of chemical threat agents or precursors could decrease the number of false alarms that are inherent to currently popular nonspecific detection techniques.

One of the limitations in the MIP development process is that phosphate derivatives readily displace monomeric ligands from the cationic center. This often destroys the luminescence properties of the MIP, which results in sensor instability and causes an issue with reliable reproducibility of results. The only way to remedy this stability issue is to make complexes with stronger binding constants between the monomeric ligand and the cationic center, which in turn would lead to decreased luminescence wavelength change upon analyte binding and would be limited to an amplitude at a given wavelength, thereby resulting in lowered sensitivity.

6.3.3 MECHANICAL SENSORS

A typical mechanical sensor is constructed of a surface-coated micro-fabricated array of silicon cantilevers. The coating material acts as the sensing surface and can be made of gold, platinum, or an organic layer. These cantilevers, when exposed, either physisorb or chemisorb the analyte, causing a surface stress-induced bending of cantilevers. Lang et al. (1999) and Baller et al. (2000) reported the detection of ethene and water vapor using such cantilever-based sensor. Typically an uncoated "reference cantilever" is included in the studies. This reference cantilever helps eliminate the environmental variables that would otherwise produce artifacts in the studies.

6.3.4 COLORIMETRIC SENSORS

Colorimetric sensors utilize a change in wavelength, either during absorption, emission, or fluorescence, as the detection method. A chemical interaction between the analyte and the sensor surface produces a change in the optical property. Several methods to detect these changes in optical properties are discussed in this section.

6.3.4.1 Fiber-Optic-Based Colorimetric Sensor

The principle of the fiber optic sensor has been previously discussed; here, its use in the colorimetric sensor will be discussed. A polymeric sensing unit is immobilized at the tip of a fiber optic cable. A unique optical signal is produced by the chemical interaction of the analyte with the polymer. The analyte vapor recognition is via a neural network information deconvolution. The information to the neural network is fed by an array of polymer-tipped optic fibers. The capture and analysis of multiple parameters, such as phase, polarity, and wavelength, contributes to discrimination even between structurally related compounds. In an elegant modification, Dickenson et al. (1996) used Neil red polymer dye that was immobilized on the tip of a fiber as a chemical recognition surface. The polymeric dye was synthesized with varying physicochemical properties, for example, pore size, hydrophobicity, and swelling properties. Individual vapors responded differently to different regions of the polyermic dye, creating unique signals. A series of trials trained the neural network for analysis and identification. When field-tested, this "laboratory trained" neural network uniquely identified different analytes with temporal data (Dickenson et al. 1996).

6.3.4.2 Color-Indicating Tubes

Mine Safety Appliances Company (MSA/Auer, Berlin, Germany) and Draeger Safety, Inc. (Pittsburgh, PA) manufacture colorimetric detector tubes for use in the field. The detector tubes are slender glass tubes approximately 10 cm long that are filled with reagents and reagent-impregnated granular solids appropriate for the type of agent to be sampled (Haupt and Mosbach 2000). The tubes are used by manually drawing an air sample through the tube with either a bellows pump or other vacuum source. The tubes, when exposed to their specific agent, change color to indicate exposure (Haupt and Mosbach 2000). For example, the phosphoric acid ester (PAE) MSA tube is capable of detecting nerve agents at a minimum concentration of approximately 0.01 mg/m^3, which is below the immediately dangerous to life and health (IDLH) level as set forth by Army Regulation (AR) 385-61. The MSA tubes, however, do not perform well at temperatures below $-5°C$ and have been shown to produce false positives (Haupt and Mosbach 2000). There is also a large amount of subjectivity in determining the color change, or lack of change, for positive indication near the chemical threat agent threshold levels (Boyd et al. 2004).

The phosphoric acid ester Draeger tube detects nerve agents at a minimum concentration of approximately $0.01-0.03$ mg/m^3, which is below the immediately dangerous to life and health (IDLH) level (Mosbach et al. 1998). The Draeger tubes are also not known to respond to potential interferrents, and detection thresholds are not severely affected by extreme humidity and temperature conditions. However, Draeger tubes are unable to detect nerve agents at, or below, the time-weighted average-airborne exposure limit (Mosbach et al. 1998).

6.3.5 CALORIMETRIC SENSORS

These sensors measure changes in the temperature caused by heat evolution during the catalytic oxidation of combustible gases. Their principle of operation is similar to that of high-temperature-resistance thermometers where the change in temperature causes a variation in resistance of the sensor material. Typically, an encapsulated platinum coil that doubles as a thermometer and a heater is used in calorimetric sensors (Auguet et al. 2006). Other examples of sensing material include palladium and platinum black. A variation is a thin film Bi-Sb gold-junctioned thermopile sensor, formed on a kapton-coated glass plate. (Kapton is a polyimide film developed by DuPont that remains stable in a wide range of temperatures, from $-269°C$ to $+400°C$; Schreiter et al. 2006). Although results are inaccurate because of heat loss, these sensors are used in detecting combustible gases at low concentrations (Auguet et al. 2006; Schreiter et al. 2006).

6.3.6 ELECTROCHEMICAL SENSORS

Electrochemical sensors are based on basic electroanalytical principles (Nernst 1904; Buck 1981a, 1981b). Selectivity and sensitivity differs from analyte to analyte since each analyte interacts differently with the electrochemical cell. Variables that are usually measured for sensing purposes include voltage (potentiometic), current (amperometic), resistance (conductometric), or capacitance.

Potentiometric sensors measure the potential difference generated at the electrode/electrolyte interface of the electrochemical cell caused by the change in the chemical concentration on the equilibrium due to presence of analyte molecules. A conductometric sensor measures the change in conductivity (or resistance) in an electrochemical cell, while a typical amperometric sensor measures the change in current. Electrochemical sensors can use inert electrodes made of Au or Pt as an active electrode. The other half of the cell is a reference electrode for completing the electrochemical measurement. Some active electrodes incorporate specific, functional membranes that may provide ion selectivity, ion permeability, or ion exchange properties. Active electrode sensors include many of the ion-selective electrode sensors. These principles have been well documented in the literature (Nernst 1904; Buck 1981a, 1981b).

6.3.7 Surface Acoustic Wave–Based Sensors

Surface acoustic wave (SAW) technology employs a pair of microsensors that respond to changes in the mass of the surface coatings resulting in vibration frequency changes when a vapor sample flows over them (Wohltjen et al. 1991). Typically, a diaphragm pump draws a sample through a concentrator to collect and concentrate a vapor sample. The concentrated sample is then thermally desorbed. The desorbed sample passes over the sensors, causing vibration frequency changes, and a microcomputer analyzes the sensor responses. The SAW MiniCAD MKII from MicroSensors Systems, Inc. (Bowling Green, KY) has demonstrated chemical threat agent vapor detection at high concentrations (ppm level) over long (> 5 min.) response times (Ho et al. 2003). Humidity changes do not cause adverse effects on the SAW detectors; however, high temperatures are known to affect the instrument's ability to detect agents, and cold temperatures affect the instrument's ability to recover from agent exposures (Ho et al. 2003).

6.4 BIOLOGICAL THREAT AGENT SENSORS

Technologies and strategies that are currently being used to sense biological threat agents, whether *in vivo* or *in vitro* in various environments, generally utilize some means of isolating and detecting immunologic, serologic, or nucleic factors associated with a generic or specific pathogen type. The focus of this section is on the use of nucleic acid fragment (DNA or RNA) detection technologies available.

6.4.1 DNA-Based Biological Agent Sensors

Deoxyribonucleic acid (DNA) forms the genetic code of all living organisms. Assays based on DNA recognition are becoming increasingly common for detecting pathogenic microorganisms and viruses. There are numerous techniques and methodologies that have been developed and utilized that exploit various serological and immunological factors of the host organism to detect and identify pathogens, many requiring technical procedures such as culturing or plating laboratory techniques. These methods have limitations that include the need for skilled personnel, specialized equipment, lengthy preparation and analysis times, and large quantities of expensive reagents. Hence, detection techniques often require culture times that are too lengthy or costly to be effective for use in very time-sensitive situations. Current research focus and developmental efforts have been directed toward reducing this time lag and the costs involved.

6.4.2 APPROACHES TO DNA SENSORS

Currently there are several techniques that have been developed that selectively and effectively couple unique biochemical reactions to a physical signal transducer. This transducer converts the biochemical signal into an electrical signal that can be monitored and measured, thereby indicating a response. Those DNA sensors that have shown definite promise are discussed briefly below.

6.4.2.1 Electrochemical Methods

The electrochemical method relies on the electroactive properties of a material that is physically attached to a single-stranded DNA (ssDNA) that acts as the receptor for the presence of a single-stranded complementary DNA (cDNA). The ssDNA is tethered to an electrode and the ssDNA hybridization with the cDNA produces a change in the electrical property of the tethered material that is then converted to an analytical signal. Several methods have been reported using this technique. Millan et al. (1994) explored the voltametric detection of ssDNA and cDNA hybridization on the surface of glass carbon and carbon paste electrodes; however, this required a very high ionic strength solution for successful detection. Wang et al. (1996) pioneered several alternative techniques, and chronopotentiometric stripping analysis (a method for trace metal analysis) was successfully employed by them to detect DNA hybridization. The same group later demonstrated the detection of small molecules and radiation-damaged DNA via a competitive binding format against an electroactive intercalant (Wang et al. 1996).

Napier et al. (1997) employed an alternative strategy in which they monitored the onset of a catalytic current that occurs when an electroactive metallo-intercalator binds to guanine moieties. Korri Youssoufi et al. (1997) presented an alternate approach by using DNA strands as molecular wires. Oligonucleotides were tethered via a polypyrrole to these electrodes. The potential at which oxidation of polypyrrole took place was monitored. This oxidation was in relation to the tethered DNA strand. An elegant ferrocene-modified oligonucleotide sensor has been reported by Ihara et al. (1997) that utilizes a sandwich-type sensing scheme. The target nucleic acid is annealed between a capture probe and a reporter oligonucleotide that is ferrocence tagged (Ihara et al. 1997).

6.4.2.2 Acoustic Wave Devices

Acoustic wave devices, also known as shear mode resonating quartz crystal microbalances, employ an AT-cut quartz crystal operating in a shear mode for transduction of ssDNA-cDNA hybridization events. This transduction is monitored as mass change

(Ito et al. 1996; Caruso et al. 1997). These experiments also provide additional information on interfacial microviscosity and charge distribution upon hybridization. Wang et al. (1996) also reported success using this method, reporting crystal coverage of up to 87% by DNA at 25°C, and a fast response time of 1–3 min. for complete evolution of the signals. They were able to show a high degree of selectivity, including detecting a single base-pair mismatch.

6.4.2.3 Optical Sensors Based on Surface Plasmon Resonance (SPR)

Surface plasmons are also known as surface plasmon polaritons (polaritons are quasiparticles resulting from strong coupling of electromagnetic waves with an electric or magnetic dipole-carrying excitation, propagated parallel along a metal/dielectric or metal/vacuum interface; Brockman et al. 2000). Since the wave is on the boundary of the metal and the external medium (air or water, for example), these oscillations are very sensitive to any change of this boundary, such as the adsorption of molecules (typically nucleic acids), to the metal surface. Nucleic acids act as a dense, but optically transparent medium. The sensitivity of the surface plasmon waves to changes to the nucleic acid layer (adsorption, desorption or recombination) can be exploited for signal transduction. The signal is detected by monitoring the alterations in the interfacial dielectric constant resulting from mass change at the surface. Detection within the picomolar range has been reported when coupled with an interferometric technique (Kim et al. 2007). This technique essentially allows detection of binding at the surface in real time, without the use of fluorescent tags or radio labels (Brockman et al. 2000). Commercial instruments are available from Biacore AB (GE Healthcare, Waukesha, WI).

6.4.2.4 Luminescence Techniques

The most sensitive of all DNA detection capabilities are the luminescence techniques. Zeptomole (zmol = 10^{-21} units) level detection of radiation-induced DNA damage has been reported using luminescence techniques (Le et al. 1998). Optical fiber incorporated fluorimetric sensors have become the focus of attention in recent times primarily due to the multiplexing capabilities. This technique was first reported by Graham et al. (1992). In their work, a single-stranded nucleic acid sequence was tethered to an optical fiber via a gluteraldehyde linkage. Binding of a prelabeled DNA strand was monitored by the fluorescence intensity of the fluorescein moiety, resulting in a nanomolar detection limit within a detection period of minutes.

Although technology has greatly improved over the years, there are still major challenges to overcome. While most biological sensor technologies emphasize detecting actual pathogens and their composite materials, a sensor for the detection of biological and chemical agent precursors capable of operating effectively in diverse environmental conditions does not currently exist.

6.5 SENSOR RESEARCH WITHIN THE ZUMWALT PROGRAM

6.5.1 ANALYTE SELECTION

Any precursor to be used in experiments should not be toxic, thereby minimizing the need for personal protective equipment. The production of a number of nerve agents involves the use of a robust and simple, relatively nontoxic, phosphonic acid precursor. This precursor is readily available and can be stored for extended periods of time without degradation. These precursors are soluble in organic solvents for ease of handling through the chemical production process of the final product, usually a nerve agent (Graupner 2003). A precursor that was chosen for study within the Admiral Elmo R. Zumwalt, Jr. National Program for Countermeasures to Biological and Chemical Threats (the Zumwalt Program) is phosphonic acid dimethylester or dimethyl hydrogen phosphite (DMHP). Dimethyl hydrogen phosphite is a versatile reactive intermediate used for chemical syntheses and is necessary for the production of a number of nerve agents, including tabun, sarin, soman, and VX, and it met the precursor guidelines described by Graupner (2003).

6.5.2 MOLECULARLY IMPRINTED POLYMERS

The basic strategy of the molecular imprinting process begins with the self-assembly formation of a prepolymeric complex (PPC) between the template molecule and monomeric ligands with complementary functionality (Maeda and Bartsch 1998). To generate a luminescent signal capable of detecting organic molecules, most MIPs utilize polyvalent cations that bind to electronegative functional groups. Many MIP applications use lanthanide ions, such as Eu^{3+}, which is complexed with coordinating ligands and allows broadband excitation in the near UV range (Boyd et al. 2004), and this formed the basis of the research into luminescent functional PPC within the Zumwalt Program.

Research within the Zumwalt Program discovered that when a PPC was exposed to the DMHP, the self-assembly process between the complexes, and subsequent displacement of water from Eu^{3+}, enhanced emission of the 610-nm spectral line and indicated the presence of the analyte of interest. Sensitivity assessments of the PPC for DMHP were conducted with both liquid and vapor samples. The minimum detection limit (MDL) for liquid samples was found to be 1.67 μg with an average complexation time of 9 sec; the linear dynamic range was 62.1 to 186 μg ±5%. The MDL for the PPC when exposed to DMHP vapor was determined to be 859 ng. The limits with vapor were thought to improve because of the DMHP complexing with the PPC over time. Response time for the complexation was determined to be 9 sec. The entire response region for the system when exposed to DMHP vapor in air was found to be logarithmic and to have a dynamic range of ±5% over a narrow mass range of 2.5 to 7.8 μg. To fully characterize the capabilities of the functional PPC, it was exposed to multiple organophosphate compounds. This allowed discernment of other possible analytes that could serve as template molecules for fully functioning, molecularly imprinted polymers prior to the addition of a cross-linking agent. Liquid organophosphates were chosen to represent several broad categories of compounds: dichlorvos for halogenated compounds, phosdrin for esterified compounds, dicrotophos for aminated compounds, and malathion for dithio compounds.

Figure 6.1 provides a graphical representation of the complexation results. Dicrotophos resulted in a larger increase in response signal than DMHP, while phosdrin was comparatively lower. It appears that the tertiary amine on dicrotophos may add stability to the PPC, which is in stark contrast to the ester of phosdrin. Therefore,

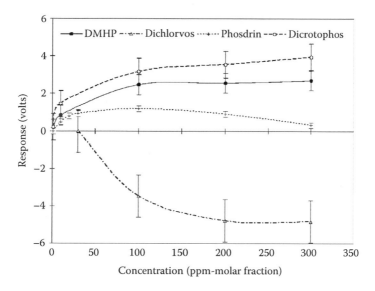

FIGURE 6.1 Exposure-response of a sensor based on functional luminescent prepolymeric complexes to increasing concentrations of dimethyl hydrogen phosphate (DMHP), dichlorvos, phosdrin, and dicrotophos, with best-fit curves shown.

dicrotophos appears to have the potential to make an excellent candidate for a molecularly imprinted polymer using this PPC. Dicrotophos is an analog for some of the more widespread chemical threat agents, such as VX, and thus the ability to incorporate it as the template molecule is beneficial for future chemical threat agent sensors. A large, decreasing response caused by exposure of the system to dichlorvos is also clearly illustrated. This indicates that a chloride group, and possibly any halide, will not show an increase in the signal when bound to the complex, and conversely may decrease the response to zero, which may limit the overall usefulness of this PPC for making a MIP for the detection of phosgene and a number of other chemical threat agents.

6.5.3 Liquid Crystal Sensors

The ultimate objective of our current sensor technology is to provide quantitative information about an analyte of interest. In most practical situations, however, a simple diagnostic sensor, as opposed to an expensive, labor-intensive quantitative sensor, would provide all the information needed for most decision-making purposes in a field situation. For example, a diagnostic chemical threat agent sensor with a detection limit lower than the lowest concentration where an airborne chemical threat agent would cause harm to humans would provide war fighters sufficient information and time to make decisions on whether to redeploy or continue on course. With this in mind, sensor research within the Zumwalt Program currently involves a concentrated effort on developing a liquid crystal (LC) diagnostic sensor.

It was hypothesized that by synergistically combining the chemical detection and signal transduction into one, significant advantages could be obtained in terms of specificity, signal strength, and usability of the device. To achieve this, the chemical that recognizes the target analyte must also generate a signal. Designer molecules, when existing in a liquid crystalline state, have the ability to do both. The liquid crystalline state, also known as the fourth state of matter (solid, liquid, and gas being the other three), exists between liquid and solid crystal, under certain conditions of temperature and pressure. This state of matter displays some characteristics similar to both liquids (e.g., flow, viscosity) and crystals (e.g., molecular order, birefringence). Temperature is a significant parameter that controls the physical state of thermotropic liquid crystals (Chandrasekhar 1992). With increasing temperature, a molecule existing in crystal state (optically anisotropic state) transitions into a liquid crystalline state, which is still optically anisotropic. In a simple case of solid to liquid crystalline to isotropic state, the compound stays in the solid state until the temperature rises and provides sufficient energy to disrupt the crystal-like ordering, at which point it transitions into the liquid (or optically isotropic) state. This sequential change of state can be easily monitored using polarized light (Chandrasekhar 1992).

Polarized light interacts with the optically anisotropic state to produce birefringence. A change in birefringence is observed if the molecules within the sample undergo a change in ordering (crystal or liquid crystal states only) or transition within the state. Such phase transitions can be brought about by a change in temperature or by other environmental factors, such as the presence of an external molecule that breaks the molecular order of the phase (Chandrasekhar 1992). A molecule that would show a change in birefringence within a phase due to the presence of an external molecule (analyte) would satisfy the needs for a dual-purpose chemical (Chandrasekhar 1992).

Selectivity of the analyte was achieved by receptor-like molecular recognition (Shah and Abbott 1999, 2001). The receptor-LC pair was so designed that they would bind to each other via a hydrogen (H) bonding linkage. An aspect of H-bonding that has significant advantage for the LC sensor is its directionality and nature of constantly breaking and re-forming (Pauling 1940). When a molecule approaches the junction of the receptor-LC, it competes for the receptor binding spot with the LC molecule. In the presence of the target analyte, the stronger molecular recognition pattern overwhelms the weaker H-bonding and the LC molecule is displaced. This displacement of the LC molecule causes a change in the alignment of the LC layer and the change in alignment is sufficient to produce a strong, visually observable (to the unassisted eye) change in birefringence. Working on these principles, a sensor with a detection limit in low ppb range has been achieved.

Portability of a sensor is of prime importance to meet the goal of protecting the citizen population. A field-deployable, molecularly imprinted polymer–based sensor with a high selectivity toward precursors of chemical threat agents was designed and developed under the Zumwalt sensor research. Developing further on the principles of portability, the program envisaged a diagnostic sensing technology. The premise was, in the case of an event, an unambiguous and simple yes or no answer is more valuable in saving lives than a detailed quantitative analysis of the toxin released. Research within the Zumwalt Program has demonstrated the proof of principle for one such diagnostic sensor system. Currently, the diagnostic sensor system is capable of detecting a nerve-agent precursor in low ppb levels.

6.6 FUTURE RESEARCH NEEDS AND RECOMMENDATIONS

The protection of defense and civilian populace against the use of chemical and biological threat agents by early detection of the event is the goal of the Zumwalt sensor research. The design and construction of a diagnostic sensor and demonstration of its proof of principle forms just the first step toward meeting the goal. The sensor's specificity to one kind of precursor, however, limits its extensive usage. Developing a broad-based detection technology that will be capable of detecting multicomponent threat release is a research goal of prime importance. Equally important is conversion of the available technology to an actual field-based prototype.

6.7 CONCLUSIONS

Sensor technology has come a long way since miners used a canary to alert them to excessive methane accumulation in their vicinity. Modern times have seen a revolution in every aspect of sensing technology and the detection limits have reduced to levels where the technology is matching nature's detection abilities. Development of fiber optics and its innovative adaptation to sensing methods has played a significant role in this revolution. Mankind, on the other hand, has also discovered and developed chemicals that are constantly pushing the toxicity limits. This, in turn, is spurring on the need for further innovation.

REFERENCES

Auguet, C., Seguin, J.L., Martorell, F., Moll, F., Torra, V., and Lerchner, J., 2006. Identification of micro-scale calorimetric devices, *J. Thermal Anal. Calorimetry*, 86, pp. 521–529.

Baller, M.K., Lang, H.P., Fritz, J., Gerber, C., Gimzewski, J.K., Drechsler, U., Rothuizen, H., Despont, M., Vettiger, P., Battiston, F.M., Ramseyer, J.P., Fornaro, P., Meyer, E., and Güntherodt, H.J., A cantilever array-based artificial nose, *Ultramicroscopy*, 82, pp. 1–9.

Boyd, J.W., 2004. Strategic sensor development for near-real-time detection of a chemical warfare agent precursor. Ph.D. Diss. Texas Tech. University, Lubbock.

Boyd, J.W., Cobb, G.P., Southard, G.E., and Murray, G.M., 2004. Development of molecularly imprinted polymer sensors for chemical warfare agents, *Johns Hopkins APL Tech. Digest*, 25, pp. 44–49.

Brockman, J.M., Nelson, B.P., and Corn, R.M., 2000. Surface plasmon resonance imaging measurements of ultrathin organic films, *Annu. Rev. Phys. Chem.*, 51, pp. 41–63.

Buck, R.P., 1981a, Electrochemistry of ion-selective electrodes, *Sensors Actuators*, 1, pp. 197–206.

Buck, R.P., 1981b, Electronic semiconducting oxides as pH sensors, *Sensors Actuators*, 1, pp. 137–146.

Caruso, F., Rodda, E., Furlong, D. F., Niikura, K., and Okahata, Y., 1997. Quartz crystal microbalance study of DNA immobilization and hybridization for nucleic acid sensor development, *Anal. Chem.*, 69, pp. 2043–2049.

Chandrasekhar, S., 1992. *Liquid Crystals*, Cambridge University Press, Cambridge, U.K.

Di Natale, C., Macagnano, A., Repole, G., Saggio, G., D'Amico, A., Paolesse, R., and Boschi, T., 1998. The exploitation of metalloporphyrins as chemically interactive material in chemical sensors, *Mater. Sci. Eng. C-Biomimetic Mater. Sensors Systems*, 5, pp. 209–215.

Dickinson, T.A., White, J., Kauer, J.S., and Walt, D.R., 1996. A chemical-detecting system based on a cross-reactive optical sensor array, *Nature*, 382, pp. 697–700.

Fergus, J.W., 2007. Solid electrolyte based sensors for the measurement of CO and hydrocarbon gases, *Sensors Actuators B-Chem.*, 122, pp. 683–693.

Freedonia Group Research, 2005. Chemical sensors to 2009: demand and sales forecasts, market share, market size, market leaders, study 2005. December, available at http://www.freedoniagroup.com/Chemical-Sensors.html (accessed October 11, 2007).

Freund, M.S. and Lewis, N.S., 1995. A chemically diverse conducting polymer-based electronic nose. *Proc. Natl. Acad. Sci. USA*, 92, pp. 2652–2656.

Graham, C.R., Leslie, D., and Squirrell, D.J., 1992. Gene probe assays on a fiberoptic evanescent wave biosensor, *Biosensors Bioelectronics*, 7, pp. 487–493.

Haupt, K. and Mosbach, K., 2000. Molecularly imprinted polymers and their use in biomimetic sensors, *Chem. Rev.*, 100, pp. 2495–2504.

Ho, C.K., Lindgren, E.R., Rawlinson, K.S., McGrath, L.K., and Wright, J.L., 2003. Development of a surface acoustic wave sensor for *in situ* monitoring of volatile organic compounds, *Sensors*, 3, pp. 236–247.

Hunter, G.W., Neudeck, P.G., Chen, L.Y., Knight, D., Liu, C.C., and Wu, Q.H. 1995. Silicon carbide-based detection of hydrogen and hydrocarbons, in *Silicon Carbide and Related Materials*, IOP Publishing., Bristol, U.K., pp. 817–820.

Hunter, G.W., Neudeck, P.G., Gray, M., Androjna, D., Chen, L.Y., Hoffman, R.W., Liu, C.C., and Wu, Q.H., 1999. SiC-based gas sensor development, in *Silicon Carbide and Related Materials*, Parts 1 and 2, Trans Tech Publications, Zurich-Uetikon, pp. 1439–1442.

Hunter, G.W., Neudeck, P.G., Gray, M., Androjna, D., Chien, L.Y., Hoffman, R.W. Jr., Liu, C.C., and Wu, Q.H., 2000. SiC-based gas sensor development, *Mater. Sci. Forum* 338–342, pp. 1439–1422.

Ihara, T., Nakayama, M., Murata, M., Nakano, K., and Maeda, M., 1997. Gene sensor using ferrocenyl oligonucleotide, *Chem. Commun.*, pp. 1609–1610.

Institute of Medicine and National Research Council, 1999. *Chemical and Biological Terrorism: Research and Development to Improve Civilian Medical Response*, National Academy Press, Washington, DC, p. 15.

Ito, K., Hashimoto, K., and Ishimori, Y., 1996. Quantitative analysis for solid-phase hybridization reaction and binding reaction of DNA binder to hybrids using a quartz crystal microbalance, *Anal. Chim. Acta*, 327, pp. 29–35.

Kim, D.K., Kerman, K., Saito, M., Sathuluri, R.R., Endo, T., Yamamura, S., Kwon, Y.S., and Tamiya, E., 2007. Label-free DNA biosensor based on localized surface plasmon resonance coupled with interferometry, *Anal. Chem.*, 79, pp. 1855–1864.

Korri Youssoufi, H., Garnier, F., Srivastava, P., Godillot, P., and Yassar, A., 1997. Toward bioelectronics: specific DNA recognition based on an oligonucleotide-functionalized polypyrrole, *J. Am. Chem. Soc.*, 119, pp. 7388–7389.

Lang, H.P., Baller, M.K., Berger, R., Gerber, C., Gimzewski, J.K., Battiston, F.M., Fornaro, P., Ramseyer, J.P., Meyer, E., and Guntherodt, H.J., 1999. An artificial nose based on a micromechanical cantilever array, *Anal. Chim. Acta*, 393, pp. 59–65.

Le, X.C., Xing, J.Z., Lee, J., Leadon, S.A., and Weinfeld, M., 1998. Inducible repair of thymine glycol detected by an ultrasensitive assay for DNA damage, *Science* 280, pp. 1066–1069.

Maeda, M. and Bartsch, R., 1998. Molecular and ionic recognition with imprinted polymers: a brief overview, in *Molecular and Ionic Recognition with Imprinted Polymers*, Bartsch, R.A. and Maeda, M., Eds., Oxford University Press, Oxford, pp. 1–8.

Mehregany, M., Zorman, C.A., Roy, S., Fleischman, A.J., Wu, C.-H., and Rajan, N., 2000. Silicon carbide for microelectromechanical systems, *Int. Mater. Rev.*, 45, pp. 85–105.

Millan, K.M., Saraullo, A., and Mikkelsen, S.R., 1994. Voltammetric DNA biosensor for cystic-fibrosis based on a modified carbon-paste electrode, *Anal. Chem.*, 66, pp. 2943–2948.

Mosbach, K., Haupt, K., Liu, X.C., Cormack, P., and Ramström, O., 1998. Molecular imprinting: *status artis et quo vadere?* in *Molecular and Ionic Recognition with Imprinted Polymers*, Bartsch, R.A. and Maeda, M., Eds., Oxford University Press, Oxford, pp. 29–48.

Napier, M.E., Loomis, C.R., Sistare, M.F., Kim, J., Eckhardt, A.E., and Thorp, H.H., 1997. Probing biomolecule recognition with electron transfer: electrochemical sensors for DNA hybridization, *Bioconjugate Chem.*, 8, pp. 906–913.

Nernst, W., 1904. Theorie der Reaktionsgeschwindigkeit in heterogenen Systemen, *Z Phys. Chem.*, 47, pp. 52–54.

Pauling, L., 1940. *The Nature of the Chemical Bond and the Structure of Molecules and Crystals: An Introduction to Modern Structural Chemistry*, Cornell University Press, Ithaca, NY.

Roy, S., McIlwain, A.K., DeAnna, R.G., Fleischman, A.J., Burla, R.K., Zorman, C.A., and Mehregany, M., 2000. SiC resonant devices for high Q and high temperature applications, Proceedings of the Hilton Head Solid State Sensor and Actuator Workshop, Hilton Head, SC, p. 22.

Schreiter, M., Gabl, R., Lerchner, J., Hohlfeld, C., Delan, A., Wolf, G., Bluher, A., Katzschner, B., Mertig, M., and Pompe, W., 2006. Functionalized pyroelectric sensors for gas detection, *Sensors Actuators B-Chem.*, 119, pp. 255–261.

Shah, R. and Abbott, N.L., 1999. Using liquid crystals to image reactants and products of acid-base reactions on surfaces with micrometer resolution, *J. Am. Chem. Soc.*, 49, pp. 11300–11310.

Shah, R. and Abbott, N.L., 2001. Principles for measurement of chemical exposure based on recognition-driven anchoring transitions in liquid crystals, *Science*, 293, pp. 1296–1299.

U.S. Army Center for Health Promotion and Preventive Medicine (USCHPPM), 2004. Chemical exposure guidelines for deployed military personnel, available at http://chppm-www.apgea.army.mil/tg.htm (accessed October 12, 2007).

U.S. Department of Defense, Army Medical Research Institute of Chemical Defense (USAMRICD), 1995. *Medical Management of Chemical Causalities Handbook*, 2nd ed., available at http://www.brooksidepress.org/Products/OperationalMedicine/DATA/operationalmed/Manuals/RedHandbook/001TitlePage.htm (accessed October 11, 2007).

Walt, D.R., 1998. Fiber optic imaging sensors, *Acc. Chem. Res.* 31, pp. 267–278.

Wang, J., 2002. Electrochemical nucleic acid biosensors, *Anal. Chim. Acta*, 469, pp. 63–71.

Wang, J., Cai, X.H., Rivas, G., Shiraishi, H., Farias, P.A.M., and Dontha, N., 1996. DNA electrochemical biosensor for the detection of short DNA sequences related to the human immunodeficiency virus, *Anal. Chem.*, 68, pp. 2629–2634.

Wernette, D.P., Swearingen, C.B., Cropek, D.M., Lu, Y., Sweedler, J.V., and Bohn, P.W., 2006. Incorporation of a DNAzyme into Au-coated nanocapillary array membranes with an internal standard for Pb(II) sensing, *Analyst*, 131, pp. 41–47.

Wohltjen, H., Jarvis, N.L., and Lint, J.R., 1991. Surface acoustic-wave (SAW) chemical microsensors and sensor arrays for industrial-process control and pollution prevention, *Abstr. Papers Am. Chem. Soc. (ENVR.)* 201, pp. 63–66.

Wolfbeis, O.S., 2000. Fiber optic chemical sensors and biosensors, *Anal. Chem.*, 72, pp. 81R–89R.

Yeh, T.S., Chu, C.S., and Lo, Y.L., 2006. Highly sensitive optical fiber oxygen sensor using Pt(II) complex embedded in sol-gel matrices, *Sensors Actuators B: Chem.*, 119, pp. 701–707.

Zemel, J.N., 1990. Microfabricated non-optical chemical sensors, *Rev. Sci. Instrum.*, 61, pp. 1579–1606.

7 Phage Display and Its Application for the Detection and Therapeutic Intervention of Biological Threat Agents

Joe A. Fralick, Prabhjit Chadha-Mohanty, and Guigen Li

CONTENTS

7.1 INTRODUCTION

Man has used biological weapons since the dawn of civilization, often for the purpose of warfare or assassination. Chapter 1 provides an overview of documented historical events involving the use of biological pathogens and toxins as weapons during warfare. A more exhaustive history of biological warfare is provided by Smart (1997) and Frischknecht (2003).

The history of biological warfare is truly a fascinating and harrowing topic; because of the potential horrors of germ warfare, two international treaties, one in 1925 and a second in 1972, have outlawed the use of biological weapons. In 1972 more than 100 nations signed the Biological and Toxin Weapons Convention (BWC), which prohibits the stockpiling or possession of biological agents except for "prophylactic, protective, or other peaceful purposes." Unfortunately, however, there is no clear protocol for enforcing such a treaty, and Iraq, among other signature countries, experimented with and amassed weapons that are specifically outlawed by the BWC, leading to a preemptive war.

Recently, individuals and rogue groups have become involved in the use of biological and chemical weapons for acts of terrorism (Purver 2002). In 1995, the Aum Shinrikyo cult conducted multiple attacks on the Tokyo subway system with the nerve agent sarin, injuring thousands and killing 12 (Cole 1996; Olson 1999). However, sarin was not the only weapon that the Aum Shinrikyo cult had obtained. They had unsuccessfully attempted terrorist attacks in Japan in the early 1990s with anthrax spores and botulinum toxin, and had attempted to obtain the Ebola virus from an outbreak in Zaire in 1993 (Atlas 2001; Leitenberg 2001). More recently, "weaponized" anthrax spores were sent through the U.S. Postal Service, resulting in the deaths of five individuals.

Because of the relative ease by which they can be manufactured and their relatively low cost, some types of biological weapons have become a "poor man's/cult's/nation's" weapon of mass destruction (WMD) and have led to experimentation and stockpiling of such weapons by some countries. This, coupled with well-organized terrorist groups capable of producing biological and chemical threat agents, has led to a rapid rise in government-sponsored research aimed at detection, prevention, and treatment for such agents. New technologies to meet these goals are clearly needed. The thrust of the Zumwalt Program has been to investigate the impact and means of countering biological and chemical weapons used against the war fighters of the United States of America.

7.2 PEPTIDE-BASED COUNTERMEASURES

A vast number of bioassays and biosensors depend on antibodies as recognition reagents, and human antibodies are popular candidates for the development of therapeutics (Zafir-Lavie et al. 2007; Filpula 2007). However, this is changing. Peptides are beginning to be utilized in detection-based formats (i.e., as capture and detection elements in biosensors) (Goldman et al. 2000) and as viable alternatives to biopharmaceuticals (Ladner et al. 2004). Small peptides (10–50 amino acids) have multiple advantages over antibodies for such uses. They are more robust (i.e., are stable at room temperature); have higher activity per mass (e.g., 15–60-fold, assuming a peptide of 10 to 50 amino acids); have better cell, organ, or tumor penetration (Chen and Harrisoon 2007); are more amenable to many diagnostic formats; may be less immunogenic; and have a lower manufacturing cost. Furthermore, most of the past limitations of peptides have been removed by new technologies, so that peptides now face no more hurdles than do antibodies in the development of pharmaceuticals (Ladner et al. 2004). However, perhaps one of the most important aspects of peptide-based detection or therapeutics for detecting and countering biological threat agents is that they can be developed in a much shorter time at much less expense than antibodies.

Phage display offers a powerful tool to identify partners of protein-protein interactions. The development of peptide therapeutics from display phage technology also offers advantages over the development of small molecule therapeutics. For example, the size of the peptide libraries (> 10^{10} different peptide sequences) from which they are selected is much greater than the small molecule libraries (~ 10^5 different species of small molecules), providing a much greater diversity from which to select for higher affinity and specificity. Peptides are also often more effective than small molecules when the therapeutics are called upon to block a protein-protein interaction (Ladner et al. 2004). Peptides can also be derivatized to a greater extent than small molecules. Finally, peptides can be manufactured to contain specific sequences for organ-specific or cell-specific delivery (i.e., sequences identified by *in vivo* display library selection by targeting specific organs, tissues, and cells) (Sergeeva et al. 2006; Ladner et al. 2004).

In this chapter we describe the use of a combinatorial approach employing display phage technology for the development of peptides that can replace antibodies for the detection, and therapeutic intervention, of biological threat agents.

7.3   COMBINATORIAL CHEMISTRY AND DISPLAY TECHNOLOGY

Combinatorial chemistry represents a new branch of biological chemistry that incorporates both molecular biology and immunology disciplines. Its goal is to use high-throughput screening of a very large collection of compounds or entities to find the few that have selective activities of interest. The overall approach integrates several drug-discovery disciplines, including synthetic and computational chemistry, analytical methodologies, molecular modeling, and high-throughput screening, and has the potential of identifying novel reagents that can be of value for both diagnostic and therapeutic uses. One of the more powerful technologies in the field of combinatorial chemistry is phage display.

7.3.1   PHAGE DISPLAY METHODOLOGY

In 1985, George Smith demonstrated that foreign proteins could be displayed on the surface of filamentous bacteriophage, M13, as a genetic fusion with the gene encoding for the capsid protein pIII (Smith 1985) (Figure 7.1). Because the displayed peptide is encoded in the viral genome, it was recognized that display technology could provide a powerful tool for selecting and evolving peptide ligands from large combinatorial peptide libraries (Cortese et al. 1996; McLafferty et al. 1993; Ladner 1995;

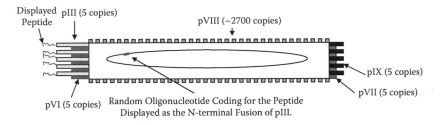

FIGURE 7.1   (See color insert following page 46.) M13 display phage. Cartoon of an M13 filamentous bacteriophage depicting the five capsid proteins (pIII, pVI, pVII, pVIII, and pIX), the single-stranded DNA genome, and the random oligonucleotide, cloned into the 5' end of gIII, coding for the sequence of the displayed peptide.

Hosse et al. 2006; Ja et al. 2005). Historically, peptides were displayed as *N*-terminal fusion peptides with either capsid protein pIII or pVIII (Figure 7.1). However, more recently, employing phagemids (i.e., plasmids carrying phage genes, peptides, and polypeptides) have been displayed from each of the five M13 coat proteins, but not all five on the same phage, suggesting greater flexibility in genetically modifying phage particles and the development of display phage with binding affinities for more than a single target (Russel et al. 2004).

The methods of phage display technology are based on the general scheme of making a large peptide display library (often involving a repertoire of 10^9 or greater) and putting it though repeated iterations of selection and amplification to select for those peptide ligands that have high affinity for the selected target (Figure 7.2). In selection, those ligands with desired properties (e.g., high affinity) are preferentially separated from the remainder of the library. In amplification, the relatively few selected ligands are copied to form a new generation (i.e., evolution). In some protocols, mutation (Yu and Smith 1996) or "exon shuffling" (Fisch et al. 1996) provides an increased combinatorial repertoire and allows a global search for ligands not

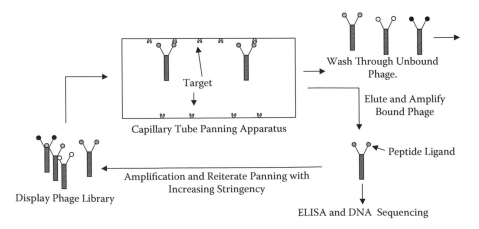

FIGURE 7.2   (See color insert following page 46.) Affinity selection. Cycles of selection for phage displayed peptides (display phage) that bind to target ligands.

included in the initial library. As stated above, the power of phage display technology is that it creates a physical linkage between a selectable function (i.e., binding of the displayed peptide to a target) and the DNA encoding that function (i.e., DNA sequence coding for the displayed peptide).

Phage displayed peptide libraries can be used to isolate peptides that bind with high specificity and affinity to virtually any target. As a result, such peptides can be used as reagents to understand molecular recognition. The latest applications of this technology include but are not exclusive of:

1. Functional selections of peptide ligands to probe regulatory networks (Venkztesh et al. 2002)
2. Epitope mapping used in vaccine design and development (Lesinksi and Westerink 2001)
3. Mapping of protein-protein contacts (Felici et al. 1991)
4. Identification of protease substrates (Hong and Boulanger 1995)
5. Receptor antagonists (Doorbar and Winter 1994)
6. Peptide mimics of nonpeptide ligands (Devlin et al. 1990)
7. Protein-DNA interactions (Bulyk et al. 2001)
8. Ligand fishing and protein engineering by directed evolution (O'Neil and Hoess 1995)
9. Toxin neutralization (Mourez et al. 2001)
10. Organ targeting in vivo (Duerr et al. 2004; Lee et al. 2002)
11. Selection of peptides that bind metals for nanowire production (Whaley et al. 2000; Mao et al. 2004)

This technology has revolutionized the means for studying molecular binding and ligand-ligand interactions.

7.3.2 AFFINITY SELECTION

Selection is determined by a peptide's affinity for a target that has usually been affixed to a solid support such as a microplate well or a magnetic bead. In a process called "panning/biopanning," display phage libraries, with large diversities, are added to the target, allowed to bind, and then the unbound phage are washed away, often with buffers of different stringencies. The attached phage are recovered, propagated in bacteria (*Escherichia coli*), and then further enriched through repeated rounds of selection and amplification with increasing stringency conditions (Figures 7.2–7.4). As the selection is reiterated, the library is appreciably reduced in complexity and those phage displaying the binding ligands are characterized using affinity analysis (e.g., enzyme-linked immunosorbent assay or ELISA), and DNA sequencing.

The logic of affinity maturation/evolution calls for low stringency and high yield in the early rounds of selection. This is because the initial library has many clones, each of which is represented by few phage particles. Hence, if the yield for a binding clone is not high in the first round of selection, that clone has a good chance of being lost. In later rounds, however, stringency can be increased in order to select for the most fit (i.e., tightest) binder. Stringency can be varied by a variety of factors such as:

Peptide Display Phage

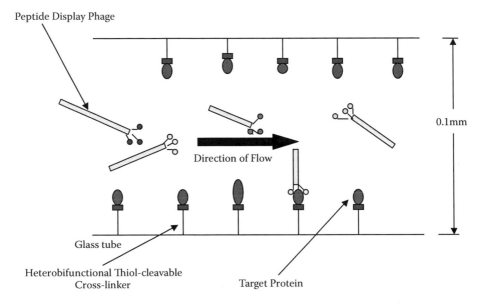

Direction of Flow

0.1mm

Glass tube

Heterobifunctional Thiol-cleavable
Cross-linker

Target Protein

FIGURE 7.3 **(See color insert following page 46.)** Capillary tube panning apparatus. A cartoon of our panning apparatus indicating the luminal surface of the glass capillary tube, the target affixed to the tube via a thiol-cleavable cross-linkage agent, and the display phage flowing through the tube.

1. The presence of nonionic detergents, which decreases nonspecific hydrophobic binding (e.g., Tween 20)
2. Temperature
3. Time allotted for attachment (selects for ligands with different attachment constants)
4. Time allotted for detachment or elution time (selects for ligands with different detachment constants)
5. Concentration of target
6. Competition of binding (e.g., addition of antibodies against target molecules)
7. The number of rounds of biopanning

to mention but a few. It should be noted, however, that there is a limit to stringency. There is always a background yield of nonspecifically bound phage; if stringency is set too high, the yield of specifically captured phage will fall below the background of nonspecifically bound phage, and hence the power of discrimination in favor of high affinity is lost. In practice, it is advisable to explore a range of stringency conditions in the final rounds of selection. Those rounds whose yields are close to background (i.e., no target) are too stringent to be useful.

7.3.3 Characterization of Selected Ligands

The aim of affinity selection is to identify high-affinity ligands to specific targets. It is, however, conceivable that a phage clone might be selected on the basis of slight growth advantages or other subtle traits unrelated to affinity. For this reason, it is

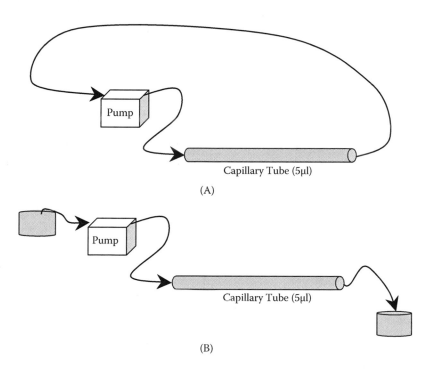

FIGURE 7.4 Affinity selection with the capillary panning apparatus. (a) To select peptides that bind to target, the display phage library was circulated through capillary tube using peristaltic pump. (b) Unbound phage were removed by washing for prescribed periods.

important to evaluate independently the affinity of the peptides displayed by the selected phage.

Although numerous methods for estimating affinity are available, there are three different methods that have been found to allow peptides to be studied in the form of phage-borne peptides, without having to synthesize them chemically. These are:

1. Phage capture assay/phage ELISA
2. Western blot analysis
3. Inhibition ELISA

In phage capture assay, selected phage clones are characterized with respect to their ability to recognize their target in a microplate based ELISA (Yu and Smith 1996) employing anti-M13 phage antibodies (i.e., phage ELISA) or plaque-forming units (pfu) (i.e., phage capture assay). In general, high yield in this assay is expected to correlate with high affinity between the receptor and the phage-borne ligand. The results provide a basis for choosing a smaller number of clones for more definitive analyses such as the inhibition ELISA. Western blot analysis can also determine relative affinities. We have found that only a very high-affinity display phage can be detected in Western blot analysis. Furthermore, in the case of a multimeric target, Western blot analysis can determine which component (i.e., protomer) of the multimeric protein is being recognized by the peptide ligands of the phage as well as any conformational requirements.

A primary consideration in screening a peptide clone is to demonstrate that the interaction of interest (target binding) can be blocked with the natural ligand. The inhibition ELISA determines the affinity of a test peptide by measuring the ability of various concentrations of the target in solution to competitively inhibit binding of a receptor to an immobilized ligand. If the strength of binding (or affinity) is high enough and the dissociation constant low enough ($K_D < 1\ \mu M$), peptide ligands can be analyzed in the form of whole virons (display phage). The virons to be used are first purified by CsCl density gradient centrifugation (Smith and Scott 1993) and the procedure is as described by Yu and Smith (1996). This is the most complex and time-consuming analysis of putative peptide ligands, and the procedure is designed so that rates of inhibition are determined and theoretical inhibition curves can be calculated for the determination of the "theoretical dissociation constant" (K_D) of the inhibitor for the receptor (i.e., K_D is a useful measure to describe the strength of binding or affinity between receptors and their ligands).

7.3.4 DNA Sequencing

Sequencing the DNA of the selected display phage provides the sequences of the peptides displayed on the winning phage. Often, in a successful screening, there is a strong consensus in the amino acid sequences of the displayed peptide (i.e., putative binding motifs). At times, a given display phage clone or putative binding motif may predominate in the final panning, although with larger libraries there tends to be greater variation within the consensus boundaries, and there exists further subgrouping within the selected pools of sequences.

7.3.5 Panning Apparatus

During studies with phage display technology a novel, closed capillary system for panning experiments was developed (Figure 7.3). The apparatus and method developed involves tethering target protein to the luminal surface of a capillary tube ($5\ \mu l$–$100\ \mu l$) via a water-soluble, heterobifunctional, thiol-cleavable cross-linkage agent and passing the display phage libraries through this tube with the aid of a peristaltic pump. This procedure enables panning against very small quantities of a target, with very precise washing and eluting procedures in the safety of a closed system (Figure 7.4). It also allows elution of the target together with bound display phage by cleaving the cross-linkage agent with a mild reducing agent. This method is more gentle than traditional acid (pH 2.2) elution (Menendez et al. 2001), thus enabling isolation of high-affinity ligands that may be inactivated by acid denaturation. Finally, the apparatus is very amenable for use in a high throughput format and could be used to concentrate an analyte for its detection.

7.4 SELECTION FOR HIGH-AFFINITY CHOLERA TOXIN BINDING PEPTIDE LIGANDS

Cholera is a life-threatening diarrheal disease caused by a potent enterotoxin secreted by *Vibrio cholerae*, appropriately called cholera toxin (CT). Fatality rates from cholera are fairly high (Faruque et al. 1998; Kaper et al. 1994) and there is a growing concern for the potential use of *V. cholerae* as a biological threat agent for warfare and terrorism (Category B list of diseases and agents listed by the U.S. Centers for Disease Control and Prevention (CDC) as potential bioterror threats, March 2005). If used as such, *V. cholerae* would be capable of causing epidemic-type disease, incapacitating a large number of individuals and creating social chaos.

The specific identification of toxins usually involves molecular recognition using complementary biomolecules. Hence, antibodies have been widely used for the detection of CT with excellent sensitivity. Recently, a method employing immobilized antibodies onto CdSe-ZnS quantum dots has provided a fluoroimmunoassay for CT with a detection limit of 10 ng/mL (Goldman et al. 2004). Other antibody detection formats have included antibody-based microarrays (Delehanty and Ligler 2002; Taitt et al. 2002; Rucker et al. 2005), surface plasmon resonance (SPR) (Choi et al. 1998; Jyoung et al. 2006), as well as SPR in combination with an ion-selective field-effect transistor device (Zayats et al. 2006). If one could replace the antibodies used in these detection assays with small peptide ligands, the devices would cost less to produce, be more robust, have much longer shelf lives, and would not require refrigeration or lyophilization for storage or transport.

7.4.1 SELECTION OF PEPTIDE LIGANDS THAT BIND CT

To identify peptide ligands that bind to cholera toxin with high affinity, a commercially available peptide display library (PH.D.-12 library, New England BioLabs) was employed along with the capillary tube panning apparatus described above. The PH.D.-12 library is a dodecapeptide,12-amino acid, phage display library that consists of random 12-mers fused to the *N*-terminal of pIII (minor coat) protein of the filamentous coliphage M13KE (see Figure 7.1). The stringency of the selection for binding was increased with increasing rounds of panning. This involved decreasing the allotted binding times, to select for ligands with rapid K_{on}, (a constant defining the attachment of the ligand with the target); increasing the Tween-20 concentrations in the wash buffer, to decrease nonspecific hydrophobic binding; and increasing the wash volume/time to select for ligands with a slow K_{off} (a constant defining the removal of the ligand with the target).

Following panning, the highest binders were identified by ELISA and their displayed peptide sequence determined. Analysis of the highest binders revealed two putative binding motifs. Interestingly, the phage with the highest affinity, CT7, displayed a peptide that contained both motifs.

7.4.2 CHARACTERIZATION OF SELECTED DISPLAY PHAGE/PEPTIDE LIGANDS

The disassociation constant (K_d) is a useful measure that describes the strength of binding between receptors and their ligands. The "theoretical K_d" (Yu and Smith, 1996) for CT7 phage binding to CT was determined to be 150 pM (Figure 7.5), indicating a very high affinity. Most monoclonal antibodies have K_d values in the nM to low μM range. The above results demonstrate that one can identify peptide ligands from a phage display library that bind with as high or higher affinity than do many antibodies.

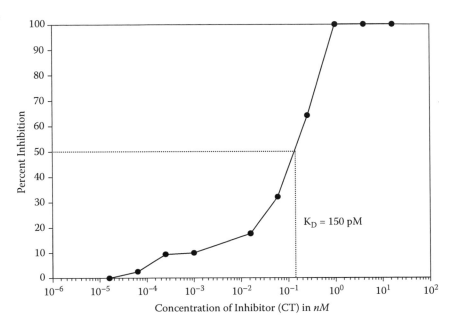

FIGURE 7.5 Determination of dissociation equilibrium constant. Dissociation equilibrium constants (K_d) were determined by the method described by Yu and Smith (1996). This method is based on assays that measure the ability of ligands in solution to inhibit the binding of biotinylated receptors to immobilized ligand. The slopes ($\Delta A_{405-490}$)/min. for each well were determined for test and standard curve wells. A regression curve of biotinylated bovine serum albumin (BSA) concentration vs. slope was generated. The concentration of the biotinylated target (Y) that had bound to the immobilized phage was determined by extrapolating from the standard curve using the slopes of the competition wells ($\Delta A_{405-490}$)/min.]. Y_{max} was the concentration of the bound target when no inhibitor was present. The percent inhibition was calculated by the formula: % inhibition = $(Y_{max} - Y)/Y_{max} \times 100$. Theoretical inhibition curves were generated by plotting % inhibition vs. concentration of inhibitor.

To further examine the feasibility of replacing an antibody with a display phage/peptide ligand, the CT7 was compared with an anti-CT monoclonal antibody (CT-mAb) for the detection of CT. The results are given in Figure 7.10. It can be seen that both the CT-mAb and CT7 can detect CT in the fM–nM range. However, stoichiometrically, four times as many CT-mAb molecules were used in the ELISA compared with the phage probe. Considering this, CT7 may be the better detection agent of the two.

Cholera toxin is a donut-shaped toxin composed of two different subunits (protomers), five copies of the B-subunit, which form a ring, and the A-subunit that sits in the ring. Immunoblot analysis, employing CT7 phage as a detection probe, was used to examine the localization of the binding of CT7 to cholera toxin. When probed with CT7 and anti-M13 horseradish peroxidase (HRP)-tagged antibody, only the pentameric form of subunit B was detected (Figure 7.6), indicating that the CT7 ligand recognizes a conformational epitope on the pentameric B-subunit ring.

To determine if the free peptide ligand can bind CT, small peptides (6–12-mers) carrying the different putative binding motifs in different orientations or locations

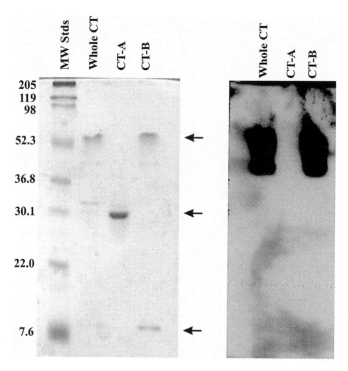

FIGURE 7.6 **(See color insert following page 46.)** Western blot analysis of CT with display phage CT7. Cholera toxin and its subunits were separated by nonreducing SDS-PAGE and transferred to a nitrocellulose membrane. CT7 and anti-M13 HRP-tagged Ab were used to probe and develop the blot. The arrows indicate the single CT-B subunit (bottom arrow), CTA (middle arrow), and pentameric CT-B (top arrow). Molecular weight standards are in kilodaltons (kDa).

were synthesized and tested for their ability to compete with CT7 for binding CT by competition ELISA (Figure 7.7). It can be seen from these results that all of the synthesized peptides can compete with display phage CT7 for binding CT and that the most effective peptide is #3, which contained the 12-mer sequence of phage CT7. In this competitive ELISA, the peptides competed with 10^{11} CT7 phage for binding to CT. It shows a 90% inhibition at 0.5 nM. This represents approximately 3×10^{10} peptide molecules, indicating that the free peptide competes very well with phage CT7 for the binding of CT. These results indicate that our 12-amino-acid peptide can bind to CT with high affinity and confirms our hypothesis that peptide ligands could replace antibodies in biothreat agent detection systems.

7.5 DEVELOPMENT OF PEPTIDE LIGANDS FOR THE NEUTRALIZATION OF CT INTOXICATION

Cholera toxin is an A-B type ADP-ribosylating toxin. The structural genes for CT are encoded within a lysogenic phage, CTxφ integrated into the large chromosome of *V. cholerae* (Moss and Vaughn 1988; Waldor and Mekalanos 1996). This toxin is

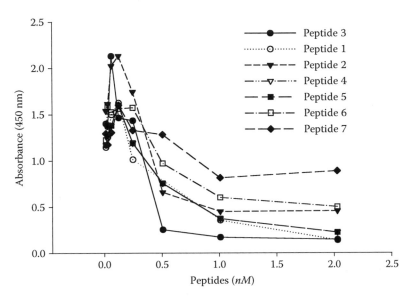

FIGURE 7.7   Inhibition ELISA of CT7 binding CT by peptide ligands. Peptide dilutions of the indicated peptides were incubated with 10^{11} pfu of CT7 display phage and then added to microtiter wells containing 1 μg of CT and incubated at 370°C for 1 h. The wells were washed six times and the amount of CT7 determined by HRP-tagged anti-M13 Abs.

an oligomeric protein consisting of a single A (enzymatic) subunit (CT-A) and five identical B (binding) subunits (CT-B) (Moss and Vaughn 1988). The five B subunits are arranged to form a pentameric ring with the A subunit occupying the central channel of this donutlike structure (Moss and Vaughn 1988; Mekalanos et al. 1979). Intact A subunit is not enzymatically active but must be nicked to produce fragments CTA1 and CTA2, which are held together via a disulfide bond. This nicking apparently occurs after the toxin is secreted from the cell. CT enters the intestinal epithelial cell as a stably folded protein by binding a lipid-based membrane receptor, ganglioside G (M1). It is thought that this receptor sorts the toxin into lipid rafts and a trafficking pathway to the endoplasmic reticulum, where the toxin unfolds and transfers its enzymatic subunit, CTA1, to the cytosol, through a poorly understood translocation pathway (Lencer 2001; Lencer et al. 1995, 1999; Plemper and Wolf 1999; White et al. 1999; Wolf 1998). Once in the cytosol, CTA1 is able to stimulate adenylate cyclase activity by catalyzing the ADP-ribosylation of the Gsα subunit of the GTP-binding regulatory protein (Moss and Vaughn 1988) resulting in elevated levels of cyclic adenosine monophosphate (cAMP). The consequence of this and perhaps other effects, such as increased synthesis of prostaglandins by the intoxicated cell, is the excessive accumulation of salt and water in the intestinal lumen, a process that leads to electrolyte imbalance, hypovolemic shock, and sometimes death. Cholera infection may be acquired through consumption of *Vibrio cholerae*–contaminated food or drink.

7.5.1 *In Vitro* Neutralization of CT with Display Phage

As described above, intoxication of intestinal cells with CT results in elevated levels of cAMP. Using this phenomenon, we have examined the ability of the high affinity display phage CT7, CT28, and CT32 to neutralize the cholera-toxin- induced increase in cAMP levels in monolayers of SW480 colorectal adenocarcinoma epithelial cells (ATCC CCL-228). A commercially available kit (R & D Systems, Minneapolis, MN) was used to determine the cAMP levels. Figure 7.8 presents the concentration-dependent neutralization of CT-induced cAMP levels by display phage and demonstrates that phage CT7, whose peptide ligand contains both CT binding motifs, provided the greatest neutralization. An irrelevant display phage (9A) showed no effect on the ability of CT to elicit cAMP production in this assay.

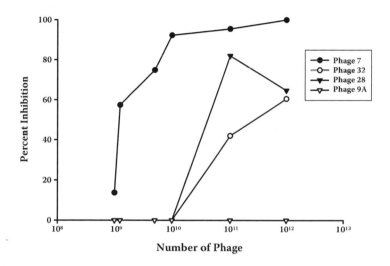

FIGURE 7.8 *In vitro* neutralization of cholera toxin activity with display phage. Monolayers of SW480 colorectal adenocarcinoma epithelial cells (ATCC CCL-228) cells were first seeded in six-well tissue culture plates (10^6/well; Corning Costar) and grown in minimal essential medium (MEM; GIBCO BRL) containing 5% fetal calf serum and 1X antibiotics at 37°C, 5% CO_2 to near confluence. Prior to adding cholera toxin to wells, cells were incubated in MEM 1% FCS containing $1mM$ isobutyl-1-methylxanthine (Sigma) for 1 hour at 37°C, 5% CO_2. 1 μg cholera toxin was activated with 2 ng trypsin in 0.2M NaCl, 0.05M Tris-HCl (pH 7.5), 0.001M EDTA, 0.003M NaN_3 (TEAN) for 30 minutes at 37°C. Varying numbers of Phage 7, 28, and 32 were incubated with 1 μg cholera toxin after trypsinization for 30 minutes at 37°C. Cholera toxin and phage were then incubated with cells for 2 h at 37°C, 5% CO_2. After incubation, the cells were transferred to 2 ml microcentrifuge tubes and centrifuged for 10 minutes at 14,000 g at RT. The pellets were washed with cold phosphate buffered saline (PBS) and the intracellular cAMP was extracted with 0.4 ml 0.1N HCl. Following incubation at RT for 20 minutes, cAMP levels were measured using an ELISA based low-pH cAMP kit (R & D Systems, Minneapolis, MN). Values were determined by using the formula: % inhibition = {(cAMP CT. alone – cAMP CT + Phage 7)/cAMP CT alone × 100}.

These results indicated that 10^{10} CT7 particles were capable of neutralizing 1 μg cholera toxin activity in this assay. One μg of cholera toxin is equivalent to ~10^{12} molecules. The inhibition levels mediated by phage CT28 and CT32 were approximately tenfold lower than those of phage CT7. It seems likely that since CT7 (and presumably CT28 and CT32) binds to the B-subunit of CT (Figure 7.6), they may be preventing CT from binding to its cell receptor and thus preventing entrance into the SW480 colorectal adenocarcinoma epithelial cells.

7.5.2 IN VIVO NEUTRALIZATION OF CT WITH CT7 PHAGE AND PEPTIDES

To determine if CT7 or peptides containing the two putative CT binding motifs could neutralize CT *in vivo*, we have examined their ability to neutralize CT in the intestine of a mouse employing the ileal loop assay (Figure 7.9). CT (2 mg) and display phage (10^{10}) or CT (1 mg) plus peptides (10^{10}) were injected into the lumen of the loop (generated by ligations with silk suture). Fluid accumulation in the loops at the end of a 4 h incubation period was used as a measure of CT activity in the mouse small intestine. This was determined by the ratio of the weight of the loop (mg) divided by the length of the loop (cm). The values are the average of six mice. The three peptides employed in this study contained the sequences of CT28 (#1), of CT32 (#2), or of CT7 (#3). We have found that all three of these peptides as well as display phage CT7 are effective in inhibiting CT activity in the intestine of a mouse. These

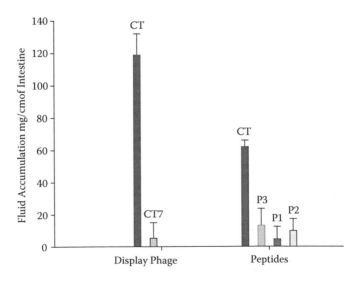

FIGURE 7.9 *In vivo* neutralization of cholera toxin activity with display phage CT7 and corresponding peptides in the intestine of a mouse. Display phage, CT7 (10^{10} pfu), or free peptides (10^{10}) containing peptide sequences from CT28 (#1), CT32 (#2), or CT7 (#3) were injected with activated CT (trypsinization for 30 min. at 37°C) into the lumen of a mouse ileal loop (generated by ligations with silk suture). Fluid accumulation in the loops at the end of a 4 h incubation period was determined by the ratios of the weight of the loop (mg) divided by the length of the loop (cm).

results further support our hypothesis that peptide ligands are capable of replacing antibodies as a therapeutic for biothreat agents.

7.6 DEVELOPMENT OF AN ANTIEPITOPIC DISPLAY PHAGE

A peptide that binds the paratope of an antibody is called an epitopic mimic or mimotope (i.e., it is an amino acid sequence that mimics the structure of an epitope that the antibody was raised against). Therapeutically mimotopes have been used as peptide vaccines (Wang and Yu 2004; Riemer et al. 2005; Li et al. 2006). For example, Wu and others (2001) isolated antibotulinum neurotoxin type A (BoNT/A) monoclonal antibodies that neutralized this toxin. Then, employing phage display technology, they isolated peptide ligands that mimicked the epitope recognized by these neutralizing monoclonal antibodies and used them to successfully vaccinate mice against BoNT/A intoxication.

Peptide mimotopes could also be used to develop simulants of epitopes recognized by antibodies used in a variety of detection systems. Such simulants would enable testing of those detection devices without personal exposure to toxins or highly pathogenic agents that require biological safety level III containment. For this purpose we have identified epitopic mimics against monoclonal antibodies used to detect *Yersinia pestis* (unpublished results) in an immunoassay format. However, a more intriguing use of mimotopes would be to develop a peptide that binds to the mimotope (i.e., an antiepitopic peptide [AEP]). Theoretically AEPs will mimic the antibodies used to generate the mimotope. Hence, such peptides could be used to replace antibodies that have been developed for a variety of diagnostic and therapeutic uses and would recognize the same epitopes, as do the antibodies used to generate them. For reasons already stated, such peptides would provide important detection and therapeutic reagents.

Attempts to develop a protocol for selecting antiepitopic peptides have employed an anti-ricin toxin (RT) polyclonal antibody: the strategy being to select for display phage from a peptide display phage library that bound to the paratope of an anti-RT antibody (i.e., ricin mimotope) and then use this phage to select for a second display phage that can bind to the mimotope peptide displayed on the phage. The selections involved subtractive panning against irrelevant antibodies (Abs) of the same isotype (rabbit immunoglobulin, IgG) in the first selection (mimotopes) and against irrelevant display phage in the second selection (antiepitopic display phage).

7.6.1 SELECTION OF RICIN MIMOTOPE

Rabbit IgG was affixed to the luminal surface of a capillary panning apparatus and the PH.D.-12 peptide display library was circulated through this tube to remove those display phage, which bound to the surface of rabbit IgG. We then used this subtractive library and our capillary panning apparatus to pan against polyclonal rabbit IgG antiricin Ab. This entailed decreasing the time of exposure of our library to the target Abs, increasing the wash time, increasing the Tween 20 concentrations, and

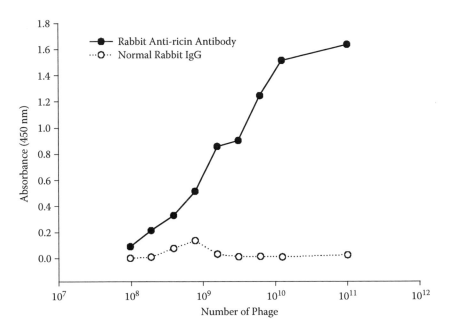

FIGURE 7.10 RT11 binds to antiricin antibody. RT11 display phage were added at the indicated concentrations (pfu) to micotiter wells to which either anti-RT Abs or irrelevant rabbit IgG was affixed (1 µg/well). The wells were incubated at 370°C for 1 h. The wells were washed six times and the amount of display phage, RT11, determined by HRP-tagged anti-M13 Abs.

decreasing the target concentrations (antiricin Abs) with successive pannings. Following the last panning, we ran an ELISA to identify the high binders. Interestingly, we found a single, dominant display phage that binds an antiricin antibody, which we refer to as RT11. To further characterize this phage we compared its binding to the antiricin Ab with that of an irrelevant phage (has no display peptide). The results are given in Figure 7.10. It can be seen that RT11 binds the antiricin Ab with high affinity. The above results may suggest that the polyclonal mixture of anti-RT Abs recognizes a dominant epitope on RT.

To further characterize display phage RT11, we used it as an analyte in a Western blot against the antiricin Ab in its native (Figure 7.11A) and in its reduced (Figure 7.11B) state (heavy and light chains). The assumption is that if R11 binds to the surface of the Ab, one would expect it to recognize both the native and reduced Ab subunits (i.e., heavy or light chains). If, however, it binds to the paratope of the antibody, it would require a specific conformation of the antibody (both heavy and light chains) and therefore should only bind the native Ab. It can be seen from this Western that the latter result is the case. This indicates that we have identified a peptide ligand that has a high affinity for the native anti-RT Ab, putatively as an epitope mimic or "mimotope." Such a mimotope could be used as a stimulant for any detection format employing this antiricin antibody. It can also be used in the development of an antiepitopic peptide ligand that would mimic the antiricin Ab.

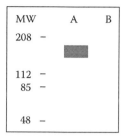

FIGURE 7.11   Western blot analysis of anti-RT with display phage, R11. Anti-RT Abs in their native (A) or reduced (B) state were separated by nonreducing SDS-PAGE and transferred to a nitrocellulose membrane. RT11 display phage and anti-M13 HRP-tagged Abs were used to probe and develop the blot. Molecular weight standards (MW Stds) are in kilodaltons (kDa).

7.6.2   Selection of Anti-R11 Peptide Ligand

Assuming that R11 represents a ricin mimotope, we proceeded to select for a second display phage that would bind the displayed mimotope with high affinity. To do this, we employed a PH.D.-12 display phage library (see above for description) and conducted a subtractive panning against an irrelevant display phage (i.e., M13 phage without a displayed peptide) to remove any peptide ligands that bound to any of the M13 capsid proteins. Those display phage from the subtractive library were then used to pan against display phage RT11. This entailed stringency selection utilizing the same stringency conditions used to select for the ricin mimotope. Following our last panning we conducted a phage ELISA from 40 randomly selected plaques. We found nine display phage that showed high affinity to R11, six of which had identical peptide sequences (putative binding motifs). Further ELISA analysis determined that the display phage with the shared sequence (i.e., six of the nine) had the highest affinity for the peptide displayed on RT11.

7.6.3   Detection of Ricin Toxin with RT37

To compare the putative antiepitopic RT display phage with the anti-RT antibody, we have examined their ability to bind to varying concentrations of RT. The amount of anti-RT antibody was maintained at a constant concentration of 124.5 ng ($\sim 5 \times 10^{10}$ antibody molecules) and the number of phage was keep constant at 1×10^{10} particles. Phage 9A is an irrelevant display phage that does not display a peptide and RT37 was one of the six high-affinity, RT11 binding phage (i.e., antiepitopic phage). The results are given in Figure 7.12. It can be seen that RT37 can detect ricin with a similar efficiency as the polyclonal antiricin antibodies used to develop this phage. To our knowledge, this is the first demonstration of an antiepitopic display phage (antibody mimic). The potential of such mimics is huge. Using such technology, one could develop small, robust, inexpensive peptides to replace existing therapeutic and diagnostic antibodies.

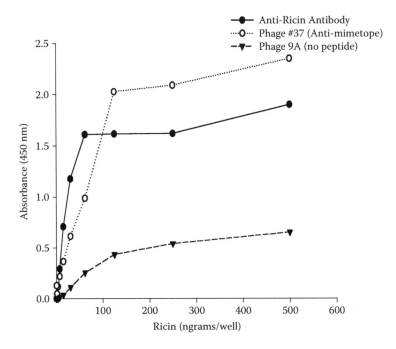

FIGURE 7.12   Comparison of RT37 with antiricin Abs. for binding ricin. The indicated display phage (10^{10} pfu) or anti-RT Abs (5×10^{10}) was added to microtiter wells to which the indicated amount of RT had been affixed. The wells were incubated at 370°C for 1 h. The wells were then washed six times and the amount of display phage were determined with anti-M13 HRP-tagged Abs and the amount of anti-RT Abs determined by HRP-tagged goat anti-rabbit IgG.

7.7   DISCUSSION AND APPLICATIONS

Phage-displayed peptide libraries can be used to isolate peptides that bind with high specificity and affinity to virtually any target, including metals (Whaley et al. 2000; Mao et al. 2004), carbohydrates (Devlin et al. 1990), and proteins (Sergeeva et al. 2006). These display phage/peptide ligands can then be used as reagents to understand molecular recognition, as lead molecules in drug design, as minimized mimics for receptors, and as replacements for antibodies used in biological threat agent detectors and biopharmaceuticals.

In this chapter we have demonstrated proof of principal for the use of display phage technology in the detection and treatment of potential biothreat agents (cholera and ricin toxins). We have described the development of a novel capillary panning apparatus that can be used for very precise, high throughput, safe panning of very small amounts of toxins. Employing this capillary panning apparatus, we have been able to select for display phage that can bind toxins (CT and RT) with affinity, similar to, or better than antibodies raised against these toxins. We have also demonstrated that some of these phage and their displayed peptides are effective at neutralizing their toxin target both *in vitro* (i.e., tissue culture) and *in vivo* (i.e., in a living animal). Finally, we have developed a method for the selection of display phage, and presumably peptide ligands, that can target the same epitopes as to which the antibodies used to develop them, bind (i.e., antiepitopic display phage/peptides). We have also used this strategy to develop an anti-CD4 receptor mimic (unpublished data). This latter process holds great promise in the development of peptide aptamers against specific antigens on highly dangerous agents without exposure to such agents. Theoretically it would also allow for the replacement of antibody biopharmaceuticals with small peptide pharmaceuticals. The advantages of such peptides over antibodies have been previously discussed.

7.8 CONCLUSIONS

As so amply stated by Eubanks and others (2007), "There is a growing need for technological advancements to combat agents of chemical and biological warfare, particularly in the context for the deliberate use of chemical and/or biological warfare agent by a terrorist organization." While antibody-based technology has been the mainstay for the detection of chemical and biological agents (Eubanks et al. 2007; Lim et al. 2005), there are other powerful technologies for the development of high-affinity analytes that could be developed for these purposes. One very powerful tool for such technology is phage display. Techniques for the development of phage display libraries with very large diversities (10^{11}) are now available (Sidhu et al. 2000), and improved screening approaches applying high stringency selection have identified peptide ligands with higher affinities toward target molecules than most antibodies (Dennis et al. 2000; and this chapter). Phage display technology has also been used to develop peptides as therapeutic alternatives to antibodies (Sidhu and Sachdev 2000; Mullen et al. 2006) and this technology has the important advantage over antibody and small molecule therapeutics with respect to development time (weeks or months versus months or years). This could be very important if a biological threat agent were used for which we had no diagnostic/therapeutic information. Future studies will involve applying this powerful technology toward the development of peptide-specific formats for the detection of biological threat agents via sensor systems, as well as the development of pharmaceuticals for their prophylaxis/treatment.

ACKNOWLEDGMENTS

We would like to acknowledge Dr. Catherine S. McVay who conducted many of these experiments and Dr. Xiaoyi Yao for her expert advice in phage display technology. We also thank Naomi McDonald and Joe Bass for their technical assistance.

REFERENCES

Atlas, R.A., 2001. Bioterrorism before and after September 11, *Crit. Rev. Microbiol.*, 27, pp. 355–379.
Bonnycastle, L.L.C., Menedez, A., and Scott, J., 2001. General phage methods, in *Phage Display: A Laboratory Manual*, Barbas, C.F., III, Burton, D.R., Scott, J.K., and Silverman, G.J., Eds., Cold Spring Harbor Laboratory Press, Cold Spring Harbor, NY, pp. 15.1–15.3.

Bulyk, M.L., Huang, X., Choo, Y., and Church, G.M., 2001. Exploring the DNA-binding specificities of zinc fingers with DNA microarrays, *Proc. Natl. Acad. Sci. USA*, 98, pp. 7158–7163.

Chen, L. and Harrisoon, S.D., 2007. Cell-penetrating peptides in drug development: enabling intracellular targets, *Biochem. Soc. Trans.*, 35, pp. 821–825.

Choi, K., Youn, H.J., Ha, Y.C., Kim, K.J., and Choi, J.D., 1998. An ultrasensitive DNA detection by using gold nanoparticle multilayer in nano-gap electrodes, *J. Microbiol.*, 36, pp. 43–48.

Cole, L.A., 1996. The specter of biological weapons, *Sci. Am.*, 275, pp. 30–35.

Cortese, R., Monaci, P., Luzzago, A., Santini, C., Bartoli, F., Cortese, I., Fortugno, P., Galfre, G., Nicosia, A., and Felici, F., 1996. Selection of biologically active peptides by phage display of random peptide libraries, *Curr. Opin. Biotechnol.*, 7, pp. 616–621.

Delehanty, J.B. and Ligler, F.S., 2002. A microarray immunoassay for simultaneous detection of proteins and bacteria, *Anal. Chem.*, 74, pp. 5681–5687.

Dennis, M.S., Eigenabot, C., Skelton, N.J., Ultsch, M.H., Santell, L., Dwyer, M.A., O'Connell, M.P., and Lazarus, R.A., 2000. Peptide exosite inhibitors of factor VIIa as anticoagulants, *Nature*, 404, pp. 465–470.

Devlin, J.J., Panganiban, L.C., and Devlin, P.E., 1990. Random peptide libraries: a source of specific protein binding molecules, *Science*, 249, pp. 404–406.

Doorbar, J., and Winter, G., 1994. Isolation of a peptide antagonist to the thrombin receptor using phage display, *J. Mol. Biol.*, 244, pp. 361–369.

Duerr, D.M., White, S.J., and Schluesener, H.J., 2004. Identification of peptide sequences that induce the transport of phage across the gastrointestinal mucosal barrier, *J. Virol. Methods*, 116, pp. 177–180.

Eubanks, L.M., Dickerson, T.J., and Janda, K.D., 2007. Technological advancements for the detection of and protection against biological and chemical warfare agents, *Chem. Soc. Rev.*, 36, pp. 458–470.

Faruque, S.H., Albert, M.J., and Mekalanons, J.J., 1998. Epidemiology, genetics, and ecology of toxigenic *Vibrio cholerae*, *Microbiol. Mol. Biol. Rev.*, 62, pp. 1301–1314.

Felici, F., Castagnoli, L., Musacchio, A., Jappelli, R., and Cesareni, G., 1991. Selection of antibody ligands from a large library of oligopeptides expressed on a multivalent exposition vector, *J. Mol. Biol.*, 20, pp. 301–310.

Filpula, D., 2007. Antibody engineering and modification technologies, *Biomol. Eng.*, 24, pp. 201–215.

Fisch, I., Kontermann, R.E., Finnern, R., Hartley, O., Soler-Gonzaliez, A.S., Griffiths, A.D., and Winer, G., 1996. A strategy of exon shuffling for making large peptide repertoires displayed on filamentous bacteriophage, *Proc. Natl. Acad. Sci. USA*, 93, pp. 7761–7766.

Frischknecht, F., 2003. The history of biological warfare. *EMBO Reports*, special issue, 4, pp. S47–S52.

Goldman, E.R., Clapp, A.R., Anderson, G.P., Uyeda, H.T. Mauro, J.M. Medintz, I.L., and Mattoussi, H., 2004. Multiplexed toxin analysis using four colors of quantum dot fluororeagents, *Anal. Chem.*, 76, pp. 684–688.

Goldman, E.R., Pazirandeh, M.P., Mauro, J., King, K.D., Frey, J.C., and Anderson, G.P., 2000. Phage-displayed peptides as biosensor reagents, *J. Mol. Recognit.*, 13, pp. 382–387.

Graff, C.P. and Wittrup, K.D., 2003. Theoretical analysis of antibody targeting of tumor spheroids: importance of dosage for penetration, and affinity for retention, *Cancer Res.*, 93, pp. 1288–1296.

Hong, S.S. and Boulanger, P., 1995. Protein ligands of the human adenovirus type 2 outer capsid identified by biopanning of a phage-displayed peptide library on separate domains of wild-type and mutant penton capsomers, *EMBO J.*, 14, pp. 4714–4727.

Hosse, R.J., Rothe, A., and Power, B.E., 2006. A new generation of protein display scaffolds for molecular recognition, *Protein Sci.,* 15, pp. 14–27.

Ja, W.W., Olsen, B.N., and Roberts, R.W., 2005. Epitope mapping using mRNA display and a unidirectional nested deletion library, *Protein Eng. Des. Sel.,* 18, pp. 309–319.

Jyoung, J.-Y, Hong, S., Lee, W., and Choi, J.W., 2006. Immunosensor for the detection of *Vibrio cholerae* 01 using surface plasmon resonance, *Biosens. Bioelectron.,* 21, pp. 2315–2319.

Kaper, J.B., Fasano, A., and Trucksis, M., 1994. Toxins of *Vibrio cholerae,* in *Vibrio cholerae and Cholera: Molecular to Global Perspectives,* ed. Wachsmuth, I.K., Blake, P.A. and, D.C, Olsvik, Eds., American Society for Microbiology Press, Washington, DC, pp. 145–176.

Ladner, R.C., 1995. Constrained peptides as binding entities, *Trends Biotechnol.,* 13, pp. 426–430.

Ladner, R.C., Sato, A.K., Gorzelany, J., and de Souza, M., 2004. Phage display-derived peptides as therapeutic alternatives to antibodies, *Drug Discovery Today,* 9, pp. 525–529.

Lee, L., Buckley, C., Blades, M.C., Panayi, G., George, A.J.T. and Pitzalis, C., 2002. Identification of synovium-specific homing peptides by in vivo phage display selection, *Arthritis Rheum.,* 46, pp. 2109–2120.

Leitenberg, M., 2001. Biological weapons in the twentieth century: a review and analysis, *Crit. Rev. Microbiol.,* 27, pp. 267–320.

Lencer, W.I., 1998. Ganglioside structure dictates signal transduction by cholera toxin in polarized epithelia and association with caveolae-like membrane domains, *J. Cell Biol.,* 141, pp. 917–927.

Lencer, W.I., 2001. Microbes and microbial toxins: paradigms for microbial-mucosal interactions V. *cholera:* invasion of the intestinal epithelial barrier by a stably folded protein toxin, *Am. J. Physiol. Gastrointest. Liver Physiol.,* 280, pp. G781–G786.

Lencer W.I., Hirst, T.R., and Holmes, R.K., 1999. Membrane traffic and the cellular uptake of cholera toxin, *Biochim. Biophys. Acta Mol. Cell. Res.,* 1450, pp. 177–190.

Lencer, W.I, Moe, S., Rufo, P.A., and Madara, J.L., 1995. Transcytosis of cholera toxin subunits across model human intestinal epithelia, *Proc. Natl. Acad. Sci. USA,* 92, pp. 10094–10098.

Lesinksi, G.B. and Westerink, M.A., 2001. Novel vaccine strategies to T-independent antigens, *J. Microbiol. Methods,* 47, pp. 135–149.

Li, M., Yan, Z., Han, W., and Zhang, Y., 2006. Mimotope vaccination for epitope-specific induction of anti-CD20 antibodies, *Cell Immunol.,* 239, pp. 136–143.

Lim, D.V., Simpson, J.M., Kearns, E.A., and Kramer, M.F., 2005. Current and developing technologies for monitoring agents of bioterrorism and biowarfare, *Clin. Mirobiol. Rev.,* 18, pp. 583–607.

Mao, C., Solis, D.J., Reiss, B.D., Kottmann, S.T., Sweeney, R.Y., Hayhurst, A., Georgiou, G., Iverson, B., and Belcher, A.M., 2004. Virus-based tool-kit for the directed synthesis of magnetic and semiconducting nanowires, *Science,* 303, pp. 213–217.

McLafferty, M.A., Kent, R.B., Ladner, R.C., and Markland, W., 1993. M13 bacteriophage displaying disulfide-constrained microproteins, *Gene,* 128, pp. 29–36.

Mekalanos, J.J., Collier, R.J., and Romig, W.R., 1979. Enzymic activity of cholera toxin, II, Relationships to proteolytic processing, disulfide bond reduction, and subunit composition, *J. Biol. Chem.,* 254, pp. 5855–5861.

Menendez, A., Bonnycastle, L.C.C., Pan, O.C.C., and Scott, J.K., 2001. Screening peptide libraries, in *Phage Display: A Laboratory Manual,* Barbas, C.F., III, Burton, D.R., Scott, J.K., and Silverman, G.J.,. Eds., Cold Spring Harbor Laboratory Press, Cold Spring Harbor, NY, pp. 17.1–17.32.

Moss, J., and Vaughn, M., 1988. CT and *E. coli* enterotoxins and their mechanisms of action, in *Handbook of Natural Toxins*, Vol. 4, ed. Hardegree, M.C. and Tu, A.T., Eds., Marcel Dekker, New York, pp. 39–87.

Mourez, M., Kane, R.S., Mogridge, J., Metallo, S., Deschatelets, P., Sellman, B.R., Whitesides, G.M., and Collier, R.J., 2001. Designing a polyvalent inhibitor of anthrax toxin, *Nat. Biotechnol.*, 19, pp. 958–961.

Mullen, L.M., Nair, S.P., Ward, J.M., Rycroft, A.N., and Henderson, B., 2006. Phage display in the study of infectious diseases, *Trends Microbiol.*, 14, pp. 141–147.

Olson, K.B., 1999. Aum Shinrikyo: once and future threat? *Emerg. Infect. Dis.*, 5, pp. 513–516.

O'Neil, K.T. and Hoess, R.H., 1995. Phage display: protein engineering by directed evolution, *Curr. Opin. Struct. Biol.*, 5, pp. 443–449.

Plemper, R.K. and Wolf, D.H., 1999. Retrograde protein translocation: eradication of secretory proteins in health and disease, *Trends Biol. Sci.*, 24, pp. 266–270.

Purver, R., 2002. Chemical and biological terrorism: the threat according to the open literature, Canadian Security and Intelligence Service, available at http://www.csis.scrs.gc.ca /eng/miscdocs/tabintr_e.html.

Riemer, A.B., Kurz, H., Klinger, M., Scheiner, O., Zielinski, C.C., and Jensen-Jarolim, E., 2005. Vaccination with cetuximab mimotopes and biological properties of induced antiepidermal growth factor receptor antibodies, *J. Natl. Cancer Inst.*, 97, pp. 1663–1670.

Rucker, V.C., Havenstrite, K.L. and Herr, A.E., 2005. Antibody microarrays for native toxin detection, *Anal. Biochem.*, 339, pp. 262–270.

Russel, M. et al., 2004. Introduction to phage display and phage biology, in *Phage Display: A Practical Approach*, Clackson, T. and Lowman, H., Eds., Oxford University Press, Oxford, pp. 1–26.

Sergeeva, A,, Kolonin, M.G., Molldrem, J.J., Pasqualini, R., and Arap, W., 2006. Display technologies: Application for the discovery of drug and gene delivery agents, *Adv. Drug Discovery Rev.*, 58, pp. 1623–1654.

Sidhu, S., Lowman, S.H.G., Cunningham, B.D., and Wells, J.A., 2000. Phage display for selection of novel binding peptides, *Methods Enzymol.*, 328, pp. 333–363.

Sidhu, S., and Sachdev, S., 2000. Phage display in pharmaceutical biotechnology, *Curr. Opin. Biotechnol.*, 11, pp. 610–616.

Smart, J.K., 1997. History of chemical and biological warfare: an American perspective, in *Medical Aspects of Chemical and Biological Warfare*, Sidell, F.R., Takafuji, E.T., and Franz, D.R., Eds., Borden Institute, Walter Reed Army Medical Center, Washington, DC.

Smith, G.P., 1985. Filamentous fusion phage: novel expression vectors that display cloned antigens on the viron surface, *Science*, 222, pp. 315–317.

Smith, G.P. and Scott, J.K., 1993. Libraries of peptides and proteins displayed on filamentous phage, *Methods Enzymol.*, 217, pp. 228–257.

Taitt, C.R., Anderson, G.P., Lingerfelt, B.M., Feldstein, M.J., and Ligler, F.S., 2002. Nineanalyte detection using an array-based biosensor, *Anal. Chem.*, 74, pp. 6114–6120.

Venkztesh, N., Zaltsman, Y., Somjen, D., Gayer, B., Boopathi, E., Kasher, R., Kulik, T., Katchalski-Katzir, E., and Kohen, F., 2002. A synthetic peptide with estrogen-like activity derived from phage-display peptide library, *Peptides*, 23, pp. 573–580.

Waldor, M.K. and Mekalanos, J.J., 1996. Lysogenic conversion by a filamentous phage encoding cholera toxin, *Science*, 272, pp. 1910–1914.

Wang, L.F. and Yu, M., 2004. Epitope identification and discovery using phage display libraries: applications in vaccine development and diagnostics, *Curr. Drug Targets*, 5, pp. 1–15.

Whaley, S.R., English, D.S., Hu, E.L., Barbara, P.F., and Belcher, A.M., 2000. Selection of peptides with semiconductor binding specificity for directed nanocrystal assembly, *Nature*, 405, pp. 665–668.

White, J., Johannes, L., Mallard, F., Girod, A., Grill, S., Reinsch, S., Keller, P., Tzschaschel, B., Echard, A., Goud, B., and Stelzer, E.H., 1999. Rab6 coordinates: a novel Golgi-to-ER retrograde transport pathway in live cells, *J. Cell Biol.,* 147, pp. 743–760.

Wolf, A.A, Jobling, M.G., Wimer-Mackin, S., Madara, J.L., Holmes, R.K., and Lencer, W.I., 1998. Ganglioside structure dictates signal transduction by cholera toxin in polarized epithelia and association with caveolae-like membrane domains. *J. Cell Biol.,* 141, pp. 917–927.

Wu, H.C., Yeh, C.T., Huang, Y.L., Tarn, L.J., and Lung, C.C., 2001. Characterization of neutralizing antibodies and identification of neutralizing epitope mimics on the *Clostridium botulinum* neurotoxin type A, *Appl. Environ. Microbiol.,* 67, pp. 3201–3207.

Yu, J. and Smith, G.P., 1996. Affinity maturation of phage-displayed peptide ligands, *Methods Enzymol.* 267, pp. 3–26.

Zafir-Lavie, I., Michaeli, Y., and Reiter, Y., 2007. Novel antibodies as anticancer agents, *Oncogene,* 26, pp. 3714–3733.

Zayats, M., Raitman, O.A., Chegel, V.I., Kharitonov, A.B., and Willner, I., 2002. Probing antigen-antibody binding processes by impedance measurements on ion-sensitive field-effect transistor devices and complementary surface plasmon resonance analyses: development of cholera toxin sensors, *Anal. Chem.,* 74, pp. 4763–4773.

8 Personnel Protective Fabric Technologies for Chemical Countermeasures

Seshadri S. Ramkumar, Utkarsh Sata, and Munim Hussain

CONTENTS

8.1   INTRODUCTION

Historical use of biological and chemical threat agents against military and civilian personnel, and recent concern regarding willingness to use chemical weapons among terrorist groups, have prompted the governments of many Western nations to initiate programs on countermeasures against such agents. Protecting those who protect against chemical and biological threats is an important challenge and immediate task for the military as well as the science and technology communities. One of the important aims of the chemical and biological defense program of the U.S. military is individual protection. Individual protection is important for war fighters, first responders, and civilians in the event of exposure to chemical warfare (CW) agents, biotoxins, and toxic industrial chemicals. Protective barriers are needed that will not only provide the necessary protection against threat agents but also be comfortable to wearers. Achieving enhanced protection and comfort in a single protective suit is a challenge that is yet to be solved (DoD 2000, 2006, 2007). A recent report by the U.S. Department of Defense (DoD) to the U.S. Congress has highlighted the importance of developing lightweight clothing with improved protection and enhanced strength that will also reduce the physiological burden caused by soldiers carrying heavy or bulky protective materials (DoD 2006). The overarching goal for the military is to provide maximum possible protection with the least possible logistical burden for the war fighter (Gurudatt et al. 1997; Lukey et al. 2004; DoD 2006, 2007). Thus, the primary focus of this chapter is on fiber-based protective materials that offer protection to war fighters and allow for personnel decontamination. This chapter discusses the current state of the art in protective and decontamination technologies such as sorbent fabrics, nanofibers, and particulate and nonparticulate nonaqueous decontamination materials.

8.2 SORBENT MATERIALS

Percutaneous, respiratory, and ocular protection are important thrust areas within the Chemical and Biological Defense Program (DoD 2006). Sorbent materials are widely used in providing dermal protection (Gurudatt et al. 1997; Wilusz 2007). For military applications, adsorption of toxic chemical vapors within fabric structures is one of the fundamental and essential mechanisms for chemical protective fabrics (Rivin and Kendrick 1997). Sorbent materials such as activated carbon are preferred adsorbent materials. The currently used Joint Service Lightweight Integrated Suit Technology (JSLIST) chem-bio suit consists of beaded activated carbon glued to nonwoven fabric (Gomes et al. 2007). Although JSLIST has several limitations that the U.S. Department of Defense is working to overcome, it is one of the best available technologies for individual protection against chemical warfare agents (Gomes et al. 2007). Charcoal-based chemical protective suits are also used by law enforcement personnel. A typical chemical protective suit used by a first responder community is shown in Figure 8.1, and it consists of an inner nonparticulate carbon layer that adsorbs toxic chemicals. The outer layer consists of a polyester and cotton-blended woven fabric that has a flame-retardant finish (Remploy Frontline 2007).

Conventional particulate activated carbons that have heterogeneous porous structures are generally made from raw materials such as synthetic polymers, coconut shell fibers, peat, coal, nutshells, lignite, or sawdust (Chiang et al. 2001). Three common forms of activated carbon particles are commercially available: powdered, granular, and pelleted. The use of granular activated carbons (GACs) to remove pollutant vapors and toxic gases from contaminated sites is one of the least expensive countermeasure technologies. However, toxic cleanups by GACs are not as efficient as those achieved by fibrous forms of activated carbon due to their heterogeneous pore structure and large mesopores, which lead to inferior adsorption dynamics (Foster et al. 1992; Hayes and Akamatsu 2000). A common characteristic of the aforementioned activated carbons is that they shred into loose particles when used in a clothing ensemble, resulting in recontamination. However, activated carbon in well-integrated fabric structures such as woven or nonwoven fabrics does not shred or break down into loose particles when used (Gurudatt et al. 1997; Hayes and Akamatsu 2000). Due to nonshredding and nonlinting characteristics, activated carbon in well-interlocked fabric forms is most often used in protective garments (Gurudatt et al. 1997). The fibrous form of activated carbon can be processed to make an activated carbon woven fabric or nonwoven felt. Therefore, such fabrics are much easier to handle compared to GACs. Activated carbon fabrics (ACFs) made from activated

FIGURE 8.1   (See color insert following page 46.) Typical activated carbon–based CBRN suit. (Photo courtesy of Neal Hinkle, TTUHSC.)

carbon fibers have become increasingly popular among the military as well as in commercial sectors during the past decade. Due to their increased adsorption capability and ease of handling, ACFs are the preferred material for adsorption of toxic vapors (Foster et al. 1992; Hayes 2002).

High surface area activated carbon fibers were first prepared by direct carbonization and activation of phenolic fibers in steam/CO_2 environment at temperatures around 1000°C (Economy and Lin 1976). These activated carbon fibers, manufactured in the form of a fabric, have received increased attention as adsorbents in air treatment processes. Because these fabrics are easy to handle, there is an increasing demand for them in various applications such as protective fabrics, filtration devices, odor absorbents, and for a wide range of ancillary industrial applications. The high cost of these fabrics has limited their potential use for a number of applications. High cost is also an issue for their use in military applications (Mangun et al. 1999).

ACFs with surface areas in the range of 1500–2500 m^2/g have the advantage of higher adsorption capabilities over granulated activated carbons, which have surface areas in range of 800–1200 m^2/g. It is possible to achieve such high surface areas in fabric forms that offer these superior adsorption characteristics (Hayes and Akamatsu 2000; Martin-Gullón and Font 2001). Greater surface area and specific pore volume of ACFs as compared to GACs make them superior substrates for protection against gaseous pollutants (Davis et al. 2001).

Homogeneous structural distribution of micropores from the external surface of the fiber to the core is another highly advantageous attribute of activated carbon fibers. This structural configuration is not observed in GACs, which are heterogeneous in their porous structure and contain significant numbers of mesopores (Hayes and Akamatsu 2000). Activated carbon fibers exhibit higher adsorption capacities and faster adsorption rates due to their homogeneous microporous structure (Hayes and Akamatsu 2000; Hayes 2002). A higher or faster rate of adsorption facilitates quick removal of toxic chemicals from a contaminated environment. Furthermore, a higher adsorption capacity enables a protective material to retain a greater volume or mass of chemical agent compared to a material that has lower adsorption capacity (Sata and Ramkumar 2006).

In the case of activated carbon fibers, pore size is small compared to GACs. Moreover, the external diameter of activated carbon fibers is at least two orders of magnitude smaller than GACs. A mesh of well-interconnected micropores from the core to the external surface of an activated carbon fiber reduces the intraparticulate diffusion time. This results in the superior adsorption dynamics of activated carbon fibers as compared to GACs (Hayes and Akamatsu 2000).

ACFs are available in both woven and nonwoven textile forms. Unlike the woven fabrics, which are made by weaving yarns, nonwoven fabrics are neither woven nor knitted fabrics. Nonwoven fabrics are manufactured by assembling raw fibers into organized webs that are bonded by mechanical, chemical, and thermal means. Household wipes, wound dressings, and bandages are some examples of nonwovens that are used in daily life. It is important to understand the adsorption characteristics of different types of commercially available fabrics made from activated carbon fibers. Variations or alterations in such weaving patterns of fabrics result in significant differences in adsorption potentials (Davis et al. 2001). Nonwoven and woven activated carbon fabrics have significantly different adsorption capacities. For example, when the chemical methyl ethyl ketone was used to characterize adsorption behavior, activated carbon nonwoven felt had a higher adsorption capacity as well as higher initial adsorption efficiency as compared to woven activated carbon fabrics (Davis et al. 2001). Irrespective of the fabric patterns (woven or nonwoven), the maximum adsorption efficiencies in the case of activated carbon–based protective substrates are usually observed in the initial stage of adsorption, followed by a steady-state decrease in adsorption efficiency until the fabric's maximum adsorption capacity is reached (Foster et al. 1992; Davis et al. 2001; Sata and Ramkumar 2006). The time at which the decrease in efficiency takes place depends on many factors such as the type of adsorbent, nature of adsorbate, partial pressure of adsorbate, concentration of liquid adsorbate, molecular structure, and chemical nature of adsorbate (Foster et al. 1992; Davis et al. 2001; Hayes and Akamatsu 2000; Hayes 2002). The adsorption capabilities of ACFs are dependent on the extent of activation (surface area) and precursor materials from which they are made. Adsorption by ACFs is quantified using adsorption capacity, which is the maximum amount of challenge chemical adsorbed by a unit amount of ACF at its saturation point. Results from a recent adsorption study on three different commercially available ACFs varying in their precursor material and extent of activation (surface area) demonstrate that these two factors heavily influence adsorption capacity of ACFs (Sata and Ramkumar 2006). Table 8.1 provides details of ACFs used in the study, and Figure 8.2 depicts adsorption of 0.1% w/v paraoxon in butanol on the three ACFs using a thermogravimetric balance.

TABLE 8.1
Details of Commercially Available ACFs Used in the Adsorption Study of a Challenge Chemical

ACF Type	Weight* (g/m²)	Woven Structure	Specific Surface Area (m²/g)*	Precursor Material
ACN-W	134.33 (5.92)	Nonwoven	1366	Polypropylene
ACN-K	136.11 (10.65)	Nonwoven	1500	Phenolic resin
ACN-G	182.17 (4.92)	Nonwoven	1196	Viscose rayon

* Weight of ACN-W was based on three repeats, ACN-K was based on eight repeats, and ACN-G was based on five repeats. Values within parenthesis indicate standard deviation values.

Source: Sata, U.R. and Ramkumar, S.S., 2006, Chemical warfare simulant adsorption by activated carbon nonwovens for personal protection, Proceedings of the International Nonwovens Technical Conference, September 25–28, Houston, TX.

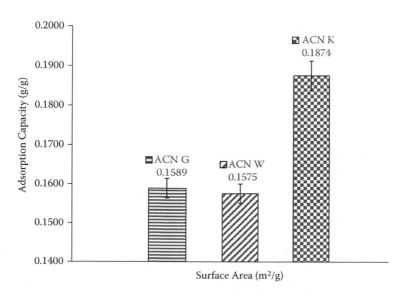

FIGURE 8.2   Adsorption capacity of ACFs with different surface areas determined using 0.1% w/v paraoxon in butanol as challenge chemical. (Sata, U.R. and Ramkumar, S.S., 2006, Chemical warfare simulant adsorption by activated carbon nonwovens for personal protection, Proceedings of the International Nonwovens Technical Conference, September 25–28, Houston, TX.)

Previous work has demonstrated the advantages of activated carbon fabrics in terms of their regeneration characteristics, cost effectiveness, and energy efficiency related to the regeneration process (Lordgooei et al. 1996). Activated carbon–based chemical warfare protective fabrics and filters are expensive, so it is important to regenerate them so that the ACFs can be reused without replacement, thereby reducing logistic burdens (Sullivan et al. 2006). In the case of activated carbon fibers, their electrical properties determine their regeneration and reuse. When electric current is passed through activated carbon fibers, the heat generated overcomes the bonding energy between adsorbed chemical molecules and the carbon fiber surface. This phenomenon, which is commonly referred to as electrothermal regeneration, is an effective method to regenerate sorbent materials for reuse (Lordgooei et al. 1996). The activated carbon fiber cloth can be electrothermally regenerated and the adsorbed chemicals can be rapidly desorbed and recovered by a simple process (Lordgooei et al. 1996). Since the surface temperature of ACFs can be controlled in this desorption process, electrothermal regeneration of ACFs is more efficient compared to conventional thermal regeneration techniques (Lordgooei et al. 1996).

In case of personal protective fabrics, the bulk of ACF-based garments often results in thermal discomfort and physiological burden to the wearer. Therefore, it is necessary to create less bulky ACFs for enhanced physical comfort in chemical protective suits (Wilusz 2007). In order to achieve maximum protection and reasonable comfort, novel fabric ensembles with or without ACFs are necessary (Wilusz 2007; Gurudatt et al. 1997). An alternative approach to the use of ACFs for chemical protection is the use of functionalized selectively permeable membranes (Wilusz 2007).

8.3 SELECTIVELY PERMEABLE MATERIALS FOR PROTECTIVE CLOTHING

At present, clothing that provides protection against skin exposure to toxic liquids, vapors, and aerosols uses either an activated carbon adsorbent in an air-permeable fabric ensemble, or an air-impermeable coated or laminated fabric ensemble. The former may allow aerosol particles to pass through and hence, cannot provide the highest level of protection. Although the latter can prevent aerosol penetration, they are uncomfortable and require cooling with prolonged use. One disadvantage of both types of fabrics is that they are bulky and heavy, which can be overcome with the use of selectively permeable membranes (SPMs) that are lightweight and can serve as efficient barriers against aerosolized agents and vapors. SPM composite fabrics allow the transport of moisture vapor from the skin to the outside environment, providing comfort to the wearer (Truong and Sarangapani 2002; Truong and Wilusz 2006). SPMs are used between outer shell fabrics and inner liner materials, replacing the core adsorbent activated carbon (Truong et al. 2006). A typical SPM-based chemical protective garment that provides protection against aerosols and vapor agents is shown in Figure 8.3. They offer a less logistic burden to war fighters as a result of their reduced weight and bulk (Truong and Wilusz 2006).

In a typical SPM fabric ensemble, gases flow into the membrane and desorb based on time, thickness of the SPM, and the concentration gradient of the chemical warfare agent according to Fick's law of diffusion (Ho and Sirkar 1992). Thus,

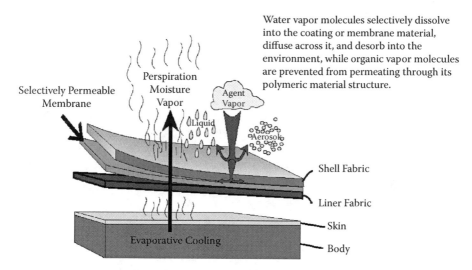

Water vapor molecules selectively dissolve into the coating or membrane material, diffuse across it, and desorb into the environment, while organic vapor molecules are prevented from permeating through its polymeric material structure.

FIGURE 8.3   Schematic diagram of a permselective composite fabric that allows the transport of inner moisture vapor to the outside environment but hinders the ingress of CW agent vapors from the outside environment to the body. (Truong, Q. and Wilusz, E., 2006, Materials technologies for next generation chemical and biological protective clothing. International Nonwovens Technical Conference, Conference Proceedings, September 25–28, Houston, TX.)

aerosolized liquids and solid particles are prevented from coming into contact with skin. However, perspiration is transported through the SPM to the external environment, providing necessary breathability. In addition, water repellent-coated shell fabrics can repel liquid agents to provide additional protection. As SPMs are nonporous, they also act as barrier materials against microorganisms (Truong and Wilusz 2006). U.S. Army Natick Soldier Research, Development and Engineering Center, located in Natick, Massachusetts, currently is conducting research to develop functionalized SPM-based protective garments that are biocidal and have self-detoxifying capabilities (Truong and Wilusz 2006). Agent-specific catalysts and antimicrobial finishing treatments can be applied to SPM protective garments to provide additional functionalities. The next-generation chemical threat agent protective clothing should be lightweight and have self-detoxifying capabilities with improved comfort (Wilusz 2007).

Ongoing SPM research focuses on properties such as weight reduction, enhanced useful life, water repellency, and impermeability to vapors, aerosols, and biotoxins (Reinartz et al. 2006). DuPont's Nafion® is a commercially available SPM. Nafion is a copolymer of tetrafluoroethylene and perfluoro [2-(fluorosulfonylethoxy)-vinyl] ether (James et al. 2000; Toshiki 1999). Although Nafion is a recognized SPM, its barrier properties for nerve agents are still under investigation (Rivin et al. 2001a). Rivin et al. (2001a) reported that Nafion in acid form is not a good barrier for the nerve agent simulant dimethyl methylphosphonate (DMMP). However, the barrier properties of Nafion can be markedly improved if there is a cation modification in the Nafion membrane (Rivin et al. 2001a; Schneider and Rivin 2004). Nafion's solubility characteristics have been the subject of interest in many studies, as it influences its barrier properties (Rivin et al. 2001b). Rivin et al. (2001b), used water and alcohols, and found that there is a marked increase in solubility of Nafion with a small amount of water in the alcohols. Rivin et al. (2001b) indicate that even at elevated temperatures, Nafion is not soluble in dry alcohols. However, in case of alcohol-water mixtures, Nafion solubilizes at temperatures above the boiling points of such mixtures. Their data suggest that water interacts more strongly with the sulfonic acid residues, whereas alcohol interacts with the fluoroether side chain, resulting in swelling of Nafion (Rivin et al. 2001b). Swelling alters the physical characteristics of Nafion and thus affects its barrier properties. Yeo and Cheng (1986) have also reported the swelling phenomenon of Nafion in hydrogen bonding solvents, providing evidence of two swelling maxima at widely separated solubility parameters (Yeo and Cheng 1986).

Wilusz et al. (2004) carried out an ion beam modification of SPMs for chemical biological protective clothing The ion beam treatment altered the diffusibility and permeability properties of SPMs (Wilusz et al. 2004). Elastomeric selectively permeable materials are also being developed at the U.S. Army Natick Soldier Center in collaboration with Innovative Chemical and Environmental Technologies Inc., located in Norwood, Massachusetts, and other industrial partners (Truong and Sarangapani 2002). Elastomeric selectively permeable materials are expected to provide acceptable barrier properties with increased comfort and reduced heat stress (Truong and Sarangapani 2002). SPM laminates are also finding use in chemical protective fabrics. Membrane Technology and Research, Inc. of Menlo Park, California, has developed a composite that has a selectively permeable material in its structure. This composite structure consists of SPMs, a sorbent layer, and a woven fabric for support and strength. A typical selectively permeable composite fabric is shown in Figure 8.4. In this structure, two identical sets of SPM composite fabric are used. The composite fabric structure consists of a sorptive layer of zeolite-loaded poly(vinyldiene fluoride) onto a woven support fabric (Wijmans and Gottschlich 1997). A thin selectively permeable polymer layer of polyvinyl alcohol is solution coated onto the sorbent layer forming the composite fabric structure. Two such composite fabrics are laminated at the selectively permeable membrane layer to form the protective fabric, which has a woven support layer on either side (Wijmans and

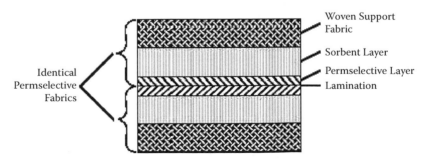

FIGURE 8.4 Design of MTR SPM protective composite fabric. (Wijmans, J.G. and Gottschlich, D.L., 1997, Protective clothing based on permselective membrane and carbon adsorption, Membrane Technology and Research, Inc.)

Gottschlich 1997). SPMs are currently generating more interest in the military due to improved barrier and comfort characteristics.

SPMs can now be found in commercial markets and specialty clothing due to their lightweight structure, liquid and aerosol repellent properties, and facilitation of moisture vapor transport. However, for military use, SPMs have limitations (Wilusz 2007). SPMs may act as liquid-repellents but may allow vapors to pass and therefore need an activated carbon layer to add extra protection capabilities. Moreover, military garments experience tremendous stress on a day-to-day basis. SPM-based ensembles are more susceptible to tearing as compared to activated carbon–based textile fabrics (Wilusz 2007). Optimizing the permselectivity of the membrane by surface modification or other such techniques is necessary to achieve a balance between comfort (e.g., moisture vapor transmission) and chemical vapor barrier properties. Furthermore, SPMs or membrane-carbon ensembles must possess acceptable mechanical strength to sustain daily military operations.

8.4 NANOFIBERS AND THEIR APPLICATIONS

Nanotechnology deals with processes and materials that are less than one billionth of a meter in at least one dimension (Subbiah et al. 2005). Nanofibers have attracted considerable attention from researchers recently due to their structural characteristics. High surface area, small pore sizes, and the possibility of producing three-dimensional structures have increased interest in nanofibers for potential applications in many different specialized areas. Nanofiber webs, due to their pore size and large surface area, may act as adsorbent materials in filters, protective masks, and clothing (Subbiah et al. 2005). The production of synthetic fibers using electrostatic forces has been in existence for more than 100 years (Subbiah et al. 2005). Electrospinning is a simple and widely used method of producing submicron-size nanofibers with diameters ranging from 100 to 500 nm (Doshi and Reneker 1995; Subbiah et al. 2005; Hussain and Ramkumar 2006). The electrospinning process uses a high-voltage electric field to produce electrically charged jets from a polymer solution or from melts, producing nanofibers upon drying. Highly charged fibers are field-directed toward an oppositely charged metal collector, which can be a flat surface or a rotating drum (Subbiah et al. 2005). In conventional spinning techniques, the fiber is subjected to a group of tensile, gravitational, aerodynamic, rheological, and inertial forces (Reneker and Chun 1996). In an electrospinning process, the spinning of fibers is achieved primarily by the tensile forces that are generated by the interaction of an induced electrical charge in a fine polymer jet and the applied electric field in the axial direction of the flow of the polymer. A typical electrospinning process is shown in Figure 8.5 and the experimental setup is shown in Figure 8.6.

A number of theoretical studies on electrospinning have been carried out to better understand initiation of the jet, and fiber deposition and collection (Reneker et al. 2000; Hohman et al. 2001a, 2001b; Shin et al. 2001a, 2001b; Yarin et al. 2001a, 2001b; Fridrikh et al. 2003). There have been efforts to characterize the structure and morphology of nanofibers as a function of process parameters and material characteristics (Subbiah et al. 2005). Production of nanofibers by electrospinning is influenced both by the electrostatic forces and the viscoelastic behavior of the

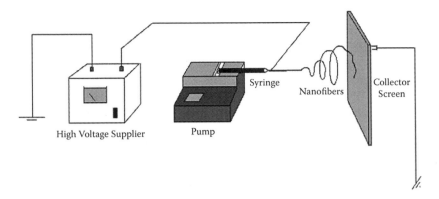

FIGURE 8.5 Schematic diagram of electrospinning process. (Mohammad, M. Hussain, PhD Dissertation, Texas Tech University. Hussain, M.M. and Ramkumar, S.S., 2006, Functionalized nanofibers for advanced applications, *Indian J. Fiber Text. Res.*, 31, 41–51.)

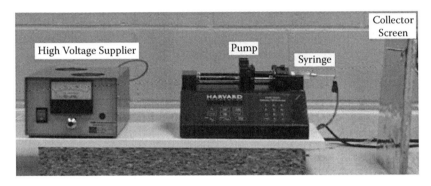

FIGURE 8.6 Experimental setup of electrospinning process.

polymer. Various theories have been proposed by different groups to explain the formation of jets and spinning of fibers (Taylor 1964; Subbiah et al. 2005; Yarin et al. 2001a, 2001b). The instability behavior of the spinning jet seems to be the reason for producing these high-surface-area, submicron-diameter fibers (Shin et al. 2001b; Subbiah et al. 2005). Although Hohman et al. (2001a), Shin et al. (2001b), and Spivak et al. (2000) have tried to model the electrohydrodynamics of the electroprocess, a generalized model for all the material systems is still elusive. This emphasizes the importance of process parameters like solution feed rate, applied voltage, nozzle-collector distance, spinning atmosphere, and material properties, including solution concentration, viscosity, surface tension, conductivity, solvent vapor pressure, and so forth, on the electrospinning process (Baumgarten 1971; Dietzel et al. 2001; Subbiah et al. 2005). A thorough review of the electrospinning process and applications of nanofibers have been documented in literature (Subbiah et al. 2005).

Although there is an abundant literature on the process-property relationships of electrospinning, limited information is available on nanofibers used as chemical protective liners. Scientists at the U.S. Army Research Development and Engineering

Command, Natick Soldier Research, Development and Engineering Center in Natick, Massachusetts, have conducted research on electrospinning of nanofibers for chemical protection (Schreuder-Gibson and Gibson 2002). Published research from the U.S. Army and other research on electrospinning have focused predominantly on the filtration and transport properties of electrospun mats (Schreuder-Gibson and Gibson 2002; Subbiah et al. 2005; Hussain and Ramkumar 2006). Gibson et al. (1999) studied the transport properties of electrospun fiber mats and concluded that nanofiber layers provide lower resistance to the moisture vapor transport than other high-performance filters. Tsai et al. (2002) developed nonwoven fibers of polyethylene oxide (PEO), polycarbonate (PC), and polyurethane (PU) by different fiber-charging methods like electrostatic spinning, corona charging, and tribocharging. They inferred that electrospun fibers have higher filtration efficiency than other nonwoven webs. PU and PC were found to have higher charge-retention capacities than electrospun PEO fibers (Tsai et al. 2002). Schreuder-Gibson and Gibson (2002) have analyzed the possibility of using thin nanofiber layers over conventionally used nonwoven filtration media for protective clothing. Polyurethane and nylon-6 nanowebs were applied over open cell foams and carbon beads and tested for airflow resistance. They concluded that the airflow resistance, filtration efficiency, and the pore sizes of nonwoven filter media could be easily altered by coatings with the lightweight electrospun nanofibers (Schreuder-Gibson and Gibson 2002). Table 8.2 provides descriptions of polymer-based electrospinning efforts and corresponding references.

8.5   FUNCTIONALIZED NANOFIBERS

Considerable amounts of research on electrospun nanofibers has focused on structure-property relationships, and filtration and transport properties (Gibson et al. 1999; Subbiah et al. 2005; Hussain and Ramkumar 2006). Only recently has attention shifted to functionalized nanofibers (Drew et al. 2003; Hussain et al. 2005). Functionalized nanofibers possess specific additives for incorporating special capabilities to nanofibers so that they can be used in a number of advanced applications, heretofore not possible (Hussain and Ramkumar 2006). These additives can include metaloxides, metal particles, enzymes, drugs, carbon nanotubes, and so on. Functionalized nanofibers find application in filters, sensors, chemical and biological protective clothing, tissue scaffolds, drug delivery, and so forth. Nanometaloxides are known to have high reactive capacities due to the presence of reactive groups and high surface areas (Rajagopalan et al. 2002). Nanocrystalline MgO has the capacity to destructively adsorb organophosphorous chemical (nerve) agents (Rajagopalan et al. 2002; Walker et al. 2002). The potential of high surface area nanowebs as chemical protective fabrics can be significantly enhanced with the reactive and adsorptive nano metaloxides. These composite webs with increased filtration and reactive efficiencies could effectively destroy chemical threat agents when used as filtration and protection media. Most recently, Drew et al. (2003) have produced polyacrylo nitrile (PAN) nanofibers using electrospinning methods and have applied thin coatings of TiO_2 and SnO_2 for catalytic activity. They immersed the electrospun webs in the coating, which then completely covered the fiber. Although they characterized the dimensions of the nanofibers before and after coating via transmission

TABLE 8.2
Electrospun Polymers and Corresponding Literature

S. No	Polymer	Solvent	Comments
1	Cellulose acetate	Acetone	Formhals attempted the electrospinning of cellulose acetate fibers in 1934 (Formhals 1934).
2	Acrylic resin (96% acrylonitrile)	DMF	Baumgarten spun acrylic fibers and studied the effect of polymer flow rate and viscosity on fiber diameter in 1971 (Baumgarten 1971). The fibers were less than 1 μm in diameter.
3	a) Polyethylene oxide b) Polyvinyl alcohol c) Cellulose acetate	a) Water/chloroform b) Water c) Acetone	Jaeger et al. studied the morphological characteristics of electrospun polymeric fibers in the diameter range of 200–800 nm (Jaeger et al. 1998).
4	a) Poly (2-hydroxy ethyl methacrylate) b) Polystyrene c) Poly (ether amide)	a) Formic acid and ethanol b) Dimethyl formamide and Diethyl formamide c) Hexa fluoro 2-propanol	Koombhongse et al. spun different fibers and reported flat ribbonlike structures and branched fibers (Koombhongse et al. 2001). The electrospun fibers were between 1 and 3 μm in diameter.
5	Polyethylene oxide (PEO)	Water	Doshi and Reneker have experimented with the spinning of PEO fibers from aqueous solutions and studied the relationship between process and solution parameters on fiber characteristics (Doshi and Reneker 1995). Electrospun fibers were about 0.05 to 5 μm in diameter
6	Polyethylene terephthalate	Mixture of dichloromethane and trifluoroacetic acid	Reneker and Chun demonstrated the spinning of polyethylene terephthalate fibers of 300 nm in diameter with cylindrical structures (Reneker and Chun 1996).
7	Polyaniline/PEO blends	Chloroform	Norris et al. produced fine fibers with desired conductivity by using Polyaniline/PEO polymeric blends (Norris et al. 2000). The fiber diameters were in the range of 950 nm to 2.1 μm.
8	Polyether urethane	Dimethyl acetamide	Wilkes observed the dependence of fiber diameter distribution and the occurrence of beaded structures on the flow rate and applied electric potential (Wilkis 2007). Fibers from 148 nm to 5 μm were obtained by Wilkes.

TABLE 8.2 (continued)
Electrospun Polymers and Corresponding Literature

S. No	Polymer	Solvent	Comments
9	Poly-L-lactide (PLLA), Polycarbonate (PC), Polyvinylcarbazole	Dichloromethane	Bognitzki et al. examined the relationship between the volatility of solvents used and the pore structure of fibers. They inferred that the solidification of fibers is controlled by onset of glass transition or by onset of crystallization (Bognitzki et al. 2001).
10	Polystyrene	Tetrahydrofuran (THF)	MacDiarmid et al. have electrospun polystyrene using THF as a solvent to produce nanofibers with a minimum diameter of 16 nm and an average diameter of 30.5 nm. They observed higher temperature in fibers due to higher conductivity (MacDiarmid et al. 2001).
11	Polybenzimidazole (PBI)	N,N-Dimethyl Acetamide (DMAc)	Kim and Reneker electrospun aromatic heterocyclic PBI polymer by electrospinning and produced birefringent fibers of approximately 300 nm in diameter (Kim and Reneker 1999).
12	Nylon 6 and Nylon 6 + montmorillonite (NLS)	1,1,1,3,3,3-hexa fluoro-2-Propanol (HFIP) and DMF	Fong et al. experimented the spinning of NLS with HFIP and DMF and observed cylindrical fibers along with some ribbon-shaped fibers with thickness of 100–200 nm and width of ~10 μm. They attributed that even though electrospinning should produce great chain alignment, rapid solvent removal inhibits perfect crystallites in the electrospun fibers (Fong et al. 2002, 42).
13	a) Polyethylene oxide (PEO) b) Polycarbonate (PC) c) Polyurethane (PU)	Isopropyl alcohol (IPA)DMF and THFDMF	Tsai et al. developed nonwoven fibers of PEO, PC, and PU by various fiber-charging methods like electrostatic spinning, corona charging, and tribocharging, and inferred that electrospun fibers have higher filtration efficiency than other nonwoven webs. PU and PC were found to have higher charge retention capacities than electrospun PEO fibers (Tsai et al. 2002). Fibers with diameter of 0.1 to 0.5 μm were produced.

(continued on next page)

TABLE 8.2 (continued)
Electrospun Polymers and Corresponding Literature

S. No	Polymer	Solvent	Comments
14.	Polyvinyl chloride	THF, DMF	Lee et al. studied the effect of volume ratio of mixed solvents on the structure and morphology of electrospun fibers (Lee et al. 2002).
15	Polyurethane	DMF	Demir et al. prepared polyurethane urea copolymer solution in DMF and observed that the average fiber diameter (AFD) increases with the solution concentration as given by AFD = (Conc.)3. A trimodal distribution of the fiber diameter has been observed with fiber diameter varying from 7 nm to 1.5 μm (Demir et al. 2002).
16	Polycaprolactone	Acetone	Reneker et al. studied the onset of the bending instability during spinning and observed the formation of a closed single- and double-loop fiber structure called "Garland." This garland structure has been observed in other copolymers like vinylidene fluoride, tetra fluoroethylene, and polyethyloazoline (Reneker et al. 2002). The fiber diameter varied from 1 μm to 1.5 μm.
17	Styrene-Butadiene-Styrene (SBS) triblock copolymer	75% THF and 25% DMF	Fong and Reneker examined the morphology of fibers with respect to microphase separation and experimented with annealing for accelerating the ordering process and stress relaxation (Fong and Reneker 1999). The electrospun fibers were around 100 nm in diameter.
18	Poly-L-lactide	Dichloromethane	Jun et al. electrospun PLA fibers and observed the cylindrical morphology of fibers with diameters ranging from 800–2400 nm (Jun et al. 2003).
19	Poly(methyl methacrylate-random) PMMA-r-TAN	Mixed solvent of toluene and DMF	Dietzel et al. produced electrospun fiber mats with specific surface chemistry from random copolymers of PMMA-r-TAN. They have demonstrated that the atomic concentration of fluorine at the surface of electrospun fibers was twice the amount seen in bulk materials (Dietzel et al. 2001). The fiber diameter was in the range of 2 μm to 300 nm.

TABLE 8.2 (continued)
Electrospun Polymers and Corresponding Literature

S. No	Polymer	Solvent	Comments
20	Polyethylene–co-vinyl acetate (PEVA), Poly lactic acid (PLA) and blend of PEVA and PLA	Chloroform	Kenawy et al. studied the potential of electrospun fiber mats as drug delivery system for the release of tetracycline hydrochloride (Kenawy et al. 2002). Electrospun PEVA + PLA blended fibers were 1–3 μm in diameter while the PLA fibers were around 3–6 μm.
21	Poly (p-phenylene terephthalamide) (PPTA) (Kevlar 49® from Dupont)	95–98 wt% sulfuric acid	Srinivasan and Reneker 1995 examined the crystal structure and morphology of the electrospun Kevlar fibers (Srinivasan and Reneker 1995). Fibers from 40 nm to a few hundreds of nanometers were produced.
22	Polyethylene terephthalate (PET), Polyethylene naphthalate (PEN)		Kim and Lee 2000 investigated the thermal properties of electrospun PET and PEN fibers made from melts (Kim and Lee 2000).
23	Silklike polymer with fibronectin functionality (SLPF)	Formic acid/ hexafluoro isopropanol	Buchko et al. used electrospinning technique to create biocompatible thin films for their use in implantable devices. They studied the morphological characteristics with regard to the changes in process and solution parameters (Buchko et al. 1999).

Source: Subbiah, T., Bhat, G.S., Tock, R.W., Parameswaran, S., and Ramkumar, S.S., 2005, Electrospinning of nanofibers, *J. Appl. Polym. Sci* 96(2), 557–569.

emission microscopy, they did not systematically measure pore size and distribution. The coating technique may affect pore size of the web, thus interfering with functionality that is possible because of large surface area. Coating metaloxides by dip immersion methods may not be a good technique and could affect the accessible surface area of nanofibers. If so, the very purpose of using electrospun nanofibers is lost (Drew et al. 2003).

Recent studies have indicated direct correlations between surface area and catalytic activity of metal oxide sensors (Li et al. 1999). Therefore, it is evident that incorporating catalytic particles by coating techniques may affect surface area and catalysis. Studies by Lee and Bhat (2003) have demonstrated that by incorporating small amounts of electrospun nanofibers in spun-bond and melt-blown nonwovens, barrier properties such as filtration efficiency and air permeability can be improved.

Other research has focused on incorporating nanoparticles like nanoclay into melt-spun polymeric fibers to enhance filtration capabilities (Bhat et al. 2006). Walker et al. (2002) have studied the incorporation of reactive materials into chemical protective fabrics for decontaminating organophosphorous chemical warfare agents such as G-agents and contact hazards such as VX and mustard gas. The G-agents are the very first category of nerve agents synthesized during the time of World War II, and VX is a highly lethal nerve agent that falls into another family of nerve agents, the V-agents, which are very persistent and cannot be washed away from surfaces very easily. Mustard gas falls under the family of sulfur mustard CW agents that can cause blisters on the skin. Nanometaloxides, such as MgO and Al_2O_3, and polyoxometallates (POMs) were incorporated into sorbent fabrics and their adsorption capabilities were evaluated using 2-chloroethyl ethylsulfide (CEES), a simulant for the CW agent mustard. Polyoxometallate incorporated microporous carbon had superior adsorption capabilities (Walker et al. 2002). In general, POM-incorporated materials provided better protection capabilities than those without POMs. The use of nanometaloxides is providing new opportunities for detoxifying toxins. As the size of metaloxides reaches the nano level, surface reactivity increases due to high concentrations of reactive edges and defect sites. High surface reactivity coupled with high surface area enhances their utility for effective detoxification of chemical threat agents (Hussain and Ramkumar 2006). Hussain et al. (2005) have conducted a study on the dispersion of nano magnesium oxides (MgO) in polyethylene oxide nanofiber webs to impart self-detoxifying capabilities to nanofibers. Results from the study (shown in Figures 8.7 and 8.8) clearly demonstrate that metaloxides can be embedded on the surface of nanofiber webs by optimizing the electrospinning process parameters such as voltage, viscosity, distance between the spinning nozzle and the collector. Figure 8.9 depicts a transmission electron microscope (TEM) image of PEO nanofibers embedded with nano MgO.

The electrospinning process was capable of incorporating the nanoparticles on the surface of the nanowebs. This is clearly evident from the regular pattern of the nodes in the transmission emission micrograph (Figure 8.8) and from the individual MgO nanoparticle embedded on the surface of a PEO nanofiber (Figure 8.9).

The mechanical properties of functionalized nanofibers can be enhanced with the addition of single-walled and multiwalled carbon nanotubes. The electrospinning process has been used successfully to embed multiwalled carbon nanotubes in PEO nanofibers (Dror et al. 2003). Composite nanofibers of polyacrylonitrile nanofibers and multiwalled carbon nanotubes have been developed by Hou et al. (2005). Multiwalled carbon nanotubes were aligned along the axes of polyacrylonitrile nanofibers, and the results of Hou et al. (2005) showed that the tensile modulus of carbon-nanotube-reinforced polyacrylonitrile fibers was enhanced by 144% and the tensile strength was enhanced by 75% (Hou et al. 2005). Ko et al. (2003) showed that the electrospinning process can be used to incorporate single-walled carbon nanotubes into nanofiber webs. The incorporation of single-walled carbon nanotubes enhanced the mechanical properties of the nanofiber web (Ko et al. 2003). Addition of materials such as carbon nanotubes into electrospun webs provides necessary space in the base structure for enhanced transport of moisture vapor. Such spacers help to increase the strength and comfort properties of chemical protective nanofibers.

FIGURE 8.7   Transmission emission microscope (TEM) image of a single polyethylene oxide nanofiber embedded with nano MgO at regular intervals. (Hussain, M.M. and Ramkumar, S.S., 2006, Functionalized nanofibers for advanced applications, *Indian J. Fiber Text. Res.*, 31, 41–51.)

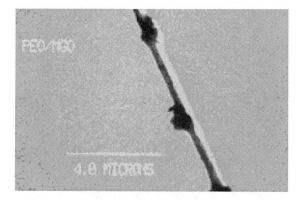

FIGURE 8.8   High-resolution TEM image of a single MgO nanoparticle embedded on the surface of a PEO nanofiber. (Hussain, M.M. and Ramkumar, S.S., 2006, Functionalized nanofibers for advanced applications, *Indian J. Fiber Text. Res.*, 31, 41–51.)

FIGURE 8.9   TEM image of PEO-MgO nanofiber showing a meshlike structure. (Mohammad, M. Hussain, PhD Dissertation, Texas Tech University.)

8.6   PERSONNEL DECONTAMINATION MATERIALS

In addition to protecting war fighters from exposure to chemical warfare agents by using protective garments, it is also critical to have materials capable of decontaminating human skin and equipment that may come into contact with skin following exposure to threat agents. Decontamination in this context essentially means removal and neutralization of a toxic agent. The most important aspect of a decontaminant is to quickly and efficiently remove toxins from the skin (Lukey et al. 2004).

Decontamination can be achieved using particulate, nonparticulate, aqueous, and nonaqueous systems. However, for personnel decontamination, it is preferable to use nonparticulate decontamination systems, as they are most convenient for end users (DoD 2004, 2005). Efficacy of nonparticulate decontamination systems may be augmented through add-on usage of aqueous systems such as Reactive Skin Decontamination Lotion (DoD 2007).

Decontamination requirements of war fighters include simplicity and speed. Therefore, decontamination processes should not be complicated and should require minimal training for their use. Any individual should be capable of using decontamination systems with relative ease following suspected exposure. The most basic method of decontamination is to use soapy water immediately after exposure (Lukey et al. 2004). However, accessibility of water stations for immediate decontamination is often not feasible for war fighters. Therefore, alternate decontamination systems with fewer logistical requirements are needed. The ideal decontamination system should be lightweight, easy to handle, portable, and should not irritate or cause allergic reactions in skin, eyes, or open wounds (DoD 2007). Such a product must not have a repulsive odor and should be safe for humans and the environment (Lukey et al. 2004). More importantly, after decontamination, such decontamination devices should not discharge vapors of the chemical agent or toxic chemicals. It should strongly retain the chemical agent under harsh conditions and render the chemical agent harmless or inert. Apart from the military environment, exposure to toxic volatile organic compounds and toxic industrial chemicals among the civilian population is a matter of concern (Lordgooei et al. 1996; Foster et al. 1992). Therefore, it is important that the next generation of decontamination systems should be highly efficient, nontoxic, and easy to use, and be cost effective so as to ensure widespread use and effectiveness.

Lukey et al. (2004) have reviewed six skin decontaminants, including: (1) hypochlorite bleach, (2) M291 decontamination kit, (3) Reactive Skin Decontamination Lotion (RSDL), (4) Sandia foam, (5) diphoterine, and (6) reactive sponge. Bleach consists of 0.5% sodium or calcium hypochlorite solution. A major disadvantage of the bleach solution is its dermal toxicity in the concentrated form (5% hypochlorite in water). However, very dilute forms of bleach (< 0.5%) are not adequately affective against chemical threat agents (Lukey et al. 2004). Furthermore, chlorine-based decontamination products are not desirable as they may corrode equipment used by war fighters (DoD 2007). The M291 decontamination kit is the current fielded personnel decontamination system used by the U.S. military. Each M291 kit is comprised of six individual decontamination pads containing a dry particulate mixture with adsorbent resin, a sulfonic acid resin, and hydroxylamine containing resin. (Lukey et al. 2004; USATDC 2006; DoD 2007). The resin can adsorb and neutralize toxic chemical threat agents (DoD 2007). The M291 kit is a requirement for all four services of the U.S. military. The disadvantage of the M291 skin decontamination kit is its particulate nature. Upon wiping, the M291 kit leaves loose particles on the skin that can cause skin irritation (Lukey et al. 2004). It is important that the M291 particulate mixture not come into contact with open wounds, eyes, mouth, and other open parts of the body (Lukey et al. 2004; USATDC 2006). Figure 8.10 depicts an opened pad of M291 decontamination kit.

Unlike the M291 decontamination kit that consists of reactive particulate resins in an applicator pad, ACF-based absorbent-adsorbent wipes are nonparticulate in nature and can be used as a dry wipe (Ramkumar 2004; Ramkumar 2005; Sata and Ramkumar 2007). This wipe has absorbent layers on top and bottom with the ACF in

FIGURE 8.10   (See color insert following page 46.) M291 applicator pad with particulate resin mixture. (Utkarsh R. Sata, PhD Dissertation, Texas Tech University.)

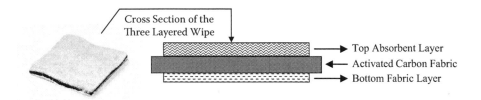

FIGURE 8.11   Three-layered dry decontamination wipe.

the core. Figure 8.11 shows the three-layered nonwoven dry decontamination wipe. The ACF adsorbs toxic vapors and helps with decontamination (Sata and Ramkumar 2006, 2007). Moreover, unlike the M291 kit, such nonparticulate materials do not leave particulate matter upon wiping a contaminated surface (Sata and Ramkumar 2007).

Reactive Skin Decontamination Lotion (RSDL) has recently been approved for procurement by the U.S. military (E-Z-EM 2007). RSDL is currently marketed by E-Z-EM, Inc. and is a patented, broad spectrum, skin decontamination lotion that is used to remove or neutralize chemical threat agents and biological warfare agents such as trichothecene mycotoxin (T2 toxin), which can cause severe skin and eye irritation. RSDL was originally developed by the Canadian Defense Research Establishment and consists of 1.25 molar potassium 2,3-butanedione monoximate in polyetheylene glycol monoethyl ethers with 10% w/v water (Sabourin et al. 2001; Lukey et al. 2004).

Sandia National Laboratory has developed two decontamination foams: (1) MDF-100 and (2) DF-200 (Lukey et al. 2004). MDF-100 consists of two components that, when mixed, generate a foam that is used for decontamination. It has been reported that MDF-100 has the ability to neutralize anthrax spores. DF-200 is a faster acting foam than MDF-100 (Lukey et al. 2004).

The U.S. Army Medical Research Institute of Chemical Defense has developed an enzyme-immobilized polyurethane sponge that can absorb chemical threat agents that can then be destroyed by enzymatic activity. These polyurethane sponges have been found to have excellent adsorption properties related to organophosphorous vapors (Lukey et al. 2004).

8.7   CONCLUSIONS

This chapter has provided a brief overview on current fiber-based technologies used for individual protection and decontamination. It is important to develop new lightweight protective materials that offer protection against chemical vapor ingress while maintaining comfort. Compared to conventional carbonaceous materials such as GACs, ACFs have several advantages, including their superior adsorption capabilities and ease of handling. Even though sorbent-fabric-based protective ensembles are heavy and bulky, it is necessary to incorporate a thin layer of ACFs in the next-generation protective garments in order to provide maximum protection. Next-generation protective garments will include SPMs and ACFs to provide the necessary protection and comfort. Furthermore, reactive particles will be incorporated in the outer shell fabric. Thus, protective fabrics will be lightweight and have self-cleaning and adsorbent capabilities. Incorporation of silver and other biocidal materials will provide biological agent countermeasure capabilities as well. The development of protective suits with both chemical and biological agent countermeasure characteristics is an immediate research and development task for the military. Similarly, the development of nonparticulate, nonaqueous reactive decontamination systems is also a priority research and development endeavor for the military.

REFERENCES

Baumgarten, P.K., 1971. Electrostatic spinning of acrylic microfibres, *J. Coll. Interface Sci.*, 36, pp. 71–79.

Bhat, G., Hegde, R., Kamath, M.G., and Deshpande, B., 2006. Nanoclay reinforced fibers and nonwovens, Proceedings of the annual conference and exhibition of the American Association of Textile Chemists and Colorists (AATCC), October 31–November 2, Atlanta, GA.

Bognitzki, M., Czado, W., Frese, T., Schaper, A., Hellwig, M., Steinhart, M., Greiner, A., and Wendorff, H., 2001. Nanostructured fibers via electrospinning, *Adv. Mater.*, 13, p. 1.

Buchko, C.J., Chen, L.C., Shen, Y., and Martin, D.C., 1999. Processing and microstructural characterization of porous biocompatible protein polymer thin films, *Polymer*, 40, pp. 7397–7407.

Chiang, Y.-C., Chiang, P.-C., and Chang, E.E., 2001. Effects of surface characteristics of activated carbons on VOC adsorption, *J. Environ. Eng.*, 127(1), pp. 54–62.

Davis, W.T., Kim, G.D., and Perry, T.C., 2001. Study of the adsorption/removal efficiency of woven and nonwoven activated carbon fabrics for MEK, *Sep. Sci. Technol.*, 36(5&6), pp. 931–940.

Demir, M.M., Yilgor, I., Yilgor, E., and Erman, B., 2002. Electrospinning of polyurethane fibers, *Polymer*, 43, pp. 3303–3309.

Dietzel, J.M., Kleinmeyer, J.D., Harris, D., and Beck-Tan, N.C., 2001. The effect of processing variables on the morphology of electrospun nanofibers and textiles, *Polymer*, 42 (1), pp. 261–272.

Doshi, J. and Reneker, D.H., 1995. Electrospinning process and application of electrospun fibers, *J. Electrostatics*, 35, pp. 151–160.

Drew, C., Liu, X., Ziegler, D., Wang, X., Bruno, F.F., Whitten, J., Samuelson, A., and Kumar, J., 2003. Metal oxide-coated polymer nanofibers, *Nano Lett.*, 3(3), pp. 143–147.

Dror, Y., Salalha, W., Khalfin, R.L., Cohen, Y., Yarin, A.L., and Zussman, E., 2003. Carbon nanotubes embedded in oriented polymer nanofibers by electrospinning, *Langmuir*, 19, pp. 7012–7020.

Economy, J. and Lin, R.Y., 1976. Adsorption characteristics of activated carbon fibers, *Appl. Polym. Symp. (New Spec. Fibers)*, 29, pp. 199–211.

E-Z-EM News Release, 2007. E-Z-EM'S RSDL Product approved for procurement by U.S. Defense Department, March 20, available at http://www.ezem.com/news/news.cfm?article_id=76 (accessed September 17, 2007).

Fong, H., Liu, W., Wang, C., and Vaia, R.A., 2002. Generation of electrospun fibers of nylon 6 and nylon 6-montmorillonite nanocomposite, *Polymer*, 43, pp. 775–780.

Fong, H. and Reneker, D.H., 1999. Elastomeric nanofibers of styrene-butadiene-styrene triblock copolymer, *J. Polym. Sci.: Part B: Polym. Phys.*, 37, pp. 3488–3493.

Formhals, A., 1934. Process and apparatus for preparing artificial threads, U.S. Patent 1975504, available at http://patft.uspto.gov (accessed September 17, 2007).

Foster, K.L., Fuerman, R.G., Economy, J., Larson, S.M., and Rood, M.J., 1992. Adsorption characteristics of volatile organic compounds onto activated carbon fibers, *Chem. Mater.*, 4(5), pp. 1068–1073.

Fridrikh, S.V., Yu, J.H., Brenner, M.P., and Rutledge, G.C., 2003. Controlling the fiber diameter during electrospinning, *Phys. Rev. Lett.*, 14(90), pp. 1–4.

Geohegan, D.B., Schittenhelm, H., Fan, X., Pennycook, S.J., Puretzky, A.A., Guillorn, M.A., Blom, D.A., and Joy, D.C., 2001. Condensed phase growth of single-wall carbon nanotubes from laser annealed particulates, *Appl. Phys. Lett.*, 78, pp. 3307–3309.

Gibson, P.W., Schreuder-Gibson, H.L., and Rivin, D., 1999. Electrospun fiber mats: transport properties, *Am. Inst. Chem. Eng. J.*, 45(1), pp. 190–195.

Gomes, C.A., Lee, Y., Puglia, J.P., Brown, P.J., and Marcus, R.K., 2007. Nano grooved fibers used in self-detoxifying fabrics, AATCC Symposium, Innovations in Nanotechnologies, Composites and Sports/Military Materials, April 11–12, Atlanta, GA.

Gurudatt, K., Tripathi, V.S., and Sen, A.K., 1997. Adsorbent carbon fabrics: new generation armor for toxic chemicals, *Defense Sci. J.*, 47(2), pp. 239–250.

Hayes, J.S., Jr., 2002. Nanostructure of activated carbon fibers and kinetics of adsorption/desorption, *Proceedings of the Air and Waste Management Associations*, 95th annual conference and exhibition, June 23–27, Baltimore, MD, pp. 689–710.

Hayes, J.S., Jr., and Akamatsu, M., 2000. Activated carbon fiber solvent recovery systems *Proceedings of the Air and Waste Management Associations*, 93rd annual conference and exhibition, June 18–22, Salt Lake City, UT, pp. 2447–2466.

Ho, W.S.W., and Sirkar, K.K. (Eds.). 1992. *Membrane Handbook*. Van Nostrand Reinhold: New York, NY.

Hohman, M.M., Shin, M., Rutledge, G.C., and Brenner, M.P., 2001a, Electrospinning and electrically forced jets: I, stability theory, *Phys. Fluids* 8(13), pp. 2201–2220.

Hohman, M.M., Shin, Y.M., Rutledge, G.C., and Brenner M.P., 2001b,. Electrospinning and electrically forced jets: II, applications, *Phys. Fluids* 8(13), pp. 2221–2236.

Hou, H.Q., Ge, J.J., and Zeng, J., 2005. Electrospun polyacrylonitrile nanofibers containing a high concentration of well-aligned multiwall carbon nanotubes, *Chem. Mater.*, 17(5), pp. 967–973.

Hussain, M.M. and Ramkumar, S.S., 2006. Functionalized nanofibers for advanced applications, *Indian J. Fiber Text. Res.*, 31, pp. 41–51.

Hussain, M.M., Subbiah, T., Guven, N., and Ramkumar, S.S., 2005. Composite nanofibers for advanced applications, Proceedings of the International Nonwoven Technical Conference, September 19–22, Saint Louis, MO.

Jaeger, R., Bergshoef, M.M., Batlle, C.M.I., Schonherr, and Vancso, G.J., 1998. Electrospinning of ultra-thin polymer fibers, *Macromol. Symp.*, 127, pp. 141–150.

James, P.J., Elliott, J.A., McMaster, T.J., Newton, J.M., Elliott, A.M.S., Hanna, S., and Miles, M.J., 2000. Hydration of Nafion studied by AFM and X-ray scattering, *J. Mater. Sci.*, 35(20), pp. 5111–5119.

Jun, Z., Hou, H., Schaper, A., Wendorff, J.H., and Greiner, A., 2003. Poly-L-lactide nanofibers by electrospinning: influence of solution viscosity and electrical conductivity on fiber diameter and fiber morphology, *e-Polymers*, paper 9. Available at http://www.e-polymers.org/journal/abstract.cfm.abstractId=401 (accessed September 17, 2007).

Kenawy, E.-R., Bowlin, G.L., Mansfield, K., Layman, J., Simpson, D.G., Sanders, E.H., and Wnek, G.E., 2002. Release of tetracycline hydrochloride from electrospun poly(ethylene-co-vinylacetate), poly(lactic acid), and a blend, *J. Controlled Release*, 81, pp. 57–64.

Kim, J., and Lee, D.S., 2000. Thermal properties of electrospun polyesters, *Polym. J.*, 32(7), pp. 616–618.

Kim, J.S. and Reneker, D.H., 1999. Polybenzimidazole nanofiber produced by electrospinning, *Polym. Eng. Sci.*, 39(5), pp. 849–854.

Ko, F., Gogotsi, Y., Ali, A., Naguib, N., Ye, H., Yang, G., Li, C., and Willis, P., 2003. Electrospinning of continuous carbon nanotube-filled nanofiber yarns, *Adv. Mater.*, 15(14), pp. 1161–1165.

Koombhongse, S., Liu, W., and Reneker, D.H., 2001. Flat polymer ribbons and other shapes by electrospinning, *J. Polym. Sci.: Part B: Polym. Phys.*, 39, pp. 2598–2606.

Lee, K.H., Kim, H.Y., La, Y.M., Lee, D.R., and Sung, N.H., 2002. Influence of mixed solvent with THF and DMF on electrospun poly(vinyl chloride) nonwoven mats, *J. Polym. Sci.: Part B: Polym. Phys.*, 40, pp. 2259–2268.

Lee, Y. and Bhat, G., 2003. Recent advances in electrospun nanofibers, Proceedings of the 12th Symposium on Processing and Fabrication of Advanced Materials, October 13–15, Pittsburgh, PA.

Li, G.J., Zhang, X.H., and Kawi, S., 1999. Relationships between sensitivity, catalytic activity and surface areas of SnO_2 gas sensors, *Sensors Actuators B.*, 60, pp. 64–70.

Lordgooei, M., Charmichael, K.R., Kelly T.W., Rood M.J., and Larson, S.M., 1996. Activated carbon cloth adsorption-cryogenic system to recover toxic volatile organic compounds, *Gas Sep. Purif.*, 10, pp. 2123–2130.

Lukey, B.J., Hurst, G.C., Gordon, R.K., Doctor, B.P., Clarkson, E., IV., and Slife, H.F., 2004. Six current or potential skin decontaminants for chemical warfare agent exposure: a literature review, in *Pharmacological Perspectives of Toxic Chemicals and Their Antidotes*, Flora, S.J.S., Romano, J.A., Baskin, S.I., and Sekhar, K., Eds., Springer-Verlag, Berlin, pp. 13–24.

MacDiarmid, A.G., Jones, W.E., Jr., Norris, I.D., Gao, J., Johnson, A.T., Jr., Pinto, N.J., Hone, J., Han, B., Ko, F.K., Okuzaki, H., and Llaguno, M., 2001. Electrostatically- generated nanofibers of electronic polymers, *Synth. Met.*, 119, pp. 27–30.

Mangun, C.L., Braatz, R.D., Economy, J., and Hall, A.J., 1999. Fixed bed adsorption of acetone and ammonia onto oxidized activated carbon fibers, *Ind. Eng. Chem. Res.*, 38(9), pp. 3499–3504.

Martin-Gullón, I. and Font, R., 2001. Dynamic pesticide removal with activated carbon fibers, *Water Res.*, 35(2), pp. 516–520.

Norris, I.D., Shaker, M.M., Ko, F.K., and MacDiarmid, A.G., 2000. Electrostatic fabrication of ultrafine conducting fibers: polyaniline/polyethylene oxide blends, *Synth. Met.*, 114, pp. 109–114.

Rajagopalan, S., Koper, O., Decker, S., and Klabunde, K.J., 2002. Nanocrystalline metal oxides as destructive adsorbents for organophosphorus compounds at ambient temperatures, *Chem. Eur. J.*, 8, pp. 2602–2607.

Ramkumar, S.S., 2004. Chemical protective composite substrate and method of producing same, U.S. Patent application 10/874793, June 23, 2004.

Ramkumar, S.S., Process for making chemical protective wipes and such wipes, U.S. Patent application 11/157124, June 20, 2005.

Reinartz, N.M., Wren, M., Kahn, R., and Howard, E., Jr., 2006. Selectively permeable membranes for chemical and biological protective clothing, Scientific Conference on Chemical and Biological Defense Research, November 13–15, Hunt Valley, MD.

Remploy Frontline, 2007. Protective suit catalogue, available at http://www.remployfrontline.com (accessed September 17, 2007).

Reneker, D.H. and Chun, I. 1996. Nanometer diameter fibers of polymers produced by electrospinning, *Nanotechnology*, 7, pp. 216–223.

Reneker, D.H., Kataphinan, W., Theron, A., Zussman, E., and Yarin, A.L., 2002. Nanofiber garlands of polycaprolactone by electrospinning, *Polymer*, 43, pp. 6785–6794.

Reneker, D.H., Yarin, A.L., Fong, H., and Koombhongse, S., 2000. Bending instability of electrically charged liquid jets of polymer solutions in electrospinning, *J. Appl. Phys.*, 9(87), pp. 4531–4547.

Rivin, D. and Kendrick, C.E., 1997. Adsorption properties of vapor-protective fabrics containing activated carbon, *Carbon*, 35(9), pp. 1295–1305.

Rivin, D., Kendrick, C., Gibson, P., Schneider, N.S., 2001a, Permselective materials for protective clothing, Proceedings of the seventh international symposium on Protection against Chemical and Biological Warfare Agents, Stockholm, Sweden, June.

Rivin, D., Kendrick, C.E., Gibson, P.W., and Schneider, N.S., 2001b, Solubility and transport behavior of water and alcohols in Nafion™, *Polymer*, 42(2), pp. 623–635.

Rivin, D., Meermeier, G., and Schneider, N.S., 2002. Simultaneous transport of water and organic molecules through a proton-conducting membrane, 23rd Army Science Conference, December 2–5, Orlando, FL.

Sabourin, C.L., Hayes T.L., and Snider T.H., 2001. A medical research and evaluation facility study on Canadian reactive skin decontamination lotion, Final report, USAMRICD, Edgewood Area, Aberdeen Proving Ground, MD.

Sata, U.R. and Ramkumar, S.S., 2006. Chemical warfare simulant adsorption by activated carbon nonwovens for personal protection, Proceedings of the International Nonwovens Technical Conference, September 25–28, Houston, TX.

Sata, U.R. and Ramkumar, S.S., 2007. New developments with nonwoven decontamination wipes, Paper presented at the International Nonwovens Technical Conference, September 24–27, Atlanta, GA.

Schneider, N.S. and Rivin, D., 2004. Interaction of dimethyl methylphosphonate with Nafion in acid and cation modifications, *Polymer*, 45, pp. 6309–6320.

Schreuder-Gibson, H. and Gibson, P., 2002. Use of electrospun nanofibers for aerosol filtration in textile structures, Proceedings of the 23rd Army Science Conference, December 2–5, Orlando, FL.

Schreuder-Gibson, H., Gibson, P., Senecal, K., Sennett, M., Walker, J., Yeomans, W., Ziegler, D., and Tsai, P.P., 2002. Protective textile materials based on electrospun nanofibers, *J. Adv. Mater.*, 34(3), pp. 44–55.

Shin, Y.M., Hohman, M.M., Brenner, M.P., and Rutledge, G.C., 2001a, Electrospinning: a whipping fluid jet generates submicron polymer fibers, *Appl. Phys. Lett.*, 8(78), pp. 1149–1151.

Shin, Y.M., Hohman, M.M, Brenner, M.P, and Rutledge, G.C., 2001b, Experimental characterization of electrospinning: the electrically forced jet and instabilities, *Polymer*, 42, pp. 9955–9967.

Spivak, A.F., Dzenis, Y.A., and Reneker, D.H., 2000. A model of steady state jets in the electrospinning process, *Mech. Res. Commn.*, 1(27), pp. 37–42.

Srinivasan, G. and Reneker, D.H., 1995. Structure and morphology of small diameter electrospun aramid fibers, *Polym. Int.*, 36, pp. 195–201.

Subbiah, T., Bhat, G.S., Tock, R.W., Parameswaran, S., and Ramkumar, S.S., 2005. Electrospinning of nanofibers, *J. Appl. Polym. Sci* 96(2), pp. 557–569.

Subbiah, T., and Ramkumar, S.S., 2004a. Composite nanofibers substrates for high efficiency aerosol filtration, *Proceedings of the 14th Annual TANDEC Conference*, University of Tennessee, Knoxville, TN, November 9–11, pp. 4.2.1–4.2.11.

Subbiah, T. and Ramkumar, S.S., 2004b, Polymeric nanofibers by electrospinning, *Proceedings of International Conference of High Performance Textiles and Apparels*, Coimbatore, India, pp. 81–90.

Sullivan, P.D., Wander, J.D., and Newsome, K.C., 2006. Electrothermal desorption of cwa simulants from activated carbon cloth, Defense Technical Information Center, http://stinet.dtic.mil/oai/oai?&verb=getRecord&metadataPrefix=html&identifier=AD A459754 (accessed September 15, 2007).

Taylor, G.I., 1964. Disintegration of water droplets in an electric field, *Proc. R. Soc., London Series A*, (280), p. 383.

Toshiki, A., 1999. Macromolecule design of permselective membranes, *Prog. Polym. Sci.*, 24, pp. 951–993.

Truong, Q. and Sarangapani, S., 2002. Development of elastomeric selectively permeable membranes for chemical/biological protective clothing, 23rd Army Science Conference, December 2–5, Orlando, FL.

Truong, Q. and Wilusz, E., 2006. Materials technologies for next generation chemical and biological protective clothing. International Nonwovens Technical Conference, Conference Proceedings, September 25–28, Houston, TX.

Tsai, P.P., Schreuder-Gibson, H., and Gibson, P., 2002. Different electrostatic methods for making electret filters, *J. Electrostatics.* 54(3–4), pp. 333–341.

U.S. Army Training and Doctrine Command (USATDC), Fort Monroe, VA, 2006. Public information, FM 3–11.5, available at https://atiam.train.army.mil/soldierPortal/atia/adlsc/view/public/22662–1/FM/3–11.5/toc.htm#toc (accessed September 17, 2007).

U.S. Department of Defense (DoD), 2000. Chemical and Biological Defense Program, Annual report to Congress, March 2000. available at http://www.acq.osd.mil/cp/cbdreports/nbc00.pdf (accessed September 15, 2007).

U.S. Department of Defense (DoD), 2001. Proliferation: threat and response, Report from the Office of the Secretary of Defense, January 2001. available at http://www.dod.mil/pubs/ptr20010110.pdf (accessed September 17, 2007).

U.S. Department of Defense (DoD), 2004. Chemical and Biological Defense Program, Annual report to Congress, May 2006. available at http://www.acq.osd.mil/cp/nbc04/2004cbrndpreport.pdf (accessed September 17, 2007).

U.S. Department of Defense (DoD), 2005. Chemical and Biological Defense Program, Annual report to Congress, March 2005. available at http://www.acq.osd.mil/cp/nbc05/cbdpreporttocongress2005.pdf (accessed September 17, 2007).

U.S. Department of Defense (DoD), 2006. Chemical and Biological Defense Program, Annual report to Congress, March 2006. available at http://www.acq.osd.mil/cp/nbc06/cbdpreporttocongress2006.pdf (accessed September 17, 2007.)

U.S. Department of Defense (DoD), 2007. Chemical and Biological Defense Program, Annual report to Congress, April 2007. available at http://www.acq.osd.mil/cp/cbdreports/cbdpreporttocongress2007.pdf (accessed September 17, 2007).

Walker, J., Schreuder-Gibson, H., Yeomans, W., Ball, D., and Hoskin, F., 2002. Development of self-detoxifying materials for chemical protective clothing, Proceedings of the Joint Service Scientific Conference on Chemical and Biological Defense Research, November 19–21, Hunt Valley, MD.

Wijmans, J.G. and Gottschlich, D.L., 1997. Protective clothing based on permselective membrane and carbon adsorption, Membrane Technology and Research, available at http://www.netl.doe.gov/publications/proceedings/97/97em/em_pdf/EMPII-5.pdf (accessed September 17, 2007).

Wilkis, G., 2007. Electrospinning, Virginia Tech Department of Chemical Engineering, available at http://www.che.vt.edu/Wilkes/electrospinning/electrspinning.html (accessed September 17, 2007).

Wilusz, E., 2007. Breaking with tradition, CBRNe World Spring 2007. available at http://www.cbrneworld.com/articles_spring_2007.html (accessed September 14, 2007).

Wilusz, E., Zukas, W.X., Fekete, Z.A., and Karasz Frank, E., 2004. Ion beam modification of permselective membranes for chemical biological protective clothing, *Polym. Preprints*, 45(1), pp. 7–8, American Chemical Society, Division of Polymer Chemistry.

Yarin, A.L., Koombhongse, S., and Reneker, D.H., 2001a, Bending instability in electrospinning of nanofibers. *J. Appl. Phys.*, 5(89), pp. 3018–3026.

Yarin, A.L., Koombhongse, S., and Reneker, D.H., 2001b, Taylor cone and jetting from liquid droplets in electrospinning of nanofibers, *J. Appl. Phys.*, 9(90), pp. 4836–4845.

Yeo, R.S. and Cheng, C.H., 1986. Swelling studies of perfluorinated ionomer membranes, *J. Appl. Polym. Sci.* 32(7), pp. 5733–5741.

9 Pathogenic and Toxic Effects of Select Biological Threat Agents

Jia-Sheng Wang

CONTENTS

9.1 INTRODUCTION

Biological threat agents are disease-causing or toxic materials that originate or are derived from naturally occurring or engineered living organisms, including bacteria, plants, rickettsiae, or viruses. These agents may be intentionally employed to cause morbidity or mortality in humans or other living organisms (Burrows and Renner 1999; Lindler et al. 2005). More extensive clarification and discussion of characteristics and identification of biological threat agents are provided in Chapter 2. Biological threat agents include a wide range of pathogenic and toxic materials produced from microorganisms that are used to intentionally interfere with the biological metabolic processes of a host. These substances work to kill or incapacitate the host and may be used to target other living organisms, including humans, other animals, and vegetation. Additionally they may be used to contaminate nonliving environments such as air, water, and soil.

Biological threat agents are an attractive weapon for use by terrorists due to their typical low visibility, high potency, accessibility, relatively low costs to produce, and relative ease of delivery (Langford 2004). For these reasons, among others, many of the diseases that may result from the use of biological threat agents considered are attractive for use by terror groups, including anthrax, brucellosis, cholera, Ebola, pneumonic plague, Q fever, smallpox, tularemia, and Venezuelan equine encephalitis. Biological toxins that might also be used as threat agents include staphylococcal enterotoxin B, botulinum toxin, ricin, and various mycotoxins (Kortepeter and Parker 1999). Developing and producing biological threat agents can be accomplished without significant high-technology resources, but finding a means of effectively disseminating these substances without degrading their infectivity is usually difficult. A potential means of biological threat agent dissemination that would likely be effective would be through the use of an aerosol generator, such as a pesticide spray apparatus or a crop-dusting airplane (Langford 2004). Another possible area of concern about terrorists using pathogens is contaminating a water or food supply (Falkenrath et al. 2001). Likewise, biological contamination of food could effectively be accomplished after processing of most food products. In such a scenario, the extent of victim exposure and infection for the biological threat agent would likely be limited to a local region; however, the fear caused would offer terrorists justification for such action (Langford 2004).

While biological threat agents, biological weapons, and biological terrorism have been well described and reviewed by many books or government publications and are introduced in Chapters 1 and 2 of this book, the focus of this chapter is to briefly review the general pathogenic effects of a few of the potential biological threat agents identified by the U.S. Centers for Disease Control and Prevention (CDC) and to briefly comment upon ongoing biotoxin research efforts related to countermeasures against such agents.

9.2 BACTERIAL AGENTS

Bacteria are unicellular, prokaryotic microorganisms of the kingdom Monera, domain Bacteria, that possess cell walls (membrane). Their chromosomes are composed of DNA, and they are capable of reproducing very rapidly via cell division. Generic classification is accomplished through description of their appearance [e.g., *cocci* (spherical), *bacilli* (rod-shaped), *strepto* (chains), or *straphylo* (clumps), etc.] (Murray et al. 2005). Various representative species of bacteria compose the most common category of biological threat agents because of their abundance in nature and easy growth (Christopher et al. 1997). Bacteria species listed in Table 2.1 of Chapter 2 are pathogenic agents known to cause human disease, and both *Bacillus anthracis* and *Salmonella* spp. have previously been used by terrorists. The causative agent for tularemia, *Francisella tularensis*, has been developed and tested as a biological weapon (Ashford 2003).

9.2.1 ANTHRAX

The pathogenic bacterium that causes the disease anthrax is *Bacillus anthracis* and it is considered a Category A potential biological terrorism agent (CDC 2007). It is a Gram-positive, encapsulated, spore-forming species and is thus highly resistant to environmental extremes of humidity and temperature. In addition to its environmental stability, *B. anthracis* is a zoonosis, is highly infectious, and may cause relatively high mortality if untreated. Zoonotic anthrax infections occur primarily in herbivores such as cattle, goats, horses, sheep, and swine and can be transmitted to humans through contact with infected animals or animal products. The mortality associated with anthrax varies with the strain of *B. anthracis*. Human infection with anthrax may occur through ingestion, inhalation, and cutaneous (skin) exposure to the spores or active form of *B. anthracis*. Utilizing the sporulated form of the bacterium, a finely milled powder may be aerosolized and inhaled. It is anticipated that a deliberate anthrax attack would likely rely on the inhalation route of exposure since it is the most deadly, as demonstrated by the anthrax attack delivered via the postal system in the fall of 2001 in the United States. As a result of those biological terrorism attacks, there was a 45% fatality rate (5 of 11) of those infected (Jernigan et al. 2002).

When anthrax spores enter the body, particularly when inhaled or ingested, they immediately become active, multiply, and release a three-part protein toxin, of which one part is deadly to humans and is referred to as the lethal factor. The lethal factor interferes with the body's ability to mount an immune response. Cutaneous (skin) infection with anthrax is rarely fatal if treated. Initial symptoms of cutaneous anthrax infection appear within days of exposure, beginning with an itchy bump that develops into a black sore, sometimes accompanied by flulike symptoms. Anthrax infection via ingestion (e.g., via contaminated food) causes food poisoning-like symptoms and may be fatal if untreated. Inhalational exposure to anthrax is the most serious form and is characterized by mild respiratory symptoms that develop into severe symptoms, breathing difficulties, shock, and eventually death if not treated. Typically symptoms appear within a week of exposure to infective *B. anthracis* but may not appear for 2 months (Inglesby et al. 2002).

There is a vaccine available for the prevention of anthrax, but it is only available to those who are at significant risk for anthrax exposure, such as military personnel and veterinarians. Postexposure treatment of individuals is with a course of antibiotics before symptoms appear if exposure is suspected, or as soon as symptoms are noted. Prompt treatment is usually effective; however, success is dependent upon exposure dosage, route of exposure, and individual susceptibility factors.

9.2.2 SALMONELLOSIS

A significant food safety threat is posed to civilian populations by specific bacterial pathogens identified as Category B threat agents such as *Salmonella* spp., *Escherichia coli* 0157:H7, and *Shigella* spp. (CDC 2007). Infection of humans and many other animals with species of salmonella bacteria causes the disease salmonellosis. The bacterium may cause symptomalogy typically associated with food poisoning, such as vomiting and diarrhea. Typhoid fever, which only infects humans, is caused by the *Salmonella typhi* serotype. Exposure to *Salmonella* spp. bacteria may occur through ingestion of contaminated foods and drinks. Intentional dissemination of *Salmonella* spp. by terrorists could be effectively accomplished by simply contaminating prepared food or drinks that are ready to be consumed. Mary Mallon, also known as Typhoid Mary, spread typhoid fever to 47 people, killing 3, through food that she handled in New York City from 1900 to 1915. As discussed in Chapter 1, the Rajneeshee religious cult caused 751 human infections by way of contaminated salad bars in several restaurants in Wasco County, Oregon, in a plot to lower voter turnout for a local political election (Christopher et al. 1997; Falkenrath et al. 2001).

There are many different subspecies or serotypes of *Salmonella* spp. capable of sickening humans (Chin and Ascher 2000), and they occur in a variety of different environments. *Salmonella* spp. may occur in water, soil, insects, factory and kitchen surfaces, animal feces, and raw meats, poultry, and seafood. The bacteria can be isolated from these sources and grown. Most typically, salmonellosis outbreaks in the United States occur because of improperly cooked food such as meat, poultry, seafood, and eggs; improperly prepared fruits, vegetables, milk, and other food and drinks; or via improperly washed hands, food contact surfaces, and utensils. There are approximately 40,000 salmonellosis cases resulting in approximately 600 deaths each year reported in the United States. Additionally, there are roughly 400 cases of typhoid fever recorded in the United States annually, but 75% of them are associated with international travel. Immediate medical attention significantly reduces mortality, but the case fatality rate is 12–30% in untreated cases (Murray et al. 2005).

Most infection from *Salmonella* spp. occur by way of the small intestines following ingestion, with an infective dose of as few as 15–20 bacterium. Inflammation follows the passage of the bacteria through the interior lining of the small intestines and into its middle layer. Prolonged untreated infection by *Salmonella* spp. can spread to the circulatory system as well as other organs, increasing disease duration and severity. Infection with *Salmonella typhi* serotype spreads from the small intestines to the bloodstream and ultimately infects the lymph nodes, liver, and spleen. Within 12–72 h following an infective exposure, symptoms of salmonellosis may include nausea, diarrhea, fever, headaches, abdominal cramps, and vomiting. Salmonellosis usually persists 4–7 days without treatment. Severe and prolonged infections often lead to severe dehydration, and potentially to spread of the infection into the bloodstream and other organs, which can cause death. A significant sequelae of salmonellosis is the development of Reiter's syndrome, characterized by joint pains, eye irritation, and painful urination that can last for months or years and lead to chronic arthritis. Typhoid fever differs from salmonellosis in that it is characterized by fever, headache, constipation, weakness, chills, muscle ache, and a rash of flat, rose-colored spots in some cases, and may persist for 3–4 weeks and can cause death due to complications (Christopher et al. 1997; Falkenrath et al. 2001).

Currently there are no commercially available vaccines for salmonellosis, but vaccines for typhoid fever are given to travelers going to countries with endemic disease. For most strains of *Salmonella* spp., antibiotics (specifically ampicillin, gentamicin, ciprofloxacin, and a combination of trimethoprim and sulfamethoxazole) are effective for both typhoid fever or severe salmonellosis; some serotypes have developed antibiotic resistance due to their use in feed animals. Treatment also includes administration of intravenous fluids and electrolytes to control severe dehydration due to diarrhea (Langford 2004).

9.2.3 TULAREMIA

The bacteria *Francisella tularensis*, which causes the disease tularemia, is a Category A biological threat agent (CDC 2007). Also known as "rabbit fever" and "deerfly fever," this zoonosis is found throughout North America, Europe, and Asia. The pathogen was initially identified in 1912 in ground squirrels in Tulare County, California, and *F. tularensis* was described as causing plaguelike illness in the ground squirrels. The pathogen may be transmitted in a number of ways, including direct contact with infective bodily fluids from an infected host (i.e., through lesions, mucous membranes, or conjunctiva), through the feeding of haematophagous arthropods (i.e., ticks, deerflies, mosquitoes, etc.), inhalation of aerosolized bacteria in water droplets, as well as through ingestion of contaminated food or water. Tularemia is not transmissible from person-to-person, so infected humans do not need to be quarantined. There were reports of waterborne outbreaks in Europe and the Soviet Union during the 1930s and 40s. There are two subspecies of *F. tularensis* that are currently recognized; Type A occurs in North America, and Type B is found in Europe. The American Type A is much more virulent than Type B. There are approximately 200 human cases of tularemia reported each year in the United States, with a case fatality rate of less than 2%. Human cases of tularemia typically occur in rural areas of the south-central and Midwestern states during the summer months, with 56% of those reported from 1990 to 2000 occurring in four states, Arkansas, Missouri, South Dakota, and Oklahoma.

The *F. tularensis* bacterium can survive for weeks at low temperatures in moist soil, water, hay, straw, and decaying plants and animals (Murray et al. 2005). In addition to its environmental stability, *F. tularensis* is considered a likely agent for bioterrorism because it can easily be aerosolized to enhance inhalation exposure, and an infective dose may be as few as 10–50 bacteria. Historically, Japan, the Soviet Union, and the United States all worked to weaponize *F. tularensis* as an offensive weapon, with the Soviet Union continuing its work into the 1990s (Dennis 2004).

Immediately following effective infection, *F. tularensis* begins pathology in the lymph nodes, lungs, spleen, liver, and kidneys, multiplying at these locations and within macrophages. Lesions develop on and inside bacteria-infected organs, causing inflammation and necrosis of surrounding tissues (Dennis 2004). Intrinsic incubation in the infected human varies, but symptoms of tularemia usually appear within 3–5 days postinfection but can appear for up to 2–3 weeks. Symptoms include fever, chills, headaches, diarrhea, muscle aches, joint pain, dry cough, sore throat, and progressive weakness. These symptoms may progress into pneumonia, chest pain, bloody sputum, difficulty breathing, and death. In the case of inhalational exposure to tularemia, such as might occur in a bioterrorist attack, symptoms are more representative of severe respiratory illness, including pneumonia and systemic infection.

Antibiotic therapy is highly effective in treating tularemia and significantly reduces fatalities, including drugs such as streptomycin, gentamicin, doxycycline, and ciprofloxacin (Langford 2004). Antibiotic treatment is generally started before confirmation. An investigational pre-exposure vaccine for tularemia was produced by the U.S. Army and is given to laboratory workers and others at high risk for infection.

9.3   VIRAL AGENTS

Viruses are a diverse class of biological threat agents, varying in size and morphology, complexity, host range, and pathologic effect on their hosts. Viruses can be generally divided based upon their genome, either RNA or DNA, which is surrounded by a protein shell. When compared with the roughly 100,000 genes in the human genome, viral genomes are very small and encode from approximately 10 to about 200 genes. Viruses, in contrast to bacterial pathogens, multiply only in living cells and are fully dependent on the host cells' energy-yielding and protein-synthesizing abilities (Murray et al. 2005). Because of the very wide range of diseases and pathologies attributed to viruses, and because there is little that can be done to interfere with the growth of viruses, many of the viral pathogens are believed to be ideally suited for use as biological terrorism agents.

9.3.1   SMALLPOX

The virus that causes smallpox, *Variola major*, is a Category A biological threat agent (CDC 2007). It is a double-stranded DNA orthopoxvirus. It is a widely held opinion among infectious disease professionals that, historically, smallpox has been responsible for more deaths than all other infectious diseases combined. After aggressive efforts by the World Health Organization to immunize the world's population, and because the virus is unable to survive for more than a week outside of its only host, humans, *Variola* was declared eradicated in 1980 (Murray et al. 2005). The smallpox vaccine can prevent or lessen the severity of the disease if administered within 96 h of exposure. The duration of effectiveness of the vaccine used to eradicate smallpox is not known, and there are reported mild to life-threatening risks are associated with the vaccine. Once a victim is symptomatic, medications and intravenous fluid can be administered to make the patient more comfortable, but there are no antivirals available for unvaccinated infected individuals (Henderson et al. 1999).

There are currently two known stocks of the virus, one at the U.S. Center for Disease Control and Prevention in Atlanta and the other at the State Research Center of Virology and Biotechnology in Koltsovo, Russia. It is widely known that from 1980 until the mid-1990s researchers in the Soviet Union developed a highly virulent strain of *Variola major* for use as a biological weapon, with a significantly reduced incubation period (Henderson and Fenner 2001).

Humans may be exposed and infected with the smallpox virus through inhalation, ingestion, or injection. It is postulated that inhalational exposure would be the most likely delivery of smallpox used by terrorists, relying upon victims to inhale *Variola* via an aerosol or through an infective individual. Effective infection with *Variola* is possible with as few as 10–100 viral particles. Once trapped in the mucous membrane, macrophages envelop the *Variola* virus particles. The virus-infected macrophages are transported to the lymph nodes, where viral replication within the macrophages occurs, and within 3–4 days the virus is released into the bloodstream, spreads throughout the body, and incubates in the bone marrow, spleen, and other organs. Typically within 12 days of initial infection victims present with flulike symptoms, including headache, fever, and fatigue, and they become infective at this stage. Over the next 2–3 weeks, the virus continues to damage the body, particularly the immune and circulatory systems. The case fatality rate for nonhemorrhagic smallpox is approximately 30%, but 2–6% of smallpox infections are classified as hemorrhagic and result in mortality rates of over 95% (Henderson 1999).

9.3.2 EBOLA

Ebola, Lassa, Machupo, and Marburg are some of the 18 different viruses that cause viral hemorrhagic fever (VHF) in humans and other primates that are classified as Category A biological threat agents (CDC 2007). Ebola is a zoonotic disease of primates and occasionally spreads to human populations that are exposed to these animals (Murray et al. 2005).

The Ebola virus increases the permeability of blood vessel membranes, which leads to hemorrhaging throughout the body, by producing specific proteins. Additionally the virus prevents the body from mounting an effective immune response. Severe pulmonary hemorrhage, gastrointestinal hemorrhage, hepatitis, or encephalitis are the causes of death in terminal infections. The case fatality rate for Ebola is 50–90% of symptomatic cases. Following infection by the Ebola virus, symptoms appear within 2–21 days and progress over the next 2 weeks until the patient dies in hypovolemic shock or in a coma. Patients who survive and recover may remain infectious even after symptoms abate. There is no effective treatment for Ebola.

As with many of the other known biological threat agents, Ebola's effective use as a weapon is totally dependent upon its delivery. Because the virus is readily transmitted human-to-human, there is clearly the potential, as with other zoonoses, that Ebola virus could be obtained from natural sources and used to infect large populations. Ebola cannot be successfully aerosolized in a dry form at this time, but the potential for aerosolization in liquid phase is speculated (Ashford 2003; Langford 2004). Ebola virus's high mortality, environmental stability, lack of an effective treatment, and its capacity to be transmitted human-to-human makes it a likely highly effective biological threat agent. It is known that investigations of methods to weaponize the virus were conducted by the former Soviet Union. There have been at least three incidents of accidental infection of research scientists with Ebola reported: England, 1976 (recovered); the United States, February 2004 (recovered); and Russia, May 2004 (died).

9.4 TOXINS

Biological toxins are substances produced or derived from living organisms, both animal and plant, such as bacteria, cyanobacteria, fungi, and some species of plants and marine fish, that cause disease pathology or other debilitating response in humans and other animals (Keeler and Tu 1983). There are a number of biological toxins that are listed as potential biological threat agents, including botulinum toxin, ricin, and staphylococcal enterotoxin B (CDC 2007; Burrows and Renner 1999; Franz 1997).

9.4.1 BOTULINUM TOXIN

A highly neuronal toxic protein produced by the soil bacteria *Clostridium botulinum* causes the paralytic disease botulism, and the toxin is considered a Category A biological threat agent (CDC 2007). Botulinum toxin is among the most lethal toxins known to man, with an approximate fatal dose being 0.4 μg in the average adult. There are seven known types of botulinum toxin, each produced by different strains of *Clostridium botulinum*, but only four of the types (A, B, E, and F) cause human botulism (Murray et al. 2005). Approximately two thirds of all *Clostridium botulinum* infections are caused by consumption of bacterial spores, which are reactivated upon entry into the body. However, ingestion of food contaminated with the toxin or bacteria is responsible for less than 25% of all cases. The most common form of ingestion infection is referred to as infant botulism because it primarily affects babies fed contaminated honey.

Mechanistically, botulinum toxin inhibits the release of the neurotransmitter acetylcholine, and thus affects the connections between nerves and the muscles they control, effectively silencing nerve action. Symptoms of toxin exposure typically appear within 18–36 h after ingestion of contaminated food, or 12–80 h after inhalation. Symptoms include slurred speech, dry mouth, blurred and double vision, respiratory distress, as well as difficulty swallowing. There is a progressive paralysis down the body symmetrically, weakening it, reducing muscle reflexes, and loss of control of the limbs. Mortality resulting from botulinum toxin is primarily attributable to paralysis of muscles required for breathing including the diaphragm. There are approximately 140 cases of botulism per year in the United States, with a few cases being fatal (Shapiro et al. 1998; Simpson 2004).

Theoretically, the toxin can be delivered to humans through a wide range of methods, including contaminating food or drink with either the toxin or the bacteria. Purified botulinum toxin is difficult to effectively aerosolize for use as an areawide weapon. Nevertheless, during 1990–1995 the Japanese terrorist cult Aum Shinrikyo attempted to spread it as an aerosol three times. It is also known that Iraq had a biological weapons program that produced a significant amount of botulinum toxin in the years leading up to the 1991 Gulf War (Christopher et al. 1997; Shapiro et al. 1998).

The administration of antitoxins that are antibodies that bind to the botulinum toxin and prevent it from affecting nerve cells is the most effective treatment postexposure. However, once the toxin is bound to cells the antitoxin is ineffective, making rapid diagnosis critical to preventing further paralysis. Due to the extensive paralysis caused by the toxin, assisted breathing, feeding, and physical therapy may be required for extended periods while new nerves connect to paralyzed muscles. Additionally, antibiotics are also administered to kill the *Clostridium botulinum* bacteria. There is currently a vaccine against botulinum toxin that is given to U.S. military troops and lab personnel at high risk of exposure (Franz 1997; Langford 2004).

9.4.2 Ricin

Phytotoxalbumin is a toxic protein derived from the beans of the castor plant (*Ricinus communis*) and is commonly called ricin. Ricin is a very potent toxin that has been used as a biological crime agent [e.g., assassination (Langford 2004)] and is considered a Category B toxin that could likely be used as a biological threat agent for bioterrorism (CDC 2007). The mechanistic action of ricin prevents protein synthesis by cells in the body, which can lead to widespread organ damage and central nervous system depression, immune system depression, liver, pulmonary, seizures, and renal failure. Initial symptoms of exposure to ricin may be fever, coughing, and gastrointestinal problems. When ingested, the toxin causes stomach irritation, gastroenteritis, bloody diarrhea, and vomiting. If inhaled, ricin causes severe lung damage, including pulmonary edema. Inhalation of the toxin may produce pathologic changes within 8 h, with severe respiratory symptoms and acute respiratory failure within 36–72 h postexposure. Phytotoxalbumin is a relatively large protein and is not easily absorbed across the skin, thus dermal exposure to ricin is of limited concern. There is not currently an available treatment or prophylaxis for ricin exposure, other than physical exclusion. Typically, if the victim of ricin exposure survives 5 days he or she has a high probability of recovery (Doan 2004).

There are approximately 91 million metric tons of castor beans processed annually in the production of castor oil worldwide; the by-products from this process are 5–10% ricin by weight. As a result of the extensive available volume of ricin produced commercially, and the relative ease and fairly low technology required to obtain the toxin, ricin's significance as a potential biological threat agent is considerable. Effective delivery or dissemination of ricin can be accomplished via inhalation, injection, or ingestion. Lyophilization (a method of freeze-drying the liquid state in a high-vacuum environment) is required to develop an aerosolized form of ricin. While the lyophilization process is not overly difficult, this step does present a technical challenge (Olsnes 2004).

9.4.3 Staphylococcal Enterotoxin B

Staphylococcal enterotoxin B (SEB) is a toxic protein produced by the bacterium *Staphylococcus aureus* with a molecular weight of 28 kilodaltons. SEB is classified as a Category B biological threat agent (CDC 2007). The toxin is highly stable and has caused many cases of food poisoning in humans (Jett et al. 2001). Outbreaks of SEB-caused food poisoning have resulted from consumption of inadequately refrigerated raw milk or cheeses, as well as through unrefrigerated cream puffs and potato salad. Additionally, SEB intoxication can also be caused by inhalation of the toxin. It is known that several countries, including the United States, have developed an aerosolized weapon form of SEB. The human staphylococcal enterotoxin B LD_{50} is ~1.7 mg/kg by inhalation, and an incapacitating inhalational dose is accomplished at only 30 ng. (LD_{50} is the amount of a material, given all at once, that causes the death of 50 percent of a group of test animals.) Following ingestion exposure, symptoms of SEB intoxication develop in 1–6 h and may include severe gastrointestinal (GI) pain, projectile vomiting, and diarrhea, or fever, chills, muscle aches, and shortness of breath when inhaled. Symptoms typically diminish within hours postexposure, but exposure to excessive doses can lead to septic shock and death if untreated. Supportive therapy with close attention to oxygenation and hydration, and in severe cases, ventilation with positive-end expiratory pressure and diuretics are recommended treatment approaches for SEB exposure (Franz 1997; Langford 2004).

Other biological toxins that are known to cause significant morbidity and mortality in humans and may be potentially used as biological terrorism weapons include aflatoxins. microcystins, saxitoxin, and T2 toxin. A brief summary of the origin and mechanism of action of these toxins is provided below.

9.4.4 AFLATOXINS

The class of biological toxins known as aflatoxins is produced by the fungi *Aspergillus flavus* and *A. parasiticus* (Ciegler et al. 1981; Cole and Cox 1981). Toxicoses in turkey poults, ducklings, and chicks were reported in Britain (Wang et al. 1998). The toxin produced by *Aspergillus flavus* called AFB_1 is the most potent and commonly occurring, has an acute toxicity in all species of mammals, birds and fish tested, and has an LD_{50} range of 0.3–9.0 mg/kg (Busby and Wogan 1984). Prolonged exposure to low level of AFB_1 leads to hepatoma, cholangiocarcinoma, hepatocellular carcinoma, and other tumors (Wang et al. 1998). The effects of acute aflatoxin intoxication include damage to both DNA and protein synthesis and enzyme activity inhibition in gastrointestinal system, mainly in the liver (Phillips et al. 2006).

9.4.5 MICROCYSTINS

The freshwater cyanobacteria *Microcystic aeruginosa* produces microcystins. This toxin has been implicated in the death of human dialysis patients (Carmichael 1988). The most common form of microcystins and presumably the biotoxin of choice to be weaponized is microcystin-LR (MCLR). Mice treated with aerosolized MCLR died within hours, and revealed an acute LD_{50} of 67 μg/kg (Anon. 1999). Mechanistically, MCLR disturbs cellular function and regulation as a strong inhibitor of protein phosphatases.

9.4.6 SAXITOXIN

The marine dinoflagellates *Gonyaulax catenella* and *G. tamarensis*, and the cyanobacteria *Anabaena circinalis* produce the toxin saxitoxin. These dinoflagellates are consumed by the Alaskan butter clam and the California sea mussel (Schantz 1986). Saxitoxin is the causative agent for paralytic shellfish poisoning and is highly toxic if ingested, with an LD_{50} calculated from shellfish poisoning in people ranging from 0.3–1.0 mg/person. Its toxicity is even more acute if aerosolized and inhaled (LD_{50} of 2 μg/kg). Mechanistically, saxitoxin is a sodium flow/sodium channel blocker. There is no therapeutic treatment for saxitoxin intoxication but supportive care, including respiratory support may limit effects (Burrows and Renner 1999; Langford 2004).

9.4.7 T2 TOXIN

The tricothecene mycotoxin T2 is produced primarily by *Fusarium sporotrichioides* and has been reported in many parts of the world (Beasley 1989; Wang et al. 1998).

T2 toxin is suspected to have been weaponized and used as a weapon agent in Laos, Cambodia, and Afghanistan (Christopher et al. 1997; Franz 1997). The toxin may be effectively delivered through the skin, by inhalation, or by ingestion. There are no antidotes effective against T2 toxin, but treatment may include supportive care, especially maintenance of electrolyte balance, so avoidance of contact is the only preventive measure.

9.5 RESEARCH RECOMMENDATIONS AND FOCUS FOR FUTURE EFFORTS

Research efforts conducted through the Admiral Elmo R. Zumwalt, Jr. National Program for Countermeasures to Biological and Chemical Threats at Texas Tech University have focused on studying combinative toxicity of biotoxins with emphasis on the newly identified food-borne and waterborne toxins. There is a historical precedence for the use of toxins as biological weapons against unprotected enemy or innocent people (Falkenrath et al. 2001; Langford 2004). During the 6th and 4th centuries BCE, Assyrians and Persians poisoned drinking water wells with ergot alkaloids, a mycotoxin. More recently trichothecene mycotoxins (T2 toxin and diacetoxysirpenol, produced by various *Fusarium* strains) were found to be the cause of the "yellow rain" episodes, which occurred in people in Afghanistan, Kampuchea, and Laos in the late 1970s. And plans were being developed for missile delivery of aflatoxin by Iraq (Langford 2004; Zilinskas 1997).

Even though for many years research efforts have been focused on the study of single toxins, or single categories of toxins, and a great deal of data regarding individual toxins has been documented, little attention has been paid in the study of combinative toxic effects of biotoxin mixtures, which may be more potent and cause more damage to human and animal health. The nature of coexistence of many types of chemicals and toxins in complex environmental samples has drawn the attention of federal governmental agencies and the scientific community in recent years. The great challenge currently facing the preventive medicine community is how to prevent chronic human diseases caused by complex toxin mixtures. Therefore, there is an urgent need for improved understanding of the mechanism of combinative toxicity of biotoxin mixtures, as well as the development of rapid and sensitive methods to detect multiple toxins in both the field and body fluids of animal and humans. Additionally, developing prevention strategies against possible use of these toxin mixtures is critical.

REFERENCES

Anon., 1999. New understanding of algae, *Environ. Health Perspect.*, 107, p. A13.

Ashford, D., 2003. Planning against biological terrorism: lessons from outbreak investigations, *Emerg. Infect. Dis.*, 9(5), pp. 515–519.

Beasley, V.R., 1989. *Trichothecene Mycotoxicosis: Pathophysiologic Effects*, Vols. I and II, CRC Press, Boca Raton, FL.

Burrows, W.D. and Renner, S.E., 1999. Biological warfare agents as threats to potable water, *Environ. Health Perspect.*, 107, pp. 975–984.

Busby, W.F. and Wogan, G.N., 1984. Aflatoxins, in *Chemical Carcinogenesis*, Searle, C., Ed., American Chemical Society, Washington, DC, pp. 945–1136.

Carmichael, W.W., 1988. Toxins of freshwater algae, in *Handbook of Natural Toxins: Marine Toxins and Venoms*, Vol. 3, Tu, A.T., Ed., Marcel Dekker, New York, NY, pp. 121–147.

Chin, J. and Ascher, M., 2000. Salmonellosis, in *Control of Communicable Diseases*, Chin, J.M. and Ascher, M., Eds., American Public Health Association, Washington, DC, pp. 440–444.

Christopher, G.W., Cieslak, T.J., Pavlin, J.A., and Eitzen, E.M., Jr., 1997. Biological warfare: a historical perspective, *J. Am. Med. Assoc.*, 278, pp. 412–417.

Ciegler, A., Burmaister, H.R., Vesonder, R.F., and Hesseltine, C.W., 1981. Mycotoxins and *N*-nitro-compounds: environmental risks, in *Mycotoxins: Occurrence in the Environment*, Shank, R., Ed., CRC Press, Boca Raton, FL, pp. 1–50.

Cole, R.J. and Cox, R.H., 1981. *Handbook of Toxic Fungal Metabolites*, Academic Press, New York, NY.

Dennis, D.T., 2004. Tularemia, in *Infectious Diseases*, Vol. II, 2nd ed., Cohen, J. and Powderly, W.G., Eds., Mosby, Edinburgh, pp. 1649–1653.

Doan, L.G., 2004. Ricin: mechanism of toxicity, clinical manifestation, and vaccine development, a review, *J. Toxicol. Clin. Toxicol.* 42, pp. 201–208.

Falkenrath, R.A., Newman, R.D., and Thayer, B.A., 2001. *America's Achilles' Heel: Nuclear, Biological, and Chemical Terrorism and Covert Attack*, MIT Press, Cambridge, MA.

Franz, D.R., 1997. Defense against toxin weapons, in *Medical Aspects of Chemical and Biological Warfare*, Sidell, F., Takafugi, E.T., and Franz, D.R., Eds., TMM Publications, Washington, DC, pp. 603–619.

Henderson, D.A., 1999. Smallpox: clinical and epidemiologic features, *Emerg. Infect. Dis.*, 5, pp. 537–539.

Henderson, D.A. and Fenner, F., 2001. Recent events and observations pertaining to smallpox virus destruction in 2002. *Clin. Infect. Dis.*, 33, pp. 1057–1059.

Henderson, D.A., Inglesby, T.V., Bartlett, J.G., Ascher, M.S., Eitzen, E., Jahrling, P.B., Hauer, J., Layton, M., McDade, J., Osterholm, M.T., O'Toole, T., Parker, G., Perl, T., Russell, P.K., and Tonat, K., 1999. Smallpox as a biological weapon: medical and public health management, Working Group on Civilian Biodefense, *J. Am. Med. Assoc.*, 281, pp. 2127–2137.

Inglesby, T.V., O'Toole, T., Henderson, D.A., Bartlett, J.G., Ascher, M.S., Eitzen, E., Friedlander, A.M., Gerberding, J., Hauer, J., Hughes, J., McDade, J., Osterholm, M.T., Parker, G., Perl, T.M., Russell, P.K., and Tonat, K., 2002. Anthrax as a biological weapon, 2002: updated recommendations for management, *J. Am. Med. Assoc.*, 287, pp. 2236–2252.

Jernigan, D.B., Raghunathan, P.S., and Bell, B.P., 2002. Investigation of bioterrorism-related anthrax, United States, 2001: epidemiologic findings, *Emerg. Infect. Dis.*, 8, pp. 1019–1028.

Jett, M., Ionin, B., and Da, R., 2001. The staphylococcal enterotoxins, in *Molecular Medical Microbiology*, Vol. 2, Sussman, M., Ed., Academic Press, San Diego, CA, pp. 1089–1116.

Keeler, R.F. and Tu, A.T., 1983. *Handbook of Natural Toxins*, Marcel Dekker, New York, NY.

Kortepeter, M.G. and Parker, G.W., 1999. Potential biological weapons threats, Special Issue, *Emerg. Infect. Dis.*, 5, pp. 523–527.

Langford, R.E., 2004. *Introduction to Weapons of Mass Destruction: Radiological, Chemical, and Biological*, Wiley-Interscience, Hoboken, NJ.

Lindler, L.E., Lebeda, F.J., and Korch, G.W., 2005. *Biological Weapon Defense: Infectious Diseases and Counter Bioterrorism*, Humana Press, Totowa, NJ.

Murray, P.R., Rosenthal, K.S., and Pfaller, M.A., 2005. *Medical Microbiology*, 5th ed., Elsevier Mosby, Philadelphia, PA, pp. 191–706.

Olsnes, S., 2004. The history of ricin, abrin and related toxins, *Toxicon*, 44, pp. 361–370.

Phillips, T.D., Afriyie-Gyawu, E., Wang, J.-S., Williams, J., and Huebner, H., 2006. The potential of aflatoxin sequestering clay, in *The Mycotoxin Factbook*, Barug, B., Bhatnagar, D., van Egmond, H.P., van der Kamp, J.W., van Osenbruggen, W.A., and Visconti, A., Eds., Wageningen Academic, Netherlands, pp. 329–346.

Schantz, E.J., 1986. Chemistry and biology of saxitoxin and related toxins, *Ann. N. Y. Acad. Sci.*, 479, pp. 15–23.

Shapiro, R.L., Hatheway, C., and Swerdlow, D.L., 1998. Botulism in the United States: a clinical and epidemiologic review, *Ann. Intern. Med.*, 129, pp. 221–228.

Simpson, L.L., 2004. Identification of the major steps in botulinum toxin action, *Annu. Rev. Pharmacol. Toxicol.*, 44, pp. 167–193.

U.S. Centers for Disease Control and Prevention (CDC), 2007. Bioterrorism agents/diseases, available at http://www.emergency.cdc.gov/agents/agentlist-category.asp (accessed October 9, 2007).

Wang, J.-S, Kensler, T.W., and Groopman, J.D., 1998. Toxicants in food: fungal contaminants, in *Current Toxicology Series: Nutrition and Chemical Toxicity*, Ioannides, C., Ed., John Wiley & Sons, New York, NY, pp. 29–57.

Zilinskas, R.A., 1997. Iraq's biological weapons: The past as future? *J. Am. Med. Assoc.*, 278, pp. 418–424.

10 Conclusions and Research Needs for the Future

Ronald J. Kendall, Galen P. Austin,
Chia-bo Chang, George P. Cobb,
Gopal Coimbatore, Stephen B. Cox,
Joe A. Fralick, Jeremy W. Leggoe,
Steven M. Presley, Seshadri S. Ramkumar,
Philip N. Smith, Jean C. Strahlendorf,
and Richard E. Zartman

CONTENTS

10.1   CONCLUSIONS

As this book, *Advances in Biological and Chemical Terrorism Countermeasures*, was being developed, the National Intelligence Council (NIC) of the United States of America issued a report in July 2007 indicating a heightened threat for terrorism in the United States and globally. Since its formation in 1973, the NIC has served as a bridge between the intelligence and policy communities of the United States and as a source of deep substantive expertise on critical national security issues (National Intelligence Council 2007). This most recent report by the NIC is a summary compilation of the intelligence assessments and projections from 16 different intelligence-gathering agencies in the United States.

The NIC stated that "The U.S. Homeland will face a persistent and evolving terrorist threat over the next three years. The main threat comes from Islamic terrorist groups and cells, especially al-Qa'ida, driven by the undiminished intent to attack the Homeland and a continued effort to adapt and improve their capabilities." Indeed, the authors believe that the threat for biological and chemical attacks in the United States and in other nations is real. This threat will not only last for 3 years but also will continue into the foreseeable future. Without question, 9/11 marked a defining point in global terrorist efforts and initiated the post–9/11 environment that

243

has resulted in significantly increased counterterrorism efforts in the United States and globally, in particular in the Western world. The July 2007 report of the NIC reads as follows,

> We assessed that greatly increased world-wide counterterrorism efforts over the past five years have constrained the ability of al-Qa'ida to attack the US homeland again and have led terrorist groups to perceive the Homeland as a harder target to strike than on 9/11. These measures have helped disrupt known plots against the United States since 9/11.

Nevertheless, the authors believe that we must proactively continue to advance our abilities and resources to provide countermeasures to biological and chemical threats as we again reiterate that the threat is real and growing.

The researchers who authored this book have focused and continue to focus their efforts on identifying novel research opportunities to meet the nation's biological and chemical threat countermeasures research and development needs as identified in a joint report by the Institute of Medicine and the National Research Council (Institute of Medicine 1999), which included:

1. Preincident communications and intelligence
2. Personal protective equipment
3. Detection and measurement of chemical and biological agents
4. Recognizing covert exposure
5. Mass-casualty decontamination and triage procedures
6. Availability, safety, and efficacy of drugs, vaccines, and other therapeutics
7. Computer-related tools for training and operation

To address the aforementioned research needs, research within the Admiral Elmo R. Zumwalt, Jr. National Program for Countermeasures to Biological and Chemical Threats has focused efforts upon four specific areas, including:

1. Modeling, simulation and visualization
2. Agent detection, remediation, and therapeutic intervention strategies
3. Sensors and personal protective devices
4. Environmental protection strategies

Although the majority of research on countermeasures to biological and chemical threats has emanated from the U.S. Army, Department of Defense funding, the application and relevance of our findings to homeland security are extremely important. In fact, our research team has concluded that better coordination is needed with military-related research in order to protect and enhance war fighters and their abilities, thereby allowing those research findings to be leveraged even further for homeland security applications. Also, we encourage an increased emphasis upon enhancement of research efforts to improve local response capabilities for biological and chemical incidents. For example, if an attack were to occur in Los Angeles, it would be very difficult for New York to assist. Therefore, Los Angeles would need to be ready to respond on its own.

In the development of this book, the authors searched widely for information specifically related to advancing training in countermeasures to biological and chemical terrorism. Indeed, few programs in the United States emphasize (in a graduate degree program in environmental toxicology) homeland security applications, with the exception of Texas Tech and a handful of other universities. With this knowledge in mind, we encourage increased emphasis on graduate education, particularly research-based programs, in countermeasures to biological and chemical terrorism, because this issue is not going away and we believe that it will be with us for the foreseeable future.

A better relationship between U.S. federal laboratories and academic institutions is critical. More joint ventures and interactions are needed to achieve synergy between academic research institutions and U.S. federal laboratories. We may have to "ruffle feathers" and increase interdisciplinary cooperation for this research enhancement. Indeed, our federal agencies should help researchers, particularly at the academic level, on commercialization and procurement issues. There are many new technologies being developed at this time where the U.S. federal government could be of great assistance in bringing commercialization support and help in transferring these to military use from the academic programs.

Even as we were developing *Advances in Biological and Chemical Terrorism Countermeasures* in 2007, in this post–9/11 environment, much of the progress being made both in the United States as well as internationally has been slow, sporadic, and relatively uncoordinated. It is essential that there be better coordination among the Department of Defense, Department of Homeland Security, Department of Health and Human Services, Department of Energy, Environmental Protection Agency, and the Food and Drug Administration, among others, with academia. All of the cognizant federal agencies have enormous resources and capabilities, and if we enhance our ability to provide overarching capabilities to interact and share science and advance new technologies to commercialization, we will be better capable of protecting the U.S. homeland, as well as advancing the global war on terror. In spite of the great threat to the United States currently, and even the issuance of the NIC document, we continue to be at a heightened state of vulnerability for a biological, chemical, or radiological attack. Nevertheless, the U.S. National Science Foundation at this time has no separate division that can support a basic research program on countermeasures to weapons of mass destruction, particularly biological and chemical materials. We suggest that this anomaly be remedied through the establishment of a comprehensive basic research program in biological and chemical threat countermeasures as soon as possible.

10.2 FUTURE RESEARCH NEEDS AND RECOMMENDATIONS

The research team that developed *Advances in Biological and Chemical Terrorism Countermeasures* concurs with the July 2007 NIC press release that the threat of a biological, chemical, nuclear, or radiological attack is real and evolving. Therefore, this should dictate a continuing vigilance by the United States and other governments globally toward counterterrorism efforts, and it will demand research and development that could provide new technologies to our first responders, leaders in homeland security, and military to thwart the efforts of terrorists to attack the United States. We believe that it is essential that there be a concerted effort between detection of these threats and our protection capabilities versus just responding. Cost-effective sensors with enhanced capabilities for detecting presences of chemical and biological material in nonideal conditions are critically needed, as well as new personal protective capabilities for our military and first responders to ensure that we can assist casualties and potentially exposed individuals without risking the lives of health-care emergency responders. Sensor-embedded protective ensembles should be developed without compromising protection and comfort.

In the development of this book, the authors recognize that the rules of engagement with biological and chemical terrorism strategies are stacked against the civilian population when we are considering a war between state-sponsored organizations versus terrorist organizations. In other words, terrorists will use any means at any level of intensity to inflict casualties and terror on the peoples of the United States or any other country who oppose their ideology. Therefore, we believe that new fields such as nanotechnology should be leveraged for new protective garments and highly mobile and applicable field-deployable sensors and materials to assist us in assessing potential biological and chemical threats.

As previously mentioned, we must stay ahead of the methodologies and technologies that terrorists may be able to access. For instance, genetically engineered organisms are a significant potential threat of the future with, perhaps, modified strains of plague and smallpox that would, with genetic engineering, be even more difficult to fight if released into a susceptible population, such as that of the United States. Therefore, there is a need to seek out new therapeutics that could address and target common pathogenic threat agents, particularly from sublethal exposures. There are many new and innovative technologies being developed that could provide therapeutics quickly that could effectively target genetically engineered organisms that may be introduced by terrorists.

One of the most daunting obstacles faced by first responders and war fighters is determining the immediate and long-term extent of the region affected by the release of a chemical or biological toxin. Providing training to personnel to understand the basic nature of dispersion processes and to understand the products provided by the modeling community is essential to enabling these personnel to react in an informed manner. Similarly, the National Research Council (NRC) has recommended that tabletop exercises need to be undertaken regularly to ensure that the modeling community understands the operational needs of first responders and war fighters (National Research Council 2003). In recognition of the limited information that will be immediately available in the wake of a release event, continuing research is needed to develop approaches by which live sensor data can be integrated to both characterize the source and improve the fidelity of dispersion predictions in real time as the release event evolves. As computational resources continue to develop, it is expected that the sophistication of meteorological and dispersion models will continue to increase to deliver progressively improving dispersion estimates. We also need to develop a simple and effective data assimilation capability of mesoscale numerical weather prediction (NWP) models based on real-time observations to provide real-time meteorological data for dispersion/diffusion computation.

For military or first responders who are assisting in homeland security efforts, we must improve our ability to protect them in their efforts to assist exposed individuals or casualties. This may be accomplished through improved barrier materials that could provide protective measures for our first responders and our military as needed, including those with self-detoxifying capabilities. In addition, combinational protective materials are needed that can be applied quickly and that have the ability to protect the wearer from both biological and chemical attacks. A nonparticulate decontamination system for personnel and sensitive equipment is needed also, including improved decontamination wipes that have the ability to decontaminate or neutralize both biological and chemical agents.

We believe that applications of research findings, particularly through applications that enhance the readiness of our war fighters, can be leveraged to address threats of agroterrorism, emerging and resurgent diseases, border security, and security at our ports of entry, including air, land, and sea. We must improve our ability to leverage our technologies and our knowledge to thwart the continuing threat of biological and chemical terrorist attacks.

We believe there needs to be a readily accessible central knowledge repository on emerging and existing biological and chemical threat agents. Even with the realization in the United States that we are under a heightened level of threat from biological, chemical, and radiological attacks, as issued by NIC, it is still very difficult to access relevant information to increase research and develop countermeasures on these threats.

We believe *Advances in Biological and Chemical Terrorism Countermeasures* is timely and highly applicable in the post–9/11 world in which we live. We believe the threat is credible but not insurmountable if we develop the technologies and cooperation that are necessary for thwarting biological and chemical attacks. Indeed, this will be an issue facing everyone for many years to come, but we have the capability, resources, and resolve as a nation to take on this threat and to overcome it. We offer *Advances in Biological and Chemical Terrorism Countermeasures* as a tool to assist us in bringing together the state-of-the-knowledge and applying new strategies to address this chemical and biological terrorist threat to the United States and to other countries.

REFERENCES

Institute of Medicine and National Research Council, 1999. *Chemical and Biological Terrorism: Research and Development to Improve Civilian Medical Response*, National Academies Press, Washington, DC.

National Intelligence Council (NIC). 2007. National intelligence estimate: the terrorist threat to the U.S. homeland, press release, 17 July, National Intelligence Council, Washington, D.C., available at http://www.dni.gov/press_release/20070717_release.pdf (accessed July 17, 2007).

National Research Council (NRC). 2003. *Tracking and predicting the atmospheric dispersion of hazardous material releases—Implications for homeland security*. National Academies Press, Washington, DC.

U.S. Department of Defense (DoD), 2006. Chemical and Biological Defense Program, annual report to Congress, March.

Glossary

1:1 clay Layered aluminosilicate having one Si tetrahedral layer and one Al octahedral layer.

2:1 clay Layered aluminosilicate having two Si tetrahedral layers and one Al octahedral layer.

Abiotic Nonliving chemical and physical factors in the environment.

Adiabatic process A process occurring without heat transfer.

Advection Transport of quantities via the bulk motion of a fluid.

Aerosol A suspension of solid or liquid particles in a gas.

Agroterrorism Covert or overt attack against agricultural food and fiber production capabilities and resources using biological or chemical agents to adulterate or limit production, distribution, and consumption of agricultural products as a means of promulgating terror and economic damage.

Air-Sea Interaction Heat and momentum exchanges between the water and air.

Aircraft Communications Addressing and Reporting System (ACARS) A meteorological observation system provided by NOAA, in which observations are taken by commercial aircraft about every 10 min.

Alpha spectrin (α-spectrin) A cytosolic protein that maintains cellular structural integrity and is susceptible to calpain and caspase cleavage.

Alveolar ducts The enlarged terminal sections of the bronchioles that branch into the terminal alveoli.

Analyte A substance or chemical that is detected by an analytical procedure.

Antigen A molecule that elicits an immune response (i.e., antibody production).

Apaf-1 Apoptotic protease activating factor 1; a cytosolic protein that forms part of the apoptosome by combining with cytochrome c and d-ATP to activate caspase-9 and participate in apoptosis.

Arthropod vector The phylum of arthropods includes insects, arachnids, and crustaceans. An arthropod vector is an arthropod, such as a mosquito, that transports pathogens from one host to another.

AT-cut quartz crystal A thin plate of quartz cut parallel to the a axis of the crystal, at an angle of $35°15'$ to the c axis (optic axis) of the crystal, and at $2°58'$ to the z plane ($01\bar{1}$-1) of the crystal.

ATP-45 Allied Technical Publication 45; a protocol for rapid estimation of zones likely to be affected by an atmospheric release of hazardous material.

Attachment constant A measure of the association of a complex.

Autophagy A Greek work derived from *auto* meaning "self" and *phagy* meaning "to eat". It represents self-cannibalism of a cell that presents as cellular degradation with numerous intracellular vacuoles.

Background information The initial estimate of the state of the atmosphere at grid points used in data analysis.

Bcl-2 protein family A family of genes and proteins that either protect against (Bcl-2 proper, Bcl-xL, Bcl-w) or initiate (Bax, BAD, Bak, and Bok) apoptosis by governing the mitochondrial outer membrane permeabilization.

Biological threat agents Microorganisms or toxins produced by living organisms that may be intentionally employed to cause morbidity or mortality in other living organisms and include bacteria, mycotoxins, rickettsia, toxins, and viruses.

Biological weapons Microorganisms or toxins produced by living organisms that have been enhanced or modified to more effectively and efficiently cause morbidity or mortality in other living organisms and include bacteria, mycotoxins, rickettsia, toxins, and viruses.

Biopanning A biological method by which one selects for only those display phage that bind to a specific target molecule.

Bioterrorism The calculated use of microorganisms or toxins produced by living organisms that may have been enhanced or modified to more effectively and efficiently cause disease, debilitate, or kill other living organisms in an attempt to intimidate or coerce a government, the civilian population, or any segment thereof, in furtherance of political, religious, or social objectives.

Birefringence Also called double refraction, is the decomposition of a ray of light into two rays (the ordinary ray and the extraordinary ray) when it passes through certain types of materials, such as calcite crystals or white carbon, depending on the polarization of the light.

Blister agents Also known as vesicants, these are chemical weapon compounds that cause severe blistering of the skin, as well as damage to the eyes, mucous membranes, respiratory tract and internal organs. This class of chemical weapon agent includes the arsenicals/Lewisites (L), phosgene oxime (CX), and sulfur mustards (HD, HN).

Blood agents Also known as cyanogens, they include arsine (SA), cyanogen chloride (CK), and hydrogen chloride (AC), and are transported in the bloodstream through the body. Blood agents do not typically affect the blood but may interrupt the production of blood components and cause toxic effect at the cellular level.

Brownian motion A random movement of microscopic particles suspended in liquids of gases resulting from the impact of molecules of the surrounding medium.

CA1 and CA3 Areas of the hippocampus in which the pyramidal neurons reside.

Capsid The protein or membrane coat that surrounds the genomic nucleic acid (DNA or RNA) of a virus. May be made up of multiple protein species.

Cation exchange capacity (CEC) Quantity of cations that an anionic colloid can hold or exchange and is expressed in terms of centimoles per kilogram (cmoles/kg).

Central limit theorem Theorem stating that the distribution of a sum of independent random variables is approximated by the normal distribution.

Centroid Center of mass; the point in a body or system of bodies at which the whole mass may be considered as concentrated.

Cerebellar granule cells A population of numerous small neurons that reside below the level of the Purkinje neurons in the cerebellar cortex. The axons of these neurons form the T-shaped parallel fibers that mediate local excitation of the Purkinje neurons.

Cerebellum Literally, "little brain," a part of the brain that lies dorsal to the pons and is intimately involved in the control of movements; damage to this area produces ataxia or motor incoordination.

Chemical terrorism The calculated use of hazardous toxic compounds or substances that may have been enhanced or modified to more effectively and efficiently debilitate or kill humans in an attempt to intimidate or coerce a government, a civilian population, or any segment thereof, in furtherance of political, religious, or social objectives.

Chemical threat agents Compounds or substances, either produced naturally or synthetically, that can cause significant morbidity or mortality in humans, as well as other organisms.

Chemical weapons Compounds or substances, either produced naturally or synthetically, that have been designed or modified to maximize exposure through delivery methodologies to cause significant morbidity or mortality in humans, as well as other organisms. These compounds or substances are generally classified as blood agents, choking agents, nerve agents, psychotic agents, and vesicants.

Choking agents Chemicals that are typically inhaled and cause tissue damage and inflict injury to the respiratory tract (especially the lungs), leading to pulmonary edema and respiratory failure. Agents include chlorine (CL), chloropicrin (PS), diphosgene (DP), and phosgene (CG).

Clay A particle less than 0.002 mm in effective diameter, (b) a soil containing a large quantity (> 40%) of clay-sized particles, (c) an inorganic particle generally as a layered aluminosilicate but may be fibrous or amorphous.

Coliphage A phage (bacterial virus) that can infect and grow on Escherichia coli cells.

Computational fluid dynamic models (CFD) Computational models of fluid flow based on numerical solution of the continuity and Navier-Stokes equations (in either instantaneous or, more commonly, some type of averaged form).

Computational instability The exponential growth of the numerical solution of the differential equation.

Concentration The quantity of material per unit volume. Concentration may be defined in terms of mass per unit volume or the number of moles of material per unit volume.

Conservative solute A dissolved substance that does not react as it is transported through a medium.

Crosslinkers Chemical agents that form covalent bonds, linking one polymer with another.

Cumulus parameterization A procedure of including the statistical effects of cumuli in a weather prediction model.

Data assimilation Inserting data into the weather prediction model to update the model state during the time integration step.

DCD Dark cell degeneration, a form of programmed cell death in which the cells display cell shrinkage and hyper-condensed cytosol and nuclei.

Decontamination Removal or neutralization of toxic agents.

Detachment constant A measure of the tendency of a complex to dissociate.

Dew-point temperature The temperature at which saturation will occur.

Diffusion/dispersion model A numerical model designed to simulate the atmospheric diffusion/dispersion processes.

Diurnally Recurring or cycling daily.

Drag coefficient Dimensionless coefficient factor representing the effect of shape and flow-field characteristics on the drag exerted on a body immersed in a flow.

Dry line A boundary between the dry and moist air masses.

Einstein summation convention A notation in which, when an index is used more than once in an equation, it is implied that the equation needs to be summed over the applicable range of indices. Also known as Einstein notation.

ELISA Enzyme-linked immunosorbent assay. A biochemical or immunological method by which an analyte is detected in a sample.

Enhancement or weaponizing Utilizing chemical, genetic, or other methods and technologies on biological pathogens, toxins, or chemical substances as a means to improve ease of delivery, longevity in the environment in which it is delivered, pathogenic or toxic effects upon the intended target population, or speed of action once within the intended target population.

Environmental compartments Those natural matrices composing the environment in which biological or chemical contaminants may occur and cause damage, including the atmosphere (air), biosphere (living organisms), hydrosphere (water), and lithosphere (soil).

Enzootic foci Locations where a disease has become endemic in a nonhuman population.

Epitope That part of an antigen that is recognized by the paratope of an antibody.

Epizootic Disease affecting many animals at once, similar to an epidemic in humans.

Equivalent aerodynamic diameter The diameter of a standard density sphere that would settle at the same rate as the actual particle.

ER Endoplasmic reticulum is an intracellular organelle responsible for protein translation, folding and transport of proteins, and serves as a storage site for calcium.

Eulerian coordinate system A coordinate system that is fixed in space.

FADD Fas-associated death domain, a protein that serves as an adaptor molecule that interacts with various cell surface receptors to mediate cell apoptotic signals via the recruitment of caspase-8.

First-order decomposition Type of reaction that depends only on the concentration of one reactant to determine the rate at which a reactant will break down into smaller components. The concentration of the dependent reactant will decrease exponentially with time.

Four-dimensional data assimilation (FDDA) A technique by which high-frequency observations are used to update the model state during the time integration step, improving the fidelity of meteorological model predictions.

Full physics A comprehensive treatment of model physics.

GABA Gamma-amino-butyric acid, an inhibitory amino acid transmitter associated with synaptic depression

GIS Geographic information system.

Glutamate Primary excitatory amino acid transmitter in the CNS associated with synaptic development, fast synaptic transmission, memory, and excitotoxicity when present in excessive amounts.

Heterobifunctional cross-linkers Cross-linkage agents that have two or more reactive groups that allow for the conjugation of macromolecules.

Hippocampus A three-layered cortical region that is instrumental in memory formation and other cognitive functions.

Horizontal resolution The horizontal grid size.

Hypovolemic shock A state of decreased blood (plasma) volume.

IgG Immunoglobulin G.

Immunofluorescence Immunohistochemistry utilizing an antibody labeled with a fluorescent molecule.

Immunohistochemistry A procedure that relies on an antibody-antigen interaction to permit visualization of tissue or cellular constituents.

Ionotropic Fast-conducting, ligand-gated ion channels in which the receptor and ion channel form an integral unit.

Isobaric surfaces Surfaces of constant pressure.

Isotype Refers to the type of chain (heavy and light) of an antibody.

Lagarangian coordinate system A coordinate system that translates with a particle of fluid.

Lateral boundary conditions The conditions specified along the outer edges of a limited-area weather prediction model.

LC Liquid crystal. Liquid crystals are substances that exhibit a phase of matter that has properties between those of a conventional liquid and those of a solid crystal. For instance, a liquid crystal (LC) may flow like a liquid but have the molecules in the liquid arranged or oriented in a crystal-like way. There are many different types of LC phases, which can be distinguished based on their different optical properties (such as birefringence).

LC_{t50} The cumulative exposure, expressed as the concentration of a chemical or biological material integrated over the time period of exposure (e.g., integrated air concentration [gram-sec]/m^3) that produces lethal effect.

Ligand (Chapter 5) Gated channel, a transmembrane plasma membrane protein that fluxes charged ions secondary to the binding of a chemical transmitter.

Ligand (Chapter 7) A molecule that binds and forms a complex with another molecule.

Lipid rafts A membrane domain that is enriched in cholesterol.

Mass-consistent wind field The wind field obeys the basic law relating the wind and mass fields.

Meso-β Scale The horizontal scale of 10 to 100 km.

Metabotropic Slow-conducting synaptic events in which the transmitter receptor and ion channel are indirectly linked via second messenger transduction processes.

Mesonet A gridded array of meteorological stations with the average spacing of the stations less than 100 km.

Mesoscale The horizontal scale of 10 to 1000 km.

Metapopulation A spatially subdivided population whose subunits (colonies) experience periodic extinction events and are linked by migration.

Microplate Also called microtiter plate, is a flat plastic plate containing rows of multiple that can be used as small test tubes.

Microflora Microscopic plant or bacterial life.

Microscale The horizontal scale of 1 km or less.

Mimotope A macromolecule that mimics the structure of an epitope of an antigen.

Model initial conditions The conditions specified at the starting time of model integration.

Nerve agents Also known as anticholinesterase agents, are chemical compounds that inhibit the ability of cholinesterase to hydrolyze acetylcholine, which is essential to mediation of neurotransmitter function in nerve impulses. Nerve agents include: VX (*O-ethyl S-(2-diisopropylaminoethyl) methylphosphonothioate*); and the G-series agents tabun (GA)–*dimethylphosphoramidocyanidate*; sarin (GB–*isopropyl methylphosphonofluoridate*; and soman (GD)–*Pinacolyl methyl phosphonofluoridate.*

Nested grid A grid system consists of fine inner mesh grids imbedded in a coarse mesh grid.

Neuropathy Any disease of peripheral nerves.

Neutral atmosphere An atmosphere in which the potential temperature is constant with respect to height. Buoyant motion in a neutral atmosphere is negligible, but vertical mixing is not suppressed. Mixing will occur throughout the full depth of a neutral layer.

NOAA National Oceanic and Atmospheric Administration.

Nonhydrostatic A condition in which the upward pressure gradient force does not equal gravity.

Nonwoven Unconventional fabrics wherein fibers are directly converted into fabrics using mechanical, chemical, and thermal methods.

Numerical weather prediction Dynamic weather prediction based on the numerical solution of physical laws.

NWS National Weather Service.

Paratope That region of an antibody that recognizes (binds) an epitope.

Peptide Short, linear polymers of alpha-amino acids formed by covalent bonds between the amino (NH2) group of one amino acid and the carboxyl (COOH) group of another.

Phage A virus that infects bacteria. The term can be singular or plural.

Phage display A high-throughput technique to identify peptides that bind to a specific target molecule.

Planetary boundary layer (PBL) The portion of the atmosphere immediately adjacent to the Earth's surface. Processes within the PBL are affected by the roughness of the planet's surface, solar heating, and evaporation. The PBL often consists of a set of sublayers, depending on the nature of the local topography.

Plaque Clear spots in a bacterial lawn caused by phage killing of bacteria as the bacterial lawn is forming.

Plaque forming unit (pfu) The number of viable phage per unit volume.

Plume The cloud of material evolving as a result of a continuous release of material into the atmosphere.

Pore water velocity Average fluid velocity inside a porous medium.

Portability A software system (e.g., weather models) that can run on various computer platforms.

Potential temperature The temperature a parcel of air would have if it were transferred adiabatically from its actual pressure at elevation to a reference sea-level pressure of 1 bar.

Protomer Consists of a single protein subunit that is assembled in a defined stoichiometry to form an oligomeric or multimeric structure.

Psychotomimetic (psychochemical) agents Pharmaceutical chemical compounds that cause symptoms similar to psychotic disorders, debilitating the victim through disorientation, confusion and hallucination.

Pyramidal neurons A primary neuron type of the hippocampus intimately involved with hippocampal physiology such as learning and memory.

Puff The cloud of material evolving as a result of a quasi-instantaneous release of material into the atmosphere.

Purkinje neurons A single row of aligned neurons that reside in the cerebellar cortex that serves as the sole inhibitory output neuron of the cerebellar cortex to the deep cerebellar nuclei.

Radiosonde A miniature radio transmitter that is carried aloft with instruments for sensing and broadcasting atmospheric conditions.

Radiological dispersal devices (RDDs) Explosive devices that disseminate radioactive materials upon detonation, colloquially known as "dirty bombs."

Radius of influence Used in obtaining gridded data. The radius is the distance from a grid point within which the observations are used to define the value at the grid point.

Real data Observed data as different from artificial data.

Real-time ability Able to accept real-time data.

Reynolds averaged Navier-Stokes (RANS) equation Equation representing the conservation of momentum in a fluid flow, subjected to a temporal or spatial averaging process in line with the approach proposed by Osborne Reynolds.

Reynolds number A dimensionless number characterizing the flow of a fluid. The Reynolds Number is defined as

$$\mathrm{Re} = \frac{\rho U D}{\mu}$$

where ρ is the density of the fluid, U is the average velocity of the fluid, D is the characteristic length of the medium that the fluid is flowing through or around, and μ is the viscosity of the fluid.

Rhizosphere The zone that surrounds the roots of plants.

Sand Particle between 2 and 0.05 mm in effective diameter, (b) a soil containing a large quantity (> 85%) of sand-sized particles.

Selectively permeable membranes Barrier materials that can let moisture vapor pass through and prevent aerosolized agents and toxic vapors from passing through.

Severe weather environment Environment favorable for developing severe weather.

Soil Unconsolidated, dynamic three-phase system comprised of solids, liquids, and gases found at the Earth's surface that serves as a medium for plant growth.

Soil texture Percentage sand, silt, and clay sized particles contained in the soil.

Solute flux density The mass flow rate of solute across a cross-section per unit area.

Sorbent materials Materials that can adsorb and hold gases, liquids, etc.

Stokes' law A set of equations that describe the settling velocities of small spherical particles in a fluid medium at low values of Reynolds' number.

Stringency Reaction conditions that dictate binding affinity.

Subgrid scale motion The scale of motion smaller than the grid size.

Synoptic forcing Continental or oceanic influences on weather patterns, such as the uplift resulting from a change in surface elevation.

Synoptic scale A horizontal scale on the order of 2000 km.

Temperature inversion In a temperature inversion, the potential temperature of the atmosphere increases with increasing height, creating a stable atmosphere in which vertical motion is suppressed.

Terrain capability Capable of treating complex terrain processes.

Terrain-following coordinates A system of vertical levels parallel to the terrain surface.

Tortuosity factor The distance a particle must travel to pass through a porous medium divided by the overall length of the medium.

Toxic Substance that causes adverse effects, morbidity, or mortality in living organisms, with either or both acute or chronic pathology; poisonous.

Toxicant Toxic substances that are produced by or are a by-product of anthropogenic activities.

Toxins Toxic substances that are produced as metabolic by-products of microorganisms, plants, and animals, or are synthetically produced, and can be used to poison other living organisms.

Turbulence closure A technique to parameterize unknown variables in turbulent flow in terms of known variables.

Turbulent kinetic energy (TKE) Average kinetic energy associated with the fluctuating component of velocity in a fluid.

Undisturbed environment Clear and calm environment, usually under a high pressure system.

Vapor pressure The pressure exerted by a vapor that is in equilibrium with its liquid form.

Velocity inflection A point at which the second derivative of the velocity with respect to height changes sign.

Voltage-gated channel A transmembrane plasma membrane protein that fluxes charged ions secondary to a change in membrane voltage.

Weaponized Aerosols in a form that optimizes the range of atmospheric transport or the potential for the aerosols to deposit within the respiratory system in a manner that maximizes their effect.

Western blot An immunological method to detect a specific protein in a given sample.

Zoonoses Diseases that normally exist and are maintained in cycles among wild animals, and whose causative pathogen may be transmitted to humans that come into contact with infected animals, or in many instances ectoparasites associated with the infected animals.

Index

A

Abbott, Shah and, studies, 174
Abiotic processes, 249
Abou-Donia studies, 140
AC, *see* Hydrogen chloride (AC)
ACARS, Aircraft Communications Addressing
    and Reporting System (ACARS)
Acetylcholine physiology, 137, *138*
Acetylcholinesterases classification, 138–139
ACF (activated carbon fabrics), 204–208
AChE, *see* Acetylcholinesterases classification
Acoustic wave devices, 170–171
Activated carbon fabrics (ACF), 204–208
Acute exposure and treatment, 139–141
Adams studies, 139
ADAPT-LODI particle dispersion model, 53
Adiabatic process, 249
Adjustment Region, 47
Advanced Regional Prediction System (ARPS),
    66
Advection
    aerosol dispersion, 35, 49
    defined, 249
    dominant transport mode, 30–31
AEC, *see* Anion exchange capacity (AEC)
Aerosols
    computational model prediction, 48–49,
        52–54
    defined, 249
    dispersion, airflows, 36–48
    fundamentals, 31–33
    Gaussian puff-based model, 56–59, *57, 59*
    inhalation properties, 33–36
    model validation field studies, 54–56
    transportation mechanisms, 33–36, *34–35*
Affinity selection, *182,* 183–184, *184–185*
Aflatoxins
    pathogenic and toxic effects, 238
    pre-9/11/2001 concerns, 109
    soil mineralogy effects, 119
    Zumwalt Program, 89–90
Agency for Toxic Substances and Disease
    Registry (ATSDR), 20
Agent Orange, 143
Aging phenomenon, 141
Agriculture, 18–19, 25–27, *26*

Agroterrorism
    defined, 249
    environmental protection assessment
        strategies, 114–115
    fundamentals, 18–19
Air, 106, 111–112
Aircraft Communications Addressing and
    Reporting System (ACARS)
    defined, 249
    dispersion predictions, 60, 61
    meteorological observations and model
        prediction combination, 62–65
Airflow dispersion, 36–48
Air-land concerns, 112
Air-Sea Interaction, 249
Akamatsu, Hayes and, studies, 204–206
Akçakaya studies, 95
Algerian troops, 136
Alibek and Handelman studies, 16
Allen studies, 35
Allied Technical Protocol-45 (ATP-45), 58, 249
Allwine and Flaherty studies, 56
Allwine studies, 56
ALOHA, *see* Computer-Aided management of
    Emergency Operations/Areal Location
    of Hazardous Atmosphere (CAMEO/
    ALOHA)
Alpha spectrin, 249
Al-Qa'ida, 9
Alveolar ducts, 249
Amherst, Jefferey (Governor General), 6
AMPA-induced excitotoxicity, 147, 149–150, 151
Analytes
    defined, 249
    selection, 171–172
Animals, *see also* Foot-and-mouth disease
    (*Rhinovirus* group) (FMD); Zoonoses
    agroterrorism concerns, 114–115
    foreign animal diseases, 26
Anion exchange capacity (AEC), 117
Anthes, Hoke and, studies, 72
Anthes studies, 71
Anthrax (*Bacillus anthracis*)
    aerosol atmospheric transport, 32
    agroterrorism concerns, 115
    antibiotic-resistant strains development, 20
    decontamination, 161